CRM Series in Mathematical Physics

More information about this series at http://www.springer.com/series/3872

Andre D. Bandrauk • Emmanuel Lorin
Jerome V. Moloney
Editors

Laser Filamentation

Mathematical Methods and Models

Springer

Editors
Andre D. Bandrauk
Faculté des Sciences
Université de Sherbrooke
Sherbrooke, QC, Canada

Emmanuel Lorin
School of Mathematics and Statistics
Carleton University
Ottawa, ON, Canada

Jerome V. Moloney
Arizona Center for Mathematical Sciences
University of Arizona
Tucson, AZ, USA

CRM Series in Mathematical Physics
ISBN 978-3-319-23083-2 ISBN 978-3-319-23084-9 (eBook)
DOI 10.1007/978-3-319-23084-9

Library of Congress Control Number: 2015951225

Springer Cham Heidelberg New York Dordrecht London

Printed on acid-free paper

Springer International Publishing AG Switzerland is part of Springer Science+Business Media (www.springer.com)

Preface

Optics and Photonics are now recognized as key enabling drivers that impact technologies in a multitude of areas such as communications, imaging, sensing, energy, and even health sciences. This recent declaration by NSF (USA) is based on the observation that optical and in particular laser technology has been developing at a steady rate. The invention of the laser in the 1960s for generating coherent sources of radiation led to the development of intense ultrafast pulses with time durations spanning femtoseconds (1 fs $= 10^{-15}$ s), the time scale of atomic (nuclear) motion, to the attosecond (1 asec $= 10^{-18}$ s), the time scale of electron motion in atoms and molecules. Intensities are also approaching or even exceeding the atomic unit (a. u.), $I_0 = 3.5 \times 10^{16}\,\mathrm{W/cm^2}$, corresponding to electric field strength $E_0 = e/a_0^2 = 5 \times 10^9\,\mathrm{V/cm}$ at the atomic unit of distance in the ground state of the H atom, $a_0 = 0.0529\,\mathrm{nm}$. Such high fields E or equivalently intensities $I = cE^2/8\pi$ produce ionized electrons with ponderomotive energies $U_p(\mathrm{eV}) = (eE)^2/(4m\omega^2) = (9.33 \times 10^{-14}) I(\mathrm{W/cm^2}) \lambda^2(\mu\mathrm{m})$, where e, m are electronic charge, mass and ω, λ are the laser frequency and wavelength. Thus, lasers at current available intensities $I = 10^{14}\,\mathrm{W/cm^2}$ and $\lambda = 800\,\mathrm{nm}$ generate free electrons with kinetic energies in the region of $\sim 10\,\mathrm{eV}$ and more. The ionization process is considered as perturbative multiphoton ionization, which for low ionization potentials I_p is transformed into a nonperturbative quantum tunnelling ionization and above barrier ionization at high intensities. For the H atom with $I_p = 13.6\,\mathrm{eV}(0.5\,\mathrm{a.\,u.})$, tunnelling ionization is defined by the Keldysh parameter $\gamma = (I_p/2U_p)^{1/2} < 1$ and above barrier ionization occurs for intensities exceeding the critical $I_c = 1.4 \times 10^{14}\,\mathrm{W/cm^2}$. Thus, at intensities $I \sim 10^{14}\,\mathrm{W/cm^2}$ one is dealing with nonlinear response of single atoms and molecules to laser fields. In macroscopic gaseous molecular media such as the atmosphere, propagation of laser pulses leads to "Laser Filamentation" due to laser-induced nonlinear refractive indices, high-order polarizations, self-focusing, and plasma generation.

This book is based on a workshop, the first of its kind ever held in Canada, and is focused on the nonlinear theoretical and mathematical problems associated with ultrafast intense laser pulse propagation in gases and in particular in air. The chapters in this book are based on lectures by invited speakers, mainly theorists and mathematicians and supplemented by active experimentalists who are acknowledged experts in this new field of nonlinear laser molecule interaction and propagation.

The co-organizers of the workshop especially thank the CRM (Centre de recherches mathématiques) for financial support and its staff for the administrative organization, NSF (USA) with a special grant and AFOSR for partial travel support, and the scientific publisher Springer for publishing the lectures of the invited speakers in this "International Year of Light." We also thank Christiane Rousseau of CRM for introducing this workshop as a new activity in the international UNESCO sanctioned program MPE (Mathematics of Planet Earth) and a new direction in research on weather and climate control via current advanced laser technologies.

Sherbrooke, QC, Canada — André D. Bandrauk
Ottawa, ON, Canada — Emmanuel Lorin
Tucson, AZ, USA — Jerome V. Moloney
June 2015

Contents

Contributors

D.R. Austin ICFO—Institut de Ciències Fotòniques, Mediterranean Technology Park, Castelldefels, Barcelona, Spain

A.D. Bandrauk Laboratoire de chimie théorique, Faculté des Sciences, Université de Sherbrooke, Sherbrooke, QC, Canada

Centre de recherches mathématiques, Université de Montréal, Montréal, QC, Canada

M. Baudelet The Townes Laser Institute, CREOL-The College of Optics and Photonics, University of Central Florida, Orlando, FL, USA

M. Baudisch ICFO—Institut de Ciències Fotòniques, Mediterranean Technology Park, Castelldefels, Barcelona, Spain

J. Biegert ICFO—Institut de Ciències Fotòniques, Mediterranean Technology Park, Castelldefels, Barcelona, Spain

ICREA—Institució Catalana de Recerca i Estudis Avançats, Barcelona, Spain

Yuri Cher Department of Mathematics, University of Toronto, Toronto, ON, Canada

See Leang Chin Centre d'optique, photonique et laser and Département de physique, de génie physique et d'optique, Université Laval, Québec, QC, Canada

A. Couairon Centre de Physique Théorique, CNRS, École Polytechnique, Palaiseau, France,

Magdalena Czubak Department of Mathematical Sciences, Binghamton University, Binghamton, NY, USA

J. Darginavičius Department of Quantum Electronics, Vilnius University, Vilnius, Lithuania

A. Dubietis Department of Quantum Electronics, Vilnius University, Vilnius, Lithuania

Éric Dumas Institut Fourier, Université Joseph Fourier, rue des Mathématiques, Saint Martin d'Hères, France

M. Durand Laboratoire d'Optique Appliquée, ENSTA ParisTech, Ecole Polytechnique, CNRS, Palaiseau, France

Townes Laser Institute. CREOL-The College of Optics and Photonics, University of Central Florida, Orlando, FL, USA

A. Durécu Onera-The French Aerospace Lab, BP 80100, 91123 Palaiseau Cedex, France

D. Faccio School of Engineering and Physical Sciences, Heriot-Watt University, Edinburgh, UK

N. Forget FASTLITE, Centre universitaire d'Orsay, Plateau du Moulon, Orsay, France

N. Garejev Department of Quantum Electronics, Vilnius University, Vilnius, Lithuania

S. Grabielle FASTLITE, Centre universitaire d'Orsay, Plateau du Moulon, Orsay, France

I. Gražulevičiūtė Department of Quantum Electronics, Vilnius University, Vilnius, Lithuania

A. Houard Laboratoire d'Optique Appliquée, ENSTA ParisTech, Ecole Polytechnique, CNRS, Palaiseau, France

M. Hemmer ICFO—The Institut de Ciéncies Fotóniques, Mediterranean Technology Park, 08860 Castelldefels, Barcelona, Spain

A. Jarnac Laboratoire d'Optique Appliquée, ENSTA ParisTech, Ecole Polytechnique, CNRS, Palaiseau, France

Jingjing Ju State Key Laboratory of High Field Laser Physics, Shanghai Institute of Optics and Fine Mechanics, Chinese Academy of Sciences, Shanghai, China

Centre d'optique, photonique et laser and Département de physique, de génie physique et d'optique, Université Laval, Québec, QC, Canada

V. Jukna Centre de Physique Théorique, École Polytechnique, Palaiseau Cedex, France

Miroslav Kolesik College of Optical Sciences, University of Arizona, Tucson, AZ, USA

Arizona Center for Mathematical Sciences, University of Arizona, Tucson, AZ, USA

David Lannes Institut de Mathématiques de Bordeaux, Université de Bordeaux & CNRS UMR, cours de la Libération, France

Ruxin Li State Key Laboratory of High Field Laser Physics, Shanghai Institute of Optics and Fine Mechanics, Chinese Academy of Sciences, Shanghai, China

K. Lim Townes Laser Institute, CREOL-The College of Optics and Photonics, University of Central Florida, Orlando, FL, USA

Peng Liu State Key Laboratory of High Field Laser Physics, Shanghai Institute of Optics and Fine Mechanics, Chinese Academy of Sciences, Shanghai, China

Y. Liu Laboratoire d'Optique Appliquée, ENSTA ParisTech, Ecole Polytechnique, CNRS, Palaiseau, France

E. Lorin School of Mathematics and Statistics, Carleton University, Ottawa, ON, Canada

Centre de recherches mathématiques, Université de Montréal, Montréal, QC, Canada

M. Lytova School of Mathematics and Statistics, Carleton University, Ottawa, ON, Canada

D. Majus Department of Quantum Electronics, Vilnius University, Vilnius, Lithuania

E. McKee Townes Laser Institute. CREOL-The College of Optics and Photonics, University of Central Florida, Orlando, FL, USA

Jerome V. Moloney College of Optical Sciences, University of Arizona, Tucson, AZ, USA

Arizona Center for Mathematical Sciences, University of Arizona, Tucson, AZ, USA

Department of Mathematics, University of Arizona, Tucson, AZ, USA

A. Mysyrowicz Laboratoire d'Optique Appliquée, ENSTA ParisTech, Ecole de Polytechnique, CNRS, Palaiseau, France

Alan C. Newell Department of Mathematics, University of Arizona, Tucson, AZ, USA

Paris Panagiotopoulos College of Optical Sciences, University of Arizona, Tucson, AZ, USA

Arizona Center for Mathematical Sciences, University of Arizona, Tucson, AZ, USA

M. Richardson The Townes Laser Institute, CREOL-The College of Optics and Photonics: CREOL and FPCE, University of Central Florida, Orlando, FL, USA

F. Silva ICFO—Institut de Ciències Fotòniques, Mediterranean Technology Park, 08860 Castelldefels, Barcelona, Spain

Catherine Sulem Department of Mathematics, University of Toronto, Toronto, ON, Canada

Jérémie Szeftel Laboratoire Jacques-Louis Lions, Université Pierre et Marie Curie, Paris, France

G. Tamošauskas Department of Quantum Electronics, Vilnius University, Vilnius, Lithuania

A. Thai ICFO—Institut de Ciències Fotòniques, Mediterranean Technology Park, 08860 Castelldefels, Barcelona, Spain

G. Valiulis Department of Quantum Electronics, Vilnius University, Vilnius, Lithuania

Tie-Jun Wang State Key Laboratory of High Field Laser Physics, Shanghai Institute of Optics and Fine Mechanics, Chinese Academy of Sciences, Shanghai, China

Centre d'optique, photonique et laser and Département de physique, de génie physique et d'optique, Université Laval, Québec, QC, Canada

Patrick Townsend Whalen College of Optical Sciences, University of Arizona, Tucson, AZ, USA

Department of Mathematics, University of Arizona, Tucson, AZ, USA

Zhizhan Xu State Key Laboratory of High Field Laser Physics, Shanghai Institute of Optics and Fine Mechanics, Chinese Academy of Sciences, Shanghai, China

Shuai Yuan Centre d'optique, photonique et laser and Département de physique, de génie physique et d'optique, Université Laval, Québec, QC, Canada

Shanghai Key Laboratory of Modern Optical System, Engineering Research Center of Optical Instrument and System, University of Shanghai for Science and Technology, Shanghai, China

State Key Laboratory of Precision Spectroscopy, East China Normal University, Shanghai, China

Heping Zeng Shanghai Key Laboratory of Modern Optical System, Engineering Research Center of Optical Instrument and System, University of Shanghai for Science and Technology, Shanghai, China

State Key Laboratory of Precision Spectroscopy, East China Normal University, Shanghai, China

Short Pulse Evolution Equation

Alan C. Newell

Abstract We introduce short pulse evolution equation (SPEE), first derived in a non-optical context in the 80s, the universal equation describing the propagation of short pulses in media which have weak dispersion in the propagation direction. We show how it connects with the first canonical examples of nonlinear wave propagation, the Korteweg–de Vries and nonlinear Schrödinger equations and argue that, in contexts for which SPEE is most useful, modifications of the latter simply do not capture the correct pulse behavior. We discuss some of SPEE's main properties and, in particular, look at its potential singular behaviors in which both the electric field gradient and its amplitude can become large. Finally, we address the practical challenge of whether very high intensity femtosecond pulses can travel significant distances in gases such as the earth's atmosphere.

1 Introduction

The main goal of my Montreal lecture was, and of this short paper is, to show that the principal propagation characteristics of short electromagnetic pulses (10–100 fs in duration) can be captured by a very simple and beautiful evolution equation which we call short pulse evolution equation (SPEE). An additional goal is to ask whether ultra short pulses might be tailored so that they reach high intensities at a specified but distant point along their trajectory paths. The ideas behind the derivation of SPEE are not new. They have their origins in the 60s when there was widespread recognition (see discussions in [1, 2]) that the propagation of waves and wave envelopes occurring in many different contexts had, because of shared symmetries, very similar propagation characteristics and could be well described by what are essentially multinomial Taylor expansions in field amplitudes and their gradients.

There are two granddaddy examples. One is the Korteweg–de Vries (KdV) equation and the other is the nonlinear Schrödinger (NLS) equation. From these starting points sprang many second generation evolution equations which added

A.C. Newell (✉)
Department of Mathematics, University of Arizona, Tucson, AZ 85721, USA
e-mail: anewell@math.arizona.edu

A.D. Bandrauk et al. (eds.), *Laser Filamentation*, CRM Series in Mathematical Physics, DOI 10.1007/978-3-319-23084-9_1

to the KdV equation the effects of diffraction or transverse dispersion (the Kadomtsev–Petviashvili or KP equation), the Benney–Roskes–Davey–Stewartson modifications of the NLS equation, and the three wave interaction (3WI) equations. A variation on the KP equation which embodies the symmetry of field reversal was originally introduced in the 80s by Kuznetsov et al. [3] in the context of sound waves in ferromagnets, elaborated on by Turitsyn and Falkovich [4] and discussed more recently in [5, 6] in the context of electromagnetic fields for which the medium crystalline structure has the centrosymmetric and isotropic properties which introduce the field reversal symmetry. This universal equation, SPEE, also describes the evolution of intense short EM pulses and those are the subject of this short paper.

Why is the study of such an equation in the context of intense ultra short pulses useful? First, it relates the behaviors of light pulses to similar behaviors which occur in different contexts and leads to a cross fertilization of understanding between waves which have similar characteristics, but arise in very different situations. The most important common ingredients are (1) weak dispersion in the direction of ray propagation, (2) almost parallel rays, and (3) weak nonlinearity in a sense we will define. Second, there is much interest in propagating high intensity laser pulses with finite energies (high intensity plus finite energy necessitates short pulse duration) through gases such as the atmosphere. There are many reasons for this, ranging from use as light bullet weapons to ionization and light emissions in the THZ ranges. For details we refer to recent pioneering works and reviews by Kosareva et al. [7], Debayle et al. [8], and Kolesik and Moloney [9]. Third, we want to develop a canonical description which allows us to describe the evolution of femtosecond pulses which range in duration from less than tens to up to a thousand cycles centered about a typical central frequency of 2.35×10^{15} rad/sec or wavelengths of 800 nm. As we shall see, such wave pulses have the potential of developing very large electric field gradients and, in regions of anomalous longitudinal dispersion, very large intensities. We want to be able to understand such interesting and potentially useful behaviors in as simple a way as possible. Fourth, whereas numerical simulations, such as UPPE [10], are extraordinarily valuable, they are more so when complemented by analytical descriptions which allow us to see which balances between ingredients produce which effects. SPEE is also useful because it allows us to suggest possible ways of using, overcoming, or circumventing some of the less desirable by-products such as ionization produced by localized high intensities. Fifth, SPEE allows us to see clearly under what circumstances, a modified NLS description of the pulse envelope, the default and knee jerk position of many in the optical sciences, is satisfactory. As we shall see, under most circumstances when intensities of ionization level obtain, NLS does not provide a uniformly valid description.

2 The Nature of SPEE

We begin by exploring and understanding the anatomies of KdV and NLS. KdV describes the deformation of each of the Riemann invariants F and G of the solution $u(x,t) = F(x-ct) + G(x+ct)$ of the wave equation (or any other linear hyperbolic system with distinct wavespeeds) under the additional influences of weak dispersion and weak nonlinearity, each of order ε, $0 < \varepsilon \ll 1$. The shapes F and G, invariant under the action of the linear wave equation, deform over long times ε^{-1} according to

$$v_\tau + 6vv_\zeta + v_{\zeta\zeta\zeta} = 0, \tag{1}$$

where $v(\zeta,\tau) = F(\zeta = x - ct, \tau = \varepsilon t)$. One can think of v as the right going wave in a narrow channel of shallow water. If weak variations with respect to a transverse direction y are included, (1) becomes (2), the KP equation,

$$\frac{\partial}{\partial\zeta}(v_\tau + 6vv_\zeta + v_{\zeta\zeta\zeta}) \pm v_{yy} = 0. \tag{2}$$

In the shallow water case (think very long oceanic waves triggered by an earthquake), the plus sign obtains. In other cases, such as ours where the weak transverse dependence is due to diffraction, the minus sign obtains, and if the transverse direction is two-dimensional, v_{yy} is replaced by $\nabla_\perp^2 v$, the two-dimensional transverse Laplacian. If the original wave equation has two speeds which are close, e.g., $c \pm \varepsilon$, then the corresponding pairs satisfy the Boussinesq equations. If the original field is constrained by the symmetry of field reversal such as in centrosymmetric crystals, then the leading order nonlinearity is cubic and (2) becomes (3), the modified KP equation (MKP).

$$\frac{\partial}{\partial\zeta}(v_\tau + 6v^2v_\zeta + v_{\zeta\zeta\zeta}) = \nabla_\perp^2 v. \tag{3}$$

(Although not the subject of this paper, it is intriguing to ask why it is that many of these universal equation representing multinomial Taylor expansions in amplitude and derivative are exactly integrable!) Equation (3) is, for all intents and purposes, SPEE. Let $\mathbf{E}(\mathbf{x} = (x,y), z, t) = \hat{e}E(\mathbf{x}, z, \tau = t - n_0 z/c)$ be linearly polarized. Then $E(\mathbf{x}, z, \tau)$ evolves along the propagation direction z as

$$\frac{\partial}{\partial\tau}\left(\frac{\partial E}{\partial z} + \frac{3\pi\hat{\chi}^{(3)}}{n_0 c}E^2\frac{\partial E}{\partial\tau} - \frac{1}{2n_0 c}\frac{\partial}{\partial\tau}\left(\hat{\chi}(0) - \hat{\chi} - \hat{\chi}\left(\mathrm{i}\frac{\partial}{\partial\tau}\right)\right)E\right) = \frac{c}{2n_0}\nabla_\perp^2 E. \tag{4}$$

If we approximate the susceptibility $\hat{\chi}(\omega)$ as $\hat{\chi}(0) + \mathrm{i}a(\omega t_0) + b(\omega t_0)^2$ (since it varies very slowly over the frequency range of interest),

$$\frac{\partial}{\partial\tau}\left(\frac{\partial E}{\partial z} + \frac{3\pi\hat{\chi}^{(3)}}{n_0 c}E^2\frac{\partial E}{\partial\tau} - \frac{bt_0^2}{2n_0 c}\frac{\partial^3 E}{\partial\tau^3} - \frac{at_0}{2n_0 c}\frac{\partial^2 E}{\partial\tau^2}\right) = \frac{c}{2n_0}\nabla_\perp^2 E. \tag{5}$$

The form (5) makes contact with earlier work as in (3) but the form (4) is more useful. In (4) and (5), τ is retarded time, $\hat{\chi}(\omega)$ is the linear susceptibility (which for air in the frequency range of interest is 5.5×10^{-4}; for future consideration $\hat{\chi}(3\omega) = 5.94 \times 10^{-4}$, $n_0 = \sqrt{1 + \hat{\chi}(0)}$, t_0 is pulse width (10–100 fs), b and a are dimensionless. The Kramers–Kronig relations demand that there is a weak absorption, namely $\hat{\chi}(\omega)$ has both a real and imaginary part, which in (5) is manifested by the Burgers' like second derivative on the left-hand side. The effects of plasma induced diffraction and absorption can and will be added later. Because $\hat{\chi}(\omega)$ is well tabulated in the literature, it is better when using numerical simulations of (4) which exchange $E(\mathbf{x}, z, \tau) = \big(1/(2\pi)\big) \int \hat{E}(\mathbf{x}, z, \omega)\mathrm{e}^{-\mathrm{i}\omega\tau}\mathrm{d}\omega$ for its Fourier transform $\hat{E}(\mathbf{x}, z, \omega)$ to use $\hat{\chi}$ directly rather than any power series representation.

The various terms in (4), (5) arise as follows: The first term $\partial^2 E/(\partial\tau\partial z)$ is simply the wave operator $\partial^2/\partial z^2 - (n_0^2/c^2)(\partial^2/\partial t^2)$ written in the longitudinal (z) and retarded time $\tau = t - (n_0/c)z$ coordinates. The second term is the nonlinear term and arises from the Kerr (cubic) dependence of the polarization on the electric field. The factor $\hat{\chi}^{(3)}$ is the Fourier transform of the nonlinear susceptibility and, in (4), is treated as constant. The nonlinearity is weak because the product $\hat{\chi}^{(3)}E^2$ is small. The balance of the first and second terms means that $E^2(\mathbf{x}, z, \tau)$ behaves as an arbitrary function f of $\tau - \big(3\pi\hat{\chi}^{(3)}/(n_0 c)\big)E^2 z$ where f is the initial shape of E^2 from which one may deduce that $(E^2)_\tau$ will become singular in finite time. The third term in (4) represents both dispersion in the direction of propagation and attenuation. Its balance with $\partial E/\partial z$ gives rise to a weak dispersion of waves with different frequencies and a Burgers' like loss but nonlinearity can suppress the dispersion (walk off in optical parlance) because, when b in (5) is negative, the balance of the first three terms gives rise to a coherent soliton in which structure all frequencies travel at the same speed. The fourth term is diffraction. When combined with focussing nonlinearity, localized objects (collapses) with both τ and x, y dependence can result.

Whereas KdV and its cousins KP, MKP, Boussinesq, and SPEE are multinomials about the origin $(0, 0, 0, 0)$ in $\partial/\partial\tau$, $\partial/\partial\zeta$, $\nabla_\perp^2$ and v (in which in (3), for example, the relative sizes are ε^3, ε, ε^4 and ε (so that all terms balance)) which describe the slow deformation of the Riemann invariant v under the joint influences of weak dispersion, nonlinearity and diffraction, the NLS equation describes the deformation of the envelope $A(x, y, z, t)$ of a strongly dispersive, weakly nonlinear (amplitude ε) and almost monochromatic (bandwidth ε) carrier wave $u = \mathrm{Re}\{A\exp(\mathrm{i}\mathbf{k}_0 \cdot \mathbf{x} - \mathrm{i}\omega(\mathbf{k}_0)t)\}$. It takes the form

$$\frac{\partial A}{\partial t} + \nabla_{\mathbf{k}}\omega \cdot \nabla A - \frac{\mathrm{i}\varepsilon}{2}\nabla^\top \Omega \nabla A - \mathrm{i}\gamma\varepsilon A^2 A^* = 0, \tag{6}$$

where $\nabla_{\mathbf{k}}\omega$ is the group velocity of the wavevector $\mathbf{k}_0$, $\nabla^{\top}, \nabla$ is the row (column) vector $(\partial/\partial x, \partial/\partial y, \partial/\partial z)$, and Ω is the dispersion tensor $\partial^2\omega/(\partial k_i \partial k_j)$, running over x, y, z. The system is strongly dispersive only when the eigenvalues of Ω are nonzero.

For example, if $\omega = \omega(k = |\mathbf{k}|)$, $\nabla_{\mathbf{k}}\omega = \omega' \hat{k}$ and $\mathbf{k}_0$ is the carrier wavevector, then $\Omega = \mathrm{diag}(\omega'/k, \omega'/k, \omega'')$, $' = \mathrm{d}/\mathrm{d}k$. If ω'' is small so that the dispersion in the direction of propagation is weak, then when we are expanding the original field u as $\mathrm{Re}\{A \exp(\mathrm{i}k_0 z - \mathrm{i}\omega_0 t)\}$ + corrections, the corrections, assuming the field reversal symmetry obtains, will contain the third harmonic $\exp(3\mathrm{i}k_0 z - 3\mathrm{i}\omega_0 t)$ with the factor $A^3/(\omega(3k_0) - 3\omega(k_0))$ and in general the $(2n-1)$th harmonic with the factor $A^{2n-1}/(\omega((2n-1)k_0) - (2n-1)\omega(k_0))$. If dispersion in the propagation direction is weak, then the denominators $\omega\big((2n-1)k_0\big) - (2n-1)\omega(k_0)$, $n = 1, 2, \ldots$ can be very small. Thus the asymptotic expansion for u will only be uniformly valid if the amplitude of A, ε, is very small indeed. If, on the other hand, $\varepsilon^2/\big(\omega(3k_0) - 3\omega(k_0)\big)$ is of order unity, as we shall see shortly it is for short pulses with intensities of most interest, even those with many oscillations under their envelopes and having relatively narrow spectra around k_0 and ω_0, the NLS description is not valid because the corrections are as large as the leading order approximation.

From the previous discussion, we draw some conclusions and introduce some important new results. It is convenient to rescale the variables

$$E \to e_0 E, \quad \tau \to t_0 \tau, \quad z \to c t_0 z, \quad x, y \to L(x, y)$$

and introduce the following length scale ratios (all of which are small)

$$\begin{aligned} \varepsilon_{\mathrm{NL}} &= \frac{c t_0}{z_{\mathrm{NL}}} = \frac{3\pi \hat{\chi}^{(3)} e_0^2}{n_0}, & \varepsilon_{\mathrm{disp}} &= \frac{c t_0}{z_{\mathrm{disp}}} = \frac{|b|}{2n_0}, \\ \varepsilon_{\mathrm{diff}} &= \frac{c t_0}{z_{\mathrm{diff}}} = \frac{(c t_0)^2}{2n_0 L^2}, & \varepsilon_{\mathrm{att}} &= \frac{c t_0}{z_{\mathrm{att}}} = \frac{a}{2n_0} \end{aligned} \tag{7}$$

so that (5) becomes ($s = \mathrm{sgn}\, b$)

$$\frac{\partial}{\partial \tau}\left(\frac{\partial E}{\partial z} + \varepsilon_{\mathrm{NL}} E^2 \frac{\partial E}{\partial \tau} - s\varepsilon_{\mathrm{disp}} \frac{\partial^3 E}{\partial \tau^3} - \varepsilon_{\mathrm{att}} \frac{\partial^2 E}{\partial \tau^2}\right) = \varepsilon_{\mathrm{diff}} \nabla_{\perp}^2 E. \tag{8}$$

Remark 1. For $1 \gg \varepsilon_{\mathrm{NL}} \gg \varepsilon_{\mathrm{disp}}, \varepsilon_{\mathrm{att}}, \varepsilon_{\mathrm{diff}}$, the principal response is to steepen all the forward facing (in τ) faces for $E > 0$ and all the backward facing ones for $E < 0$. Without regularization from dispersion and attenuation, the slopes would become infinite and thereafter the electric field multivalued. With regularization they will form shocks dominated either by diffusion (shocks are narrow discontinuities) or dispersion (shocks are wide with lots of wiggles). An initial spectrum will become broad and continuous. Credit must go to Alterman and Rauch [11] and Schaffer and Wayne [12] who, unaware of the original work of Kuznetsov et al., clearly recognized that the cubic dependence of polarization on the electric field combined with weak or essentially zero dispersion in the direction of propagation could lead to potential shocks.

I want to emphasize that, to this point, the shape $E(\mathbf{x}, z = 0, \tau = t)$ can be very short with only a few oscillations and therefore a correspondingly broad spectrum. The dynamics will steepen certain slopes and broaden others with the net effect of significantly broadening the spectrum even further. But the shape can also be envelope like with a narrow spectrum supported near some carrier frequency ω_0. It is this shape that we address with the second remark.

Remark 2. If the initial pulse has the shape of an envelope under which there are many oscillations of a basic carrier wave $\exp(\mathrm{i}\omega_0\tau)$, then a natural response is to ask if its evolution can be captured by developing an equation for the slowly varying envelope of the leading harmonic. Let us apply this idea to SPEE written in the form (4). Write E as $a\mathrm{e}^{\mathrm{i}\omega_0\tau} + a^*\mathrm{e}^{-\mathrm{i}\omega_0\tau} +$ corrections.

To leading order we find

$$a_z - \frac{\mathrm{i}\omega_0}{2n_0c}\big(\hat{\chi}(0) - \hat{\chi}(\omega_0)\big)a = 0$$

whose solution is

$$a = b\exp\frac{\mathrm{i}\omega_0}{2n_0c}\big(\hat{\chi}(0) - \hat{\chi}(\omega_0)\big)z.$$

Note that when combined with $\exp \mathrm{i}\omega_0\tau$ this gives

$$b\exp\left(\mathrm{i}\omega_0 t - \frac{\mathrm{i}n_0\omega_0}{c}z + \frac{\mathrm{i}\omega_0}{2n_0c}\big(\hat{\chi}(0) - \hat{\chi}(\omega_0)\big)z\right)$$

and the exponent is simply

$$\left(\mathrm{i}\omega_0 t - \sqrt{1 + \chi(\omega_0)}\frac{\omega_0}{c}z\right)$$

when we write $\hat{\chi}(\omega_0)$ as $\hat{\chi}(0) - \big(\hat{\chi}(0) - \hat{\chi}(\omega_0)\big)$ and expand. Now let us calculate the correction to E which reflects the third harmonic $a_3\mathrm{e}^{3\mathrm{i}\omega_0\tau}$ which will be generated by the cubic term in (4). We find

$$a_{3z} - \big(\hat{\chi}(0) - \hat{\chi}(3\omega_0)\big)a_3 = \frac{3\pi\chi^{(3)}}{n_0c}\mathrm{i}\omega_0 b^3\exp\left(\frac{3\mathrm{i}\omega_0}{2n_0c}\big(\hat{\chi}(0) - \hat{\chi}(\omega_0)\big)z\right).$$

Solving, we find that the amplitude of a_3 divided by the amplitude of b (or a) which we designate as e_0 is

$$\frac{2\pi\chi^{(3)}e_0^2}{\hat{\chi}(3\omega_0) - \hat{\chi}(\omega_0)}.$$

For air, $n_2 = \chi^{(3)}/(2n_0)$ is roughly $3 \times 10^{-23}\,\mathrm{m^2/W}$ and the intensity at which air begins to ionize is $5 \times 10^{17}\,\mathrm{W/m^2}$ (and we of course want to be able to describe the envelope at these high intensities), and $\hat{\chi}(\omega_0) \simeq 5 \times 10^{-4}$ and $\hat{\chi}(3\omega_0) - \hat{\chi}(\omega_0) = 0.4 \times 10^{-4}$ (for lower frequencies, the difference is even less) which makes the above ratio approximately 2. At the very least, we can say it is order one.

What happens therefore to such a pulse?

First we recognize that all its odd harmonics will be excited significantly. Each lobe (oscillation) in the envelope will steepen on some faces and broaden on others as discussed in Remark 1. Whereas before it reaches these high intensities, the nonlinearity can give rise to third and higher harmonics whose amplitudes are small enough so that the NLS equation will obtain (namely, in the parlance of optics, walk off will separate the first from the third harmonic), that will not be the case once the intensities increase to the levels of interest. At these levels, the harmonics will phase lock and travel together. Some investigators may try to remedy the breakdown of NLS by introducing coupled envelope equations for a, a_3, and so on but this is a clumsy representation. The simplest description is SPEE. It has the capacity to describe the collective behavior of all odd harmonics simultaneously.

Remark 3. Collapses and the initiation of local large intensities. Although the NLS equation is not the correct analytical description for pulses in media with very weak dispersion in the direction of propagation, it nevertheless serves as a useful learning tool to investigate the synergy which can occur between nonlinearity and diffraction/dispersion from which interplay the pulse intensity can become locally infinite. We return to (6), move in the group velocity frame, assume the dispersion matrix to be diagonal and positive (so that $\Omega = \mathrm{diag}(\omega'/k, \omega'/k, \omega'') > 0$), rescale x, y, z, assume A to depend only on the r coordinate in n dimensions, and allow for a nonlinearity of order $2\sigma + 1$ in amplitude to obtain

$$\frac{\partial A}{\partial t} - \mathrm{i}\nabla^2 A - \mathrm{i}\alpha (AA^*)^\sigma A = 0, \tag{9}$$

where

$$\nabla^2 A = \frac{1}{r^{n-1}} \frac{\partial}{\partial r} r^{n-1} \frac{\partial A}{\partial r}.$$

The power $N = \int_0^\infty AA^* r^{n-1}\,\mathrm{d}r$ is constant. The Hamiltonian, also a motion constant, is given by

$$H = \int \left(\frac{\partial A}{\partial r} \cdot \frac{\partial A^*}{\partial r} - \frac{\alpha}{\sigma + 1} (AA^*)^{\sigma+1} \right) r^{n-1}\,\mathrm{d}r$$

and

$$\frac{\partial A}{\partial t} = -\mathrm{i} \frac{\delta H}{\delta A^*}.$$

The well-known virial theorem gives the result that if

$$V = \int_0^\infty r^2 AA^* r^{n-1}\,\mathrm{d}r, \quad C = 2\mathrm{i}\int_0^\infty r\left(A\frac{\partial A^*}{\partial r} - A^*\frac{\partial A}{\partial r}\right) r^{n-1}\,\mathrm{d}r,$$

then

$$\frac{\mathrm{d}V}{\mathrm{d}t} = C, \quad \frac{\mathrm{d}C}{\mathrm{d}t} = \frac{\mathrm{d}^2 V}{\mathrm{d}t^2} = 8H - \frac{4\alpha}{\sigma+1}(\sigma n - 2)\int_0^\infty (AA^*)^{\sigma+1} r^{n-1}\,\mathrm{d}r. \tag{10}$$

Thus, if $\alpha > 0$ (focussing nonlinearity) and the initial state is chosen so that $H < 0$, then for $\sigma n \geq 2$, the second time derivative of a positive quantity is negative which, after a finite time, leads to the contradiction that V eventually becomes negative. Instead what happens is that at some intermediate time t_0, the analysis is invalidated by the occurrence of a singularity in A whose modulus takes the shape $\Lambda^{1/\sigma} f(\Lambda r)$ where $\Lambda(t_0 - t)$ tends to infinity as $t \to t_0$. The power or number of particles is then given by $N = \Lambda^{-(\sigma n - 2)/\sigma} N_c$, $N_c = \int_0^\infty f^2(\zeta)\zeta^{n-1}\,\mathrm{d}\zeta$, a pure number. The case $\sigma n = 2$ is called critical collapse because the number of particles or power is conserved. For $\sigma n > 2$, the collapse is supercritical and particles and power are continuously radiated from the singular region as the pulse collapses. Of course, the original derivation of NLS required weak nonlinearity so that the analytical theory breaks down before the state of infinite amplitude is reached. Nevertheless, the theory correctly predicts the local intensification of the pulse envelope.

Turitsyn and Falkovich [4] showed that a similar result obtains for SPEE when attenuation is neglected. We can write (8) as

$$\frac{\partial E}{\partial z} = \frac{\partial}{\partial \tau} \cdot \frac{\delta K}{\delta E}$$

with Hamiltonian

$$\frac{\varepsilon_{\mathrm{diff}}}{3} K = \hat{K} = \int \left(\frac{D}{2} E_\tau^2 + \frac{\varepsilon_{\mathrm{diff}}}{2}(F_x^2 + F_y^2) - \frac{\varepsilon_{\mathrm{NL}}}{12} E^4 \right) \mathrm{d}\tau\,\mathrm{d}x\,\mathrm{d}y,$$

where $F_\tau = E$ and $D = -s\varepsilon^{\mathrm{diss}}$. For $s = \operatorname{sgn} b = +1$ dispersion is normal; for $s = -1$, it is anomalous.

A straightforward computation yields

$$\frac{\partial^2}{\partial z^2}\int (x^2 + y^2) E^2\,\mathrm{d}\tau\,\mathrm{d}x\,\mathrm{d}y = 16\varepsilon_{\mathrm{diff}}\left(\hat{K} - \frac{D}{2}\int E_\tau^2\,\mathrm{d}\tau\,\mathrm{d}x\,\mathrm{d}y\right). \tag{11}$$

For the case of anomalous dispersion $D > 0$, i.e. $s = \operatorname{sgn} b = -1$, and the choice of initial conditions such that $\hat{K}$, conserved under the evolution, is negative, the intrinsically positive quantity on the left-hand side of (11) will eventually become negative. This contradiction can only be avoided if E becomes infinite at some intermediate time. Thus, in the absence of dissipation, (8) can give rise to locally collapsing filaments.

Before we discuss the nature of the collapse, let us remind ourselves that for most frequencies in air and in many gases, the dispersion in the direction of propagation, while very small, is not anomalous but normal; namely, $s = +1$. In this case, we cannot make the same conclusion that the pulse will develop arbitrarily large intensities. Indeed, previous work by us [13] in the case of NLS has shown that in the context of three-dimensional signs $(+, +, -)$ ($\omega'' < 0$ with $\alpha > 0$ means the dispersion in the propagation is normal) and $\sigma = 1$, the two-dimensional collapse with critical power $N = \Lambda^{-(\sigma n-2)/\sigma} N_c = N_c$ when $\sigma n = 2$, the collapse is arrested when the normal dispersion $\omega''(\partial^2 A/\partial z^2)$ term is turned on. The mechanism for the gradual erosion of the critical power pulse is one of four wave mixing in which, if k_0 is the carrier wavenumber under the collapsing envelope pulse, wavevectors $\mathbf{k}_1$ and $\mathbf{k}_2$ with $2k_0\hat{z} = \mathbf{k}_1 + \mathbf{k}_2$, $\omega(2k_0) = \omega(\mathbf{k}_1) + \omega(\mathbf{k}_2)$ are excited and carry the power of the pulse away from the filament. Such behavior is similar to what is experienced by a deep water wave soliton with elevation $\eta = A\exp\left(ik_0x - i\sqrt{gk_0}t\right) + (*)$, where $A = 2a\,\mathrm{sech}\,a\left(x - \frac{1}{2}\sqrt{g/k_0}t\right)$. When rescaled, the envelope A satisfies

$$\frac{\partial A}{\partial t} - \frac{i}{2}\left(\frac{\partial^2 A}{\partial x^2} - \frac{\partial^2 A}{\partial y^2}\right) - i\alpha A^2 A^* = 0$$

and is unstable to waves $\mathbf{k}_1(k_{1x}, k_{1y}), \mathbf{k}_2(k_{2x}, k_{2y} = -k_{2x})$ which travel away from the original propagation direction $\hat{x}$ and are generated by four wave mixing.

But, we should point out, in the case of 2D NLS, the collapse is critical, namely it carries just enough power to sustain collapse. Any perturbation which can remove particles or power will interrupt the collapse. For SPEE, therefore, it is worth looking at the nature of the collapse. In (8), write

$$E(x, y, z, \tau) = (Z - z)^{-\alpha} F\left(X = \frac{x}{(Z-z)^{\beta}}, Y = \frac{y}{(Z-z)^{\beta}}, T = \frac{\tau - \tau^*}{(Z-z)^{\gamma}}\right), \tag{12}$$

where Z is the position of the collapse and τ^* follows the pulse maximum. From (12), the pulse power

$$P = \int E^2\,dx\,dy\,d\tau = (Z - z)^{-2\alpha+2\beta+\gamma} \int F^2(X, Y, T)\,dX\,dY\,dT, \tag{13}$$

the latter being a calculable real number. If we balance the four terms in (8) (in the absence of dissipation), we find

$$-\alpha - \gamma - 1 = -3\alpha - 2\gamma = -\alpha - 4\gamma = -\alpha - 2\beta. \tag{14}$$

Equating terms one, two, and four (thereby ignoring dispersion), we find $\gamma = 1-2\alpha$, $\beta = 1-\alpha$, and then $P = (Z-z)^{3-6\alpha}P_c$ so that the collapse is supercritical, critical, or subcritical, depending on whether $\alpha < \frac{1}{2}$, $\alpha = \frac{1}{2}$ or $\alpha > \frac{1}{2}$. This means as the one parameter (α) family of self-similar solutions (12) progresses in z, their power has a z dependence which is not constant but evolves as $(Z-z)^{6-3\alpha}$. If we include dispersion, the parameter α is determined to be $\frac{1}{3}$ and then $\beta = \frac{2}{3}$, $\gamma = \frac{1}{3}$ and $3-6\alpha = 1$ so that, as the filament nears the point at which E becomes infinite, its power decreases linearly as $(Z-z)$.

In the final section of this short paper, we will discuss the ramifications of Remarks 1–3 in the context of ultra short light pulses. In particular, we will be asking if it might be possible to tailor a pulse so that it reaches high intensities at a specified but distant point on its trajectory path. But first, we introduce a derivation of SPEE, originally presented in [6].

3 Derivation of SPEE

The reasons for introducing a brief derivation of SPEE are: (1) To show how it arises from Maxwell's equations and therefore how to include extra effects such as charge and currents arising from plasma formation. (2) To connect with the equations such as UPPE used as a basis for most numerical simulations. (3) To make clear the origins of the various terms in SPEE and to show that the effects of the weak dependency of the electrical susceptibility on frequency, weak nonlinearity (whereas pulse intensities I are large, the nonlinear refractive index n_2 is small and therefore $n_2 I$ is still small), diffraction (for beams wide compared to the dominant wavelength) all lead to long distance modifications of the pulse shape which, under linear wave propagation, would travel unchanged. The result we strictly obtain is for a linearly polarized field $\mathbf{E} = \hat{y}E(x,z,t)$ but, with a little additional work, can be shown to be true for an $E(x,y,z,t)$ which depends on both transverse coordinates (x,y). If $\mathbf{E} = \hat{y}E(x,y,z,t)$, one needs to add a small electrical field component in the propagation direction z so as to satisfy $\nabla \cdot \mathbf{D} = 0$ or $\nabla \cdot \mathbf{D} = \rho$. So, in our derivation, diffraction is strictly $\frac{\partial^2 E}{\partial x^2}$ but in the final answer we write $\nabla_\perp^2 E = \partial^2 E/\partial x^2 + \partial^2 E/\partial y^2$.

We take as constitutive relations $\mathbf{B} = \mu\mathbf{H}$ and

$$\mathbf{D} = \varepsilon_0 \hat{y}\biggl(E + \int_{-\infty}^{t} \chi(t-\tau)E(\tau)\,\mathrm{d}\tau + \int_{-\infty}^{t} \chi^{(3)}(t-\tau_1, t-\tau_2, t-\tau_3)E(\tau_1)E(\tau_2)E(\tau_3)\,\mathrm{d}\tau_1\,\mathrm{d}\tau_2\,\mathrm{d}\tau_3\biggr). \quad (15)$$

We write the Fourier transforms of $E(x,z,t)$, the linear and nonlinear susceptibility $\chi(t)$ and $\chi^{(3)}(t_1,t_2,t_3)$ to be $e(x,z,\omega) = \int_{-\infty} E(x,z,t)\mathrm{e}^{\mathrm{i}\omega t}$, $\hat{\chi}(\omega)$ and $\hat{\chi}^{(3)}(\omega_1,\omega_2,\omega_3)$ respectively. From Maxwell's equation $\nabla \times \nabla \times \mathbf{E} = \mu(\partial^2\mathbf{D}/\partial t^2)$ and $\mu\varepsilon_0 = 1/c^2$, we obtain the exact relation

$$\frac{\partial^2 e}{\partial x^2} + \frac{\partial^2 e}{\partial z^2} + \frac{\omega^2}{c^2}(1 + \hat{\chi}(\omega))e = -\frac{2\pi\omega^2}{c^2} \int \chi^{(3)}(\omega_1, \omega_2, \omega_3) e(\omega_1) e(\omega_2) e(\omega_3) \times \delta(\omega - \omega_1 - \omega_2 - \omega_3)\, \mathrm{d}\omega_1\, \mathrm{d}\omega_2\, \mathrm{d}\omega_3, \tag{16}$$

where $\delta(x)$ is the Dirac delta function. Anticipating that we will be taking nonlinearity and diffraction to be small when compared to the linear response $(n^2(\omega)\omega^2/c^2)e$, $n^2(\omega) = 1 + \hat{\chi}(\omega)$, we write $e(r, z, \omega)$ as the sum of forward and backward fields. Let

$$e(r, z, \omega) = A(r, z, \omega)\mathrm{e}^{\mathrm{i}k_0(\omega)z} + B(r, z, \omega)\mathrm{e}^{-\mathrm{i}k_0(\omega)z}, \tag{17}$$

where $k_0^2(\omega) = (\omega^2/c^2)(1 + \hat{\chi}_r(\omega))$, $\hat{\chi}_r(\omega)$ the real part of $\hat{\chi}(\omega)$ and make the free choice of a relation between A and B to be

$$\frac{\partial A}{\partial z}\mathrm{e}^{\mathrm{i}k_0(\omega)z} + \frac{\partial B}{\partial z}\mathrm{e}^{\mathrm{i}k_0(\omega)z} = 0$$

(cf. method of variation of parameters). Then substituting (17) into (16) and adding and subtracting the free choice, we obtain the exact relations:

$$2\mathrm{i}k_0(\omega)\frac{\partial A}{\partial z} = -\frac{\mathrm{i}\omega^2}{c^2}\hat{\chi}_i(\omega)A - \frac{\mathrm{i}\omega^2}{c^2}\hat{\chi}_i(\omega)B\mathrm{e}^{-2\mathrm{i}k_0(\omega)z} - \frac{\partial^2 A}{\partial x^2} - \frac{\partial^2 B}{\partial x^2}\mathrm{e}^{-2\mathrm{i}k_0(\omega)z} - \frac{2\pi}{c^2}\omega^2 P\mathrm{e}^{-\mathrm{i}k_0(\omega)z}, \tag{18}$$

where

$$P = \int \hat{\chi}^{(3)}(\omega_1, \omega_2, \omega_3)\delta(\omega - \omega_1 - \omega_2 - \omega_3)\, \mathrm{d}\omega_1\, \mathrm{d}\omega_2\, \mathrm{d}\omega_3 \times (A(\omega_1)\mathrm{e}^{-\mathrm{i}k_0(\omega_1)z}) + B(\omega_1)\mathrm{e}^{-\mathrm{i}k_0(\omega_1)z}(\omega_1 \to \omega_2)(\omega_1 \to \omega_3), \tag{19}$$

and an equivalent equation for $\partial B/\partial z$. These equations are exact. There have been no approximations thus far. We will now ask how can we approximate solutions to (18) and the analogous equation for B if the right-hand side is small? Suppose we set $A = A_0 + A_1 + \cdots$, $B = B_0 + B_1 + \cdots$. To leading order, A_0 and B_0 will be independent of z but, in order to remove secular terms (terms growing as z) in the iterates $A_1, B_1, \ldots$, we will have to choose their slow variations $\partial A_0/\partial z$ and $\partial B_0/\partial z$ accordingly. So the first task is to identify the secular terms in A_1 and B_1. The equation for A_1 is

$$2\mathrm{i}k_0(\omega)\frac{\partial A_1}{\partial z} = -\frac{\mathrm{i}\omega^2}{c^2}\hat{\chi}_i(\omega)A_0 - \frac{\mathrm{i}\omega^2}{c^2}\hat{\chi}_i(\omega)B_0\mathrm{e}^{-2\mathrm{i}k_0(\omega)z} - \frac{\partial^2 A_0}{\partial x^2} - \frac{\partial^2 B_0}{\partial x^2}\mathrm{e}^{-2\mathrm{i}k_0(\omega)z} + P_0\mathrm{e}^{-\mathrm{i}k_0(\omega)z}, \tag{20}$$

where P_0 is P given in (19) with A replaced by A_0. Since, to leading order, A_0 and B_0 are z independent, it is clear that, by direct integration from $z = 0$ to $z = z$, the first and third terms on the right-hand side of (20) are secular whereas the second and fourth, each of whose fast dependence is $((e^{-ik_0(\omega)z} - 1)/ik_0(\omega))$ are not. The more interesting discussion involves the nonlinear term one member of which, when integrated from 0 to z, is

$$\begin{aligned}\int & \hat{\chi}^{(3)}(\omega_1, \omega_2, \omega_3) A_0(\omega_1) A_0(\omega_2) A_0(\omega_3)\\ & \times \left(\frac{\exp\big(ik_0(\omega_1) + k_0(\omega_2) + k_0(\omega_3) - k_0(\omega)\big)z\big) - 1}{i\big(k_0(\omega_1) + k_0(\omega_2) + k_0(\omega_3) - k_0(\omega)\big)}\right)\\ & \times \delta(\omega - \omega_1 - \omega_2 - \omega_3)\, d\omega_1\, d\omega_2\, d\omega_3. \end{aligned} \tag{21}$$

In order to assess whether this term gives rise to secular behavior, we must ask what is its limiting behavior as z becomes large. To answer this, we require a little mathematics. We state two important results. If $f(x)$ is an ordinary (measurable) function which is absolutely integrable and $h(x)$ is not identically zero over a finite interval in the domain of integration, we know that $\lim_{z\to\infty} \int f(x) \exp(ih(x)z)\, dx = 0$ (the Riemann–Lebesgue lemma) and that

$$\lim_{z\to\infty} \int f(x) \Delta\big(h(x)\big)\, dx = \int \left(\pi \operatorname{sgn} z \delta\big(h(x)\big) + iP\left(\frac{1}{h(x)}\right)\right) f(x)\, dx,$$

where $\Delta(h) = (e^{ihz} - 1)/ih$ and P denotes the Cauchy Principal Value. This means that as long as the amplitudes $A_0(\omega), B_0(\omega)$ are ordinary (as opposed to, say, Dirac delta) functions and are absolutely integrable, the nonlinear terms will give a bounded contribution to A_1 and be therefore nonsecular unless $h(\omega_1, \omega_2, \omega_3) = k_0(\omega_1) + k_0(\omega_2) + k_0(\omega_3) - k_0(\omega)$ is identically zero over finite regions of ω_2, ω_3 space (note: $\omega_1 = \omega - \omega_2 - \omega_3$). Namely, the nonlinear term will give no cumulative long distance effect unless h is identically zero. This occurs only when there is no (at least to leading order) longitudinal dispersion. We will comment at the end of this section, what happens if the pulse envelopes are very long and the amplitudes $A_0(\omega)$ have delta function behavior.

For now, we focus on the case of short pulses for which the Fourier transforms $e(x, z, \omega)$ are smooth in ω and for which (21) is only secular when $k_0(\omega_1) + k_0(\omega_2) + k_0(\omega_3) - k_0(\omega)$ is (almost) identically zero; namely, when $k_0(\omega) = n_0(\omega/c)$ for constant $n_0 = \sqrt{1 + \hat{\chi}_r}$. It should now also be clear that any of the other products in P_0 such as those involving $A_0(\omega_1)A_0(\omega_2)B_0(\omega_3)$ with fast z behavior $\Delta\big(k_0(\omega_1) + k_0(\omega_2) - k_0(\omega_3) - k_0(\omega)\big)$ never give rise to secular terms. Thus, the interaction between right and left going pulses is small and does not affect the deformations of either A_0 and B_0.

Having established that for short pulses, the nonlinear terms only give a cumulative long distance effect when $\hat{\chi}(\omega)$ is weakly dependent on ω and $k_0(\omega) = n_0\omega/c$ is nondispersive, we now return to (16) and write $\hat{\chi}(\omega)$ as $\hat{\chi}(\omega) - \hat{\chi}(0) + \hat{\chi}(0)$ and assume that $\hat{\chi}(\omega) - \hat{\chi}(0)$ is small compared to $\hat{\chi}(0)$. This small difference will give rise to weak dispersion and attenuation. We recall that since $E(x, z, t)$ and $D(x, z, t)$ are real, $\hat{\chi}(-\omega) = \hat{\chi}(\omega)$ so that $\hat{\chi}_r(\omega)$ is even in ω and $\hat{\chi}_i(\omega)$ is odd in ω. We will then obtain an equation (18) for $A(x, z, \omega)$ in which $k_0(\omega) = n_0(\omega/c) = \sqrt{1 + \hat{\chi}_r(0)}$ and the susceptibility deviation from $\hat{\chi}(0)$ will manifest itself as the term $\left((\hat{\chi}(0) - \hat{\chi}(\omega))/c^2\right)\omega^2 A$ on the right-hand side of (18).

When we now repeat the analysis, we find the only secular terms arising in A_1 are those for which all fast dependence on z have been removed. To suppress these secular terms in A_1, we allow A_0 to be slowly varying. The result is the equation which is the Fourier transform of SPEE;

$$\begin{aligned} 2in_0\frac{\omega}{c}\frac{\partial A_0}{\partial z} &= \frac{\hat{\chi}(0) - \hat{\chi}(\omega)}{c^2}\omega^2 A_0 - \frac{\partial^2 A_0}{\partial x^2} \\ &\quad - \frac{2\pi\omega^2}{c^2}\int \hat{\chi}^{(3)}(\omega_1, \omega_2, \omega_3)A_0(\omega_1)A_0(\omega_2)A_0(\omega_3) \\ &\quad \times \delta(\omega - \omega_1 - \omega_2 - \omega_3)\,\mathrm{d}\omega_1\,\mathrm{d}\omega_2\,\mathrm{d}\omega_3. \end{aligned} \tag{22}$$

It says, in effect, that the evolution of the right going Fourier amplitude of the electric field in frequency space travels without deformation on distances of the order of many wavelengths but is distorted over much longer distances by a combination of dispersion and attenuation, diffraction and nonlinearity.

The facts that the forward and backward going components can be separated and that $\hat{\chi}(\omega)$ is almost constant suggest that we rewrite (22) as an equation for the leading order component of the electric field

$$E_0\left(x, z, \tau = t - \frac{n_0 z}{c}\right) = \frac{1}{2\pi}\int A_0(x, z, \omega)\mathrm{e}^{-\mathrm{i}\omega(t-(n_0 z/c))}\,\mathrm{d}\omega. \tag{23}$$

We note that $-\mathrm{i}\omega A_0$ has its Fourier transform $\partial E_0/\partial\tau$. Then if we take $\hat{\chi}^{(3)}$ constant, (22) becomes

$$\frac{2n_0}{c}\frac{\partial^2 E_0}{\partial\tau\partial z} + \frac{2\pi\hat{\chi}^{(3)}}{c^2}\frac{\partial^2 E_0^3}{\partial\tau^2} + \mathrm{FT}\left\{\frac{\hat{\chi}(0) - \hat{\chi}(\omega)}{c^2}\omega^2 A_0\right\} = \nabla_\perp^2 E_0 \tag{24}$$

which is (4). We have now replaced $\partial^2 E_0/\partial x^2$ by the full horizontal Laplacian.

How might the NLS equation enter this picture? If the pulse is a long envelope, then the Fourier transform $A(x, z, \omega)$ of the right going pulse will be close to a Dirac delta function centered on the first harmonic carrier frequencies ω_0 and $-\omega_0$. In that case, the integral on the right-hand side of (19) becomes an algebraic expression with frequencies $\pm\omega_0$, $\pm 3\omega_0$ and the exponent we called $h(\omega_1, \omega_2, \omega_3)$ can become zero when $\omega_1 = \omega_0, \omega_3 = -\omega_2 = \pm\omega_0$ leading to the modal interaction term familiar in NLS, $A_0^2(\omega_0)A_0^*(\omega_0)$. But, as we have pointed out in Sect. 2, the terms $A_0^3(\omega_0)$ corresponding to the frequency $3\omega_0$ will give rise to correction A_1 which have the denominator $k_0(3\omega_0) - 3k_0(\omega_0)$ which is small. Indeed as we have already shown, for intensities of interest, the NLS for the envelope of the first harmonic will not correctly describe the field $E(x, z, \tau)$.

We next address, in the final section, what modifications and additions one must make to SPEE in order to include other effects, the most important of which will be the effect of localized plasma generation by a high intensity pulse peak on pulse propagation and absorption. Our goal is to ask whether it might be possible to tailor light pulses to reach high intensities at a specified but distant location on its trajectory path.

4 Challenges, Questions, and Conclusions

Conclusions as to whether SPEE (in forms (4), (5), (8)) in the absence of any losses first leads to infinite gradients or infinite amplitudes are not definitive in all cases. Most evidence suggests that when dispersion in the direction of propagation is normal ($s = 1$ in (8)), electric field gradients dramatically steepen whereas when dispersion is anomalous ($s = -1$ in (8)), the high local intensity singularity obtains. It is therefore important to gain a better understanding on this question.

Without wishing to overmathematize the discussion, let us make several observations. First, if we set $\hat{\chi}(0) - \hat{\chi}(\omega) = 0$ in (4), we obtain dispersionless SPEE (it is just a nonlinear wave equation) and it is easy to see that

$$\frac{\partial}{\partial z}\int (E_\tau)^2\, \mathrm{d}x\, \mathrm{d}y\, \mathrm{d}\tau = -2\varepsilon_{\mathrm{NL}}\int E^2 E_\tau^2 = -\frac{\varepsilon_{\mathrm{NL}}}{2}\int (E^2)_\tau^2\, \mathrm{d}x\, \mathrm{d}y\, \mathrm{d}\tau$$

as, when multiplied by E, the diffraction term integrates to zero. Because the right-hand side remains finite and negative, and the left-hand side is the derivative of an intrinsically positive quantity, it is hard not to believe that at some finite z the integrals must fail to exist which means that the electric field gradient becomes singular. Although diffraction can in principle conspire with nonlinearity to produce infinite amplitudes, it cannot stop gradient steepening. So what happens when we reintroduce $\hat{\chi}(0) - \hat{\chi}(\omega)$? Let us discuss this in the context of (8). The steepening gradient will be regularized either by the dispersion term $-s\varepsilon_{\mathrm{disp}}\partial^3 E/\partial\tau^3$ or by the attenuation $-\varepsilon_{\mathrm{att}}\partial^2 E/\partial\tau^2$. Let us imagine the former dominates. Then a second important observation to make is that, if we ignore transverse effects, the resulting modified KdV equation $E_z + \varepsilon_{\mathrm{NL}}E^2E_\tau - s\varepsilon_{\mathrm{disp}}E_{\tau\tau\tau} = 0$ only has coherent soliton solutions when the dispersion is anomalous ($s = -1$). When dispersion is normal, the steepening gradients do not break into solitons but into fine wiggles with time scales $\tau \approx \sqrt{\varepsilon_{\mathrm{disp}}/\varepsilon_{\mathrm{NL}}}$. This leads to spectral broadening. A third observation then is that the term $-(D/2)\int E_\tau^2\, \mathrm{d}x\, \mathrm{d}y\, \mathrm{d}\tau$ is positive in the virial theorem (11)

($D = -s\varepsilon_{\text{disp}}$) and is of order unity, independent of the size of $\varepsilon_{\text{disp}}$. The possible negativity of the Hamiltonian $\hat{K}$ is counterbalanced and the argument for the occurrences of high intensities weakened.

If attenuation, the Burgers term $\varepsilon_{\text{att}} E_{\tau\tau}$ in (8), dominates, then gradient regularization is achieved by shock creation. Shocks will be of a width (in τ) proportional to ε_{att}. Again spectral broadening will result. Again, if $s = +1$, the term $(D/\tau) \int E_\tau^2 \, dx \, dy \, d\tau$ in the virial theorem (11) counterbalances the negativity of $\hat{K}$ and the formation of large intensities.

These conclusions obtain whether or not the original pulse has few or many oscillations under its envelope. As we have already pointed out, the only difference is whether an initially broad spectrum further broadens to a continuous one or whether an initially narrow spectrum develops all its odd harmonics.

We now turn to the case of collapsing filaments in which case the pulse amplitude becomes locally large and eventually singular. Recall that the basic reason for the localization of light intensity is the presence of a self-focussing nonlinearity where the effective refractive index increases with intensity. In that case, the light rays will turn towards those x, y, τ regions of the pulse which have the largest intensity. As a result, the intensity then becomes even larger and attracts even more light rays. In 1D, this self-focussing saturates in a soliton like object. In 2D, it forms a collapsing filament which, if the envelope were to be described by NLS, has the critical amount of power required to sustain collapse. For SPEE, we do not know what happens at $D = 0$. In 3D, with $D = -s\varepsilon_{\text{disp}} > 0$, the collapse is supercritical and its power decays as $Z - z$ as the singular location $z = Z$ is approached.

We would like to address the following questions: What is the nature of the collapse and does it change in any significant way as $D \to 0$; namely, as D approaches zero through positive values? For $D = 0$, and in the absence of any kind of attenuation, which singularly comes first? What does that choice depend on? When the amplitude singularity is the first to appear either for $D > 0$ or $D = 0$, what is the nature of the regularization which occurs? If $D = 0$, and the amplitude singularity occurs, what is the nature of the collapsing filament? We have shown that in that case

$$E = (Z - z)^{-\alpha} F\left(\rho = \frac{r}{(Z - z)^{1-\alpha}}\right), \quad T = \left(\frac{\tau}{(Z - z)^{1-2\alpha}}\right)$$

with α undetermined by a simple balance argument. How is α determined then? By imposing certain restrictions on the solutions of $F(\rho, T)$ such as demanding that $F_\rho(0, T) = 0$ and $F(\rho, T)$ approaches zero without oscillations as $\rho \to \infty$ so that α is determined from a nonlinear eigenvalue problem. Is it possible that, for $D = 0$, $\alpha = \frac{1}{2}$, so that the power P in the collapsing filament does not tend to zero as $z \to Z$ but stays finite?

We now turn to the nature of the regularization when collapse events occur and the pulse locally develops very large amplitudes, large enough to ionize the gas (e.g., air) through which it is traveling. The ionization process creates a plasma in which a current is initiated and this current adds both to defocussing dispersion (acts similar to diffraction) and attenuation and power absorption. If one re-derives the evolution of E of Sect. 3 including these effects, one essentially has to replace $\varepsilon_0(\partial/\partial\tau)E^3$ by $\varepsilon_0(\partial/\partial\tau)E^3 + j$ on the left-hand side of (8) where j is the plasma current in the direction of the polarization of $\mathbf{E}$. How is j to be calculated? There is no rigorous derivation although M. Kolesik in this volume has developed an interesting approach. The efforts to date essentially use the Drude model for which

$$\frac{\partial j}{\partial t} + \gamma j = \frac{q^2}{m} E\rho, \quad \frac{\partial \rho}{\partial t} = W(I)(\rho_a - \rho) + \frac{\sigma}{U}\rho I, \tag{25}$$

where ρ is the density of electrons of charge q and mass m generated by ionization, I is electric field intensity and $W(I) \sim (I/I_c)^\alpha$, $\alpha \simeq$ 6–7 describes multiphoton ionization, and as $(\sigma/U)\rho I$ captures avalanche ionization. The collision rate γ is extremely important. The reason is its inverse γ^{-1} measures the time it takes for the plasma to form. That time is generally believed to be in the range of picoseconds (10^{-12} s). Therefore while short pulses of femtosecond duration with local intensities I in excess of the ionization threshold I_c will create plasma, that plasma will be created in the wake of the pulse.

The power containing part of the pulse will not see the plasma directly. Therefore there should be little or no defocussing effects on the lead part of the pulse. Unfortunately, it is not true that the lead part of the pulse will escape the effects of plasma absorption. The energy to ionize the gas has got to come from somewhere. For the sake of achieving high intensity pulses traveling to a distance, it would help if that energy could be provided by part of a supercritical collapsing pulse which is radiated away, or the extra power between the initial power P and the critical power in a critical collapse which is also radiated away computations to date in the case of a supercritical collapsing pulse of NLS suggests that the presence of plasma absorption attenuates the central collapsing filament and arrests the growth of large intensities.

So how might one circumvent the absorption of the plasma? I do not have what I can call a good idea as to how one might achieve this. But I do have a number of suggestions which may be worth exploring.

1. Can one find a way to use the excess power between the initial power and the critical power, or the continuously radiating power in a supercritical collapse, to balance the cost of ionization?
2. If normal dispersion keeps the pulse from collapsing and attaining locally high intensities above the ionization threshold until the pulse reaches the target, can one modify the atmosphere in the neighborhood of a target so that, effectively $\hat{\chi}(0) - \hat{\chi}(\omega)$ is zero there? The size of the region to be so affected would be the collapse time multiplied by c.
3. Depending on the initial transverse shape of the beam, it may be possible to create multiple filaments (along a ring for example). Can one overcome losses from plasma absorption by using the powers in developing filaments with subionization intensities to resupply the power of a neighboring high intensity filament. In principle, because the medium is focussing, it should be possible to do this.

Acknowledgement The author is grateful for support from The Air Force contract FA 9550-10-1-US61 and from NSF DMS 1308862.

References

1. A.C. Newell, *Solitons in Mathematics and Physics*. CBMS-NSF Regional Conference Series in Applied Mathematics, vol. 48 (SIAM, Philadelphia, 1985)
2. J.V. Moloney, A.C. Newell, *Nonlinear Optics*, 2nd edn. (Westview Press, Boulder, CO, 2001)
3. E.A. Kuznetsov, S.L. Musher, A.V. Shafarenko, JETP Lett. **37**(5), 241 (1983)
4. S.K. Turitsyn, E.G. Falkovich, Sov. Phys. JETP **62**(1), 146 (1985)
5. A.A. Balakin, A.G. Litvak, V.A. Mironov, S.A. Skobelev, J. Exp. Theor. Phys. **104**(3), 363 (2007)
6. K. Glasner, M. Kolesik, J.V. Moloney, A.C. Newell, Int. J. Optics **2012**, 868274 (2012)
7. O.G. Kosareva, W. Liu, N.A. Panov, J. Bernhardt, Z. Ji, M. Sharifi, R. Li, Z. Xu, J. Liu, Z. Want, J. Ju, X. Lu, Y. Jiang, Y. Leng, X. Liang, V.P. Kandidov, S.L. Chin, Laser Phys. **19**(8), 1776 (2009)
8. A. Debayle, L. Gremillet, L. Bergé, C. Köhler, Opt. Express **22**(11), 13691 (2014)
9. M. Kolesik, J.V. Moloney, Rep. Prog. Phys. **77**(1), 016401 (2014)
10. M. Kolesik, J.V. Moloney, Phys. Rev. E **70**(3), 036604 (2004)
11. D. Alterman, J. Rauch, Phys. Lett. A **264**(5), 390 (2000)
12. T. Schaffer, C.E. Wayne, Physica D **196**(1–2), 90 (2004)
13. G. Luther, J.V. Moloney, A.C. Newell, Physica D **74**(1–2), 59 (1996)

Variants of the Focusing NLS Equation: Derivation, Justification, and Open Problems Related to Filamentation

Éric Dumas, David Lannes, and Jérémie Szeftel

Abstract The focusing cubic NLS is a canonical model for the propagation of laser beams. In dimensions 2 and 3, it is known that a large class of initial data leads to finite-time blow-up. Now, physical experiments suggest that this blow-up does not always occur. This might be explained by the fact that some physical phenomena neglected by the standard NLS model become relevant at large intensities of the beam. Many ad hoc variants of the focusing NLS equation have been proposed to capture such effects. In this paper, we derive some of these variants from Maxwell's equations and propose some new ones. We also provide rigorous error estimates for all the models considered. Finally, we discuss some open problems related to these modified NLS equations.

1 Introduction

The cubic, focusing, nonlinear Schrödinger (NLS) equation in space dimension d is given by

$$\begin{cases} i\partial_\tau v + \Delta v + |v|^2 v = 0\,, & \tau > 0\,,\ x \in \mathbb{R}^d, \\ v(0,x) = v_0(x), & x \in \mathbb{R}^d. \end{cases} \tag{1}$$

It is a canonical model for the propagation of laser beams.

É. Dumas (✉)
Institut Fourier, Université Joseph Fourier, 100, rue des Mathématiques, BP 74, 38402 Saint Martin d'Hères, France
e-mail: Eric.Dumas@ujf-grenoble.fr

D. Lannes
Institut de Mathématiques de Bordeaux, Université de Bordeaux & CNRS UMR 5251, 351, cours de la Libération, France
e-mail: David.Lannes@math.u-bordeaux1.fr

J. Szeftel
Laboratoire Jacques-Louis Lions, Université Pierre et Marie Curie, 4, place Jussieu, 75005 Paris, France
e-mail: Jeremie.Szeftel@upmc.fr

A.D. Bandrauk et al. (eds.), *Laser Filamentation*, CRM Series in Mathematical Physics, DOI 10.1007/978-3-319-23084-9_2

From a result of Ginibre and Velo [1], (1) is locally well-posed in $H^1 = H^1(\mathbb{R}^d)$ for $d = 1, 2, 3$, and thus, for $v_0 \in H^1$, there exist $0 < T \leq +\infty$ and a unique solution $v(\tau) \in \mathcal{C}([0, T), H^1)$ to (1) and either $T = +\infty$, we say the solution is global, or $T < +\infty$ and then $\lim_{\tau \uparrow T} \|\nabla v(\tau)\|_{L^2} = +\infty$, we say the solution blows up in finite time.

The NLS equation (1) also admits the following (formal) conservation laws:

L^2-norm: $\|v(\tau)\|_{L^2}^2 = \|v_0\|_{L^2}^2$;
Energy: $E(v(\tau)) = \frac{1}{2}\int|\nabla v(\tau,x)|^2\,\mathrm{d}x - \frac{1}{4}\int|v(\tau,x)|^4\,\mathrm{d}x = E(v_0)$;
Momentum: $\mathrm{Im}(\int \nabla v(\tau,x)\overline{v(\tau,x)}\,\mathrm{d}x) = \mathrm{Im}(\int \nabla v_0(x)\overline{v_0(x)}\,\mathrm{d})$.

It is also known that a large group of symmetries leaves the equation invariant: if $v(\tau, x)$ solves (1), then $\forall(\lambda_0, \tau_0, x_0, \beta_0, \gamma_0) \in \mathbb{R}_*^+ \times \mathbb{R} \times \mathbb{R}^d \times \mathbb{R}^d \times \mathbb{R}$, so does

$$u(\tau,x) = \lambda_0 v(\lambda_0^2\tau + \tau_0, \lambda_0 x + x_0 - \beta_0 t)e^{\left(i\frac{\beta_0}{2}\cdot\left(x-\frac{\beta_0}{2}\tau\right)\right)}e^{i\gamma_0}. \tag{2}$$

The scaling symmetry $u(\tau, x) = \lambda_0 v(\lambda_0^2\tau, \lambda_0 x)$ leaves the homogeneous Sobolev space $\dot{H}^{s_c}(\mathbb{R}^d)$ invariant, where $s_c = d/2 - 1$.

Referring to conservation of the L^2-norm by the flow, (1) is said to be L^2-subcritical if $s_c < 0$, L^2-critical if $s_c = 0$, and L^2-supercritical if $s_c > 0$. Thus, (1) is L^2-subcritical if $d = 1$, L^2-critical if $d = 2$, and L^2-supercritical if $d \geq 3$. In the subcritical case, global existence (in $\mathcal{C}([0, \infty), L^2)$) holds for arbitrarily large data in L^2. It turns out that in this case, global existence (in $\mathcal{C}([0, \infty), H^1)$) also holds for arbitrarily large data in H^1, due to the conservation of mass and energy. In the critical and supercritical cases however, there exists stable finite-time blow-up dynamics. This has been known since the 60s using global obstructive arguments based on the virial identity (see, e.g., [2]).[1]

There is, however, a discrepancy between the blow-up results predicted by (1) and physical observations. Indeed, while the blow-up signifies a break-down of the solution v, physical observations show in many cases that lasers begin to focus according to the scenarios associated with (1) but depart from this behavior slightly before the focusing time. The reason advanced by physicists is that some physical phenomena that have been neglected to derive (1) become relevant at high intensities, and therefore near focusing. This phenomenon is called *filamentation*: defocusing physical phenomena are triggered at high intensities and halt the collapse of the beam. This interplay between diffraction, self-focusing, and defocusing mechanisms allows for the beam to propagate along several times the focusing distance (called Rayleigh length in optics) and the resulting structure is called *filament*.

Many variants of (1) have been derived in optics to take into account these additional physical phenomena and reproduce the filamentation mechanism. In many

[1] Much more is known about the finite-time blow-up dynamics for the focusing NLS and we refer the interested reader to [3–10] and the references therein.

cases, it is a mathematical open problem to prove whether these additional terms prevent focusing or not, and a fortiori to understand the modification of the dynamics induced by them.

Rather than adding as usual ad hoc modifications to (1) in order to take new physical effects into account, we choose here to rigorously derive such modifications from Maxwell's equations. We then comment on some of the most physically relevant open mathematical problems that these modified equations raise and that are natural milestones towards the understanding of filamentation. These variants can roughly be classified into two groups, depending on whether they take ionization processes into account or not.

Notation 1. In the brief presentation below, we denote by z the direction of propagation of the laser, by $X_\perp = (x, y)$ the transverse variables, and by $\Delta_\perp = \partial_x^2 + \partial_y^2$ the transverse Laplace operator. In dimension $d = 2$, the variable y is omitted (and hence $\Delta_\perp = \partial_x^2$), while in dimension $d = 1$, functions depend only on z (so that $\Delta_\perp = 0$).

1.1 Models Without Ionization Processes

We give below a family of variants to (1) that incorporate many physical phenomena neglected by (1). It is of course possible to look at one or several of these additional effects simultaneously. We state the equations in their most general form, starting with a family of *scalar* NLS equations, and then give the corresponding *vectorial*—and more general—form of these equations. Let us therefore consider

$$iP_2(\varepsilon\nabla)\partial_\tau v + (\Delta_\perp + \alpha_1\partial_z^2)v + i\alpha_2 v + (1 + i\varepsilon\boldsymbol{\alpha}_3\cdot\nabla)\big[\big(1 + f(\varepsilon^r|v|^2)\big)|v|^2 v\big] = 0, \quad (3)$$

where v is a complex-valued function. Here, $\varepsilon > 0$ is a (small) parameter; $P_2(\varepsilon\nabla)$ is a second (at most) order, self-adjoint, positive operator; $\alpha_1 = 0, \pm 1$; $\alpha_2 \geq 0$; $\boldsymbol{\alpha}_3 \in \mathbb{R}^d$; $f\colon \mathbb{R}^+ \to \mathbb{R}$ is a smooth mapping vanishing at the origin, and $r > 0$. The physical meaning of these terms is commented below:

1. Nonlinearity. The cubic nonlinearity in NLS corresponds to a first order approximation of the nonlinear optical phenomena. At high intensities, it is often worth including some next order terms captured here by the additional term $f(\varepsilon^r|v|^2)$. We consider here three situations:

   (a) Cubic nonlinearity: $f = 0$.
   (b) Cubic/quintic nonlinearity: $f(r) = -r$.
   (c) Saturated nonlinearity: f is a smooth function on $\mathbb{R}^+$ vanishing at the origin and such that $\big(1 + f(r)\big)r$ is bounded on $\mathbb{R}^+$ (e.g., $f(r) = -r/(1 + r)$).

2. Group velocity dispersion (GVD). The coefficient α_1 accounts for the dispersion of the group velocity and three different situations are possible:

(a) No GVD: $\alpha_1 = 0$.
(b) Anomalous GVD: $\alpha_1 = 1$.
(c) Normal GVD: $\alpha_1 = -1$.

3. Damping. The coefficient α_2 accounts for damping phenomena:

   (a) No damping: $\alpha_2 = 0$.
   (b) Damping: $\alpha_2 > 0$.

4. Off-axis variations of the group velocity. The operator $P_2(\varepsilon\nabla)$ is here to account for the fact that self-focusing pulses become asymmetric due to the variation of the group velocity of off-axis rays.[2] The operator $P_2(\varepsilon\nabla)$ is a second order, self-adjoint, and positive operator in the sense that $(P_2(\varepsilon\nabla)u, u) \geq C|u|_*^2$, with $|u|_*^2 \geq |u|_{L^2}$. The norm $|\cdot|_*$ may also control derivatives of u; we consider three cases:

   (a) No off-axis dependence: $P_2(\varepsilon\nabla) = 1$, and therefore $|\cdot|_* = |\cdot|_{L^2}$.
   (b) Full off-axis dependence: the norm $|\cdot|_*$ controls all first order derivatives, $|u|_*^2 \sim |u|_{L^2} + \varepsilon^2|\nabla u|_{L^2}$.
   (c) Partial off-axis dependence: the norm $|\cdot|_*$ controls some but not all first order derivatives. More precisely, there exist j ($j < d$) linearly independent vectors $\mathbf{v}_j \in \mathbb{R}^d$ such that $|u|_*^2 \sim |u|_{L^2} + \varepsilon^2 \sum_{k=1}^{j} |\mathbf{v}_k \cdot \nabla u|_{L^2}$.

5. Self-steepening of the pulse. The operator $(1+i\varepsilon\boldsymbol{\alpha}_3\cdot\nabla)$ in front of the nonlinearity accounts for off-axis dependence of the nonlinearity, responsible for the possible formation of *optical shocks*. Various cases are considered here:

   (a) No self-steepening. This corresponds to $\boldsymbol{\alpha}_3 = 0$ and to the usual situation where the nonlinearity does not contain any derivative.
   (b) Longitudinal self-steepening. When $\boldsymbol{\alpha}_3$ is colinear to $\mathbf{e}_z$, there is a derivative in the nonlinearity along the direction z of propagation of the laser.
   (c) Transverse self-steepening. When $\boldsymbol{\alpha}_3 \neq 0$ and $\boldsymbol{\alpha}_3 \cdot \mathbf{e}_z = 0$, there is a derivative in the nonlinearity along a direction orthogonal to the direction of propagation.
   (d) Oblique self-steepening. When $\boldsymbol{\alpha}_3$ is neither colinear nor orthogonal to $\mathbf{e}_z$.

Remark 1. The standard Schrödinger equation (1) is obtained with $P_2(\varepsilon\nabla) = 1$, $\alpha_2 = 0, \boldsymbol{\alpha}_3 = 0, f = 0$, and $\alpha_1 = 1$. Using the above terminology, it corresponds to a cubic nonlinearity, without damping terms, off-axis variation of the group velocity and self-steepening, and with anomalous GVD.

As previously said, (3) stems from a more general *vectorial* equation. For the sake of simplicity, we give here the equation corresponding to the cubic case (or

[2] This phenomenon is often referred to in optics as *space-time focusing* [11]; we do not use this terminology here because this would be misleading. Indeed, physicists usually take z as the evolution variable and treat t as a space variable. This amounts to permuting t and z in (3) and elsewhere.

$f = 0$ in (3)0):

$$iP_2(\varepsilon\nabla)\partial_\tau \mathbf{v} + (\Delta_\perp + \alpha_1\partial_z^2)\mathbf{v} + i\alpha_2\mathbf{v} + \tfrac{1}{3}(1 + i\varepsilon\boldsymbol{\alpha}_3\cdot\nabla)[(\mathbf{v}\cdot\mathbf{v})\overline{\mathbf{v}} + 2|\mathbf{v}|^2\mathbf{v}] = 0, \quad (3)_{\text{vect}}$$

where $\mathbf{v}$ is now a $\mathbb{C}^2$-valued function.

Remark 2. Equation (3) is in fact a particular case of $(3)_{\text{vect}}$ corresponding to initial data living on a one-dimensional subspace of $\mathbb{C}^2$. Indeed, if the initial condition to $(3)_{\text{vect}}$ has the form $\mathbf{v}_{|_{\tau=0}} = v^0(x)\mathbf{v}_0$ with $\mathbf{v}_0 \in \mathbb{R}^2$ and v^0 a scalar-valued function, then the solution to $(3)_{\text{vect}}$ takes the form $\mathbf{v}(\tau, x) = v(\tau, x)\mathbf{v}_0$, where v solves (3) with initial condition v^0.

1.2 *Models with Ionization Processes*

In addition to the physical phenomena taken into account in (3), it is necessary at high intensities to include ionization processes for a correct description of the laser pulse. The reason why this phenomenon is singled out here is because a system of two equations must be considered instead of the single equation (3). In the most simple case (i.e., $P_2 = 1, f = 0, \alpha_2 = 0, \boldsymbol{\alpha}_3 = 0$; for a more general model, see (50)), this system is given by

$$\begin{cases} i(\partial_t + c_g\partial_z)u + \varepsilon(\Delta_\perp + \alpha_1\partial_z^2)u + \varepsilon(|u|^2 - \rho)u = -i\varepsilon c(\alpha_4|u|^{2K-2}u + \alpha_5\rho u)\,, \\ \partial_t\rho = \varepsilon\alpha_4|u|^{2K} + \varepsilon\alpha_5\rho|u|^2, \end{cases} \quad (4)$$

with $\alpha_4, \alpha_5 \geq 0$, $c > 0$, and where ρ is the density of electrons created by ionization, while $\mathbf{c}_g = c_g\mathbf{e}_z$ is the group velocity associated with the laser pulse.

The system (4) does not directly compare to (3) and $(3)_{\text{vect}}$; indeed, (4) is written in the fixed frame of the laboratory, while (3) and $(3)_{\text{vect}}$ are written in a frame moving at the group velocity $\mathbf{c}_g = c_g\mathbf{e}_z$ and with respect to a rescaled time $\tau = \varepsilon t$. Rather than (4), the NLS equation with ionization used in the physics community is its version written in the same variables as (3) and $(3)_{\text{vect}}$. More precisely, if we set

$$u(t, X_\perp, z) = v(\varepsilon t, X_\perp, z - c_g t), \quad \rho(t, X_\perp, z) = \tilde{\rho}(\varepsilon t, X_\perp, z - c_g t),$$

and $\tau = \varepsilon t$, Eq. (4) are *approximated*[3] by the following ones,

[3]The approximation lies in the equation on ρ. In the new variables, the second equation of (4) is given by

$$\varepsilon\partial_\tau\tilde{\rho} - c_g\partial_z\tilde{\rho} = \varepsilon\alpha_4|v|^{2K} + \varepsilon\alpha_5\tilde{\rho}|v|^2.$$

In the physics literature, the term $\varepsilon\partial_\tau\tilde{\rho}$ is neglected, and this corresponds to (5).

$$\begin{cases} i\partial_\tau v + (\Delta_\perp + \alpha_1\partial_z^2)v + (|v|^2 - \tilde{\rho})v = -ic(\alpha_4|v|^{2K-2}v + \alpha_5\tilde{\rho}v), \\ -c_g\partial_z\tilde{\rho} = \varepsilon\alpha_4|v|^{2K} + \varepsilon\alpha_5\tilde{\rho}|v|^2. \end{cases} \tag{5}$$

The ionization processes taken into account by the systems (4) and (5) are:

1. Photo-ionization. This corresponds to $\alpha_4 > 0$ and $K > 0$ (K is the number of photons necessary to liberate one electron).
2. Collisional ionization. When $\alpha_5 > 0$, a term corresponding to collisional ionization is added to the evolution equation on ρ.

Remark 3. 1. The coupling with ρ can of course be added to any equation of the family (3).
2. A vectorial variant of (4) and (5) can also be derived in the same lines as $(3)_{\text{vect}}$.
3. When $\alpha_4 = \alpha_5 = 0$ and $\rho_{|_{t=0}} = 0$ (respectively $\lim_{z\to-\infty}\tilde{\rho} = 0$) one recovers (3) from (4) (respectively (5)).

The rest of the paper is as follows. In Sect. 2, we recall the Maxwell equations, we give an abstract formulation and we discuss the spaces of initial data used for the Cauchy problem. In Sect. 3, we prove our main result about the rigorous derivation of general abstract versions of the models (3), $(3)_{\text{vect}}$, and (4), (5). In Sect. 4, we analyze the role of the various parameters in (3) and (4) or (5). In particular, we consider whether they indeed prevent the breakdown in finite time or not. We also formulate a number of interesting open problems for these modified NLS equations. Finally, an appendix contains explicit computations for a physically relevant system of Maxwell's equations which show that the abstract models derived in Sect. 3 take indeed the form of (3), $(3)_{\text{vect}}$, and (4).

1.3 Notations

We denote by

- $\mathcal{F}u(\xi) = \hat{u}(\xi)$, the Fourier transform of u with respect to the space variables $x \in \mathbb{R}^d$.
- $\mathcal{F}_t u(\omega)$, the Fourier transform of u with respect to the time variable t.
- $f(D)$, the Fourier multiplier with symbol $f(\xi)$: $\widehat{f(D)u}(\xi) = f(\xi)\hat{u}(\xi)$.
- $f(D_t)$, Fourier multipliers with respect to time.
- $\Lambda = (1-\Delta)^{1/2}$, the Fourier multiplier with symbol $(1+|\xi|)^{1/2}$.

2 The Maxwell Equations and an Abstract Mathematical Formulation

2.1 The Maxwell Equations

The Maxwell equations in a non-magnetizable medium are a set of two equations coupling the evolution of the *magnetic field* $\mathtt{B}$ to the *electric induction* $\mathtt{D}$,

$$\begin{cases} \partial_t \mathtt{B} + \operatorname{curl} \mathtt{E} = 0, \\ \partial_t \mathtt{D} - \dfrac{1}{\mu_0} \operatorname{curl} \mathtt{B} = -\mathtt{J}, \end{cases} \tag{6}$$

where $\mathtt{D}$ is given in terms of the *electric field* $\mathtt{E}$ and a *polarization* $\mathtt{P}$—modeling the way the dipole moment per unit volume depends on the strength of the electric field—by the relation

$$\mathtt{D} = \epsilon_0 \mathtt{E} + \mathtt{P}, \tag{7}$$

and where we used standard notations ϵ_0 and μ_0 for the *electric permittivity* and *magnetic permeability* in vacuum. The evolution equations (6) go along with two constitutive laws,

$$\nabla \cdot \mathtt{D} = \rho, \quad \nabla \cdot \mathtt{B} = 0, \tag{8}$$

where ρ is the electric charge density.

As a consequence of the relation $\nabla \cdot \mathtt{D} = \rho$ and the second equation of (6) we get the continuity equation coupling ρ to the *current density* $\mathtt{J}$,

$$\partial_t \rho + \nabla \cdot \mathtt{J} = 0. \tag{9}$$

Introducing the speed of light in vacuum

$$c = \frac{1}{\sqrt{\epsilon_0 \mu_0}},$$

Eq. (6) can also be rewritten as a set of two evolution equations on the magnetic field $\mathtt{B}$ and the electric field $\mathtt{E}$,

$$\begin{cases} \partial_t \mathtt{B} + \operatorname{curl} \mathtt{E} = 0, \\ \partial_t \mathtt{E} - c^2 \operatorname{curl} \mathtt{B} = -\dfrac{1}{\epsilon_0} \partial_t \mathtt{P} - \dfrac{1}{\epsilon_0} \mathtt{J}. \end{cases} \tag{10}$$

In order to get a closed system of equations, we still need two physical informations:

1. A description of the polarization response to the electric field $\mathtt{E}$.
2. A description of the current density $\mathtt{J}$.

We first address the description of the polarization response in absence of current density and then proceed to describe the modification to be made when current density is included.

2.1.1 The Polarization Response to the Electric Field

Throughout this section, we assume that there is no charge nor current density ($\rho = 0, \mathtt{J} = 0$). The general case will be handled in Sect. 2.1.2 below.

There exist various ways to describe the polarization $\mathtt{P}$; we use here a simple and natural model called "nonlinear anharmonic oscillator" according to which the polarization is found by solving the second order ODE

$$\partial_t^2 \mathtt{P} + \Omega_1 \partial_t \mathtt{P} + \Omega_0^2 \mathtt{P} - \nabla V_{\mathrm{NL}}(\mathtt{P}) = \epsilon_0 b \mathtt{E}, \tag{11}$$

where $b \in \mathbb{R}$ is a coupling constant and $\Omega_0, \Omega_1 > 0$ are frequencies, and where V_{NL} accounts for nonlinear effects. When such effects are neglected, the description (11) goes back to Lorentz [12] and expresses the fact that electrons are bound to the nucleus by a spring force. Nonlinearities have been added to this description by Bloembergen [13] and Owyyoung [14] and the mathematical investigation of their influence was initiated by Donnat et al. [15, 16] (see also [17]).

Remark 4. In physics books, the polarization $\mathtt{P}$ is often sought as an expansion

$$\mathtt{P} = \epsilon_0 [\chi^1[\mathtt{E}] + \chi^2[\mathtt{E}, \mathtt{E}] + \chi^3[\mathtt{E}, \mathtt{E}, \mathtt{E}] + \cdots],$$

where the operator χ^1 is called the *linear susceptibility* of the material, while for $j > 1$, the operators χ^j are the *jth order nonlinear susceptibilities*. It is easy to check that the linear susceptibility corresponding to (11) is given by the nonlocal (in time) operator

$$\chi^1[\mathtt{E}] = \chi^1(D_t)\mathtt{E} \quad \text{with } \chi^1(\omega) = \frac{b}{\Omega_0^2 - \omega^2 + \mathrm{i}\Omega_1 \omega},$$

where we used the Fourier multiplier notation,

$$\mathcal{F}_t[\chi^1(D_t)\mathtt{E}](\omega) = \chi^1(\omega)\mathcal{F}_t\mathtt{E}(\omega).$$

Example 1. Typical examples for $V_{\rm NL}(\mathtt{P})$ are

1. Cubic nonlinearity:

$$V_{\rm NL}(\mathtt{P}) = \frac{a_3}{4}|\mathtt{P}|^4 \quad \text{and therefore} \quad \nabla V_{\rm NL}(\mathtt{P}) = a_3|\mathtt{P}|^2\mathtt{P}.$$

2. Cubic/quintic nonlinearity:

$$V_{\rm NL}(\mathtt{P}) = \frac{a_3}{4}|\mathtt{P}|^4 - \frac{a_5}{6}|\mathtt{P}|^6 \quad \text{and therefore} \quad \nabla V_{\rm NL}(\mathtt{P}) = a_3|\mathtt{P}|^2\mathtt{P} - a_5|\mathtt{P}|^4\mathtt{P}.$$

3. Saturated nonlinearity: there exists a function $v_{\rm sat}\colon \mathbb{R}^+ \to \mathbb{R}$, with $v'_{\rm sat}$ and $v''_{\rm sat}$ bounded on $\mathbb{R}^+$ and such that

$$V_{\rm NL}(\mathtt{P}) = \tfrac{1}{2}v_{\rm sat}(|\mathtt{P}|^2) \quad \text{and therefore} \quad \nabla V_{\rm NL}(\mathtt{P}) = v'_{\rm sat}(|\mathtt{P}|^2)\mathtt{P};$$

for instance, one can take

$$v_{\rm sat}(r) = \frac{a_3}{2}\frac{r^2}{1+(2a_5/3a_3)r},$$

in which case $\nabla V_{\rm NL}(\mathtt{P}) = a_3|\mathtt{P}|^2\mathtt{P} - a_5|\mathtt{P}|^4\mathtt{P} + \text{h.o.t}$, and is therefore the same at the origin as in (2) above, up to higher-order terms (seventh-order terms here).

We show in section "The Case Without Charge nor Current Density" in Appendix 1 that Maxwell's equations can be put under the following dimensionless form[4] for all the nonlinearities considered in Example 1,

$$\begin{cases} \partial_t \mathtt{B} + \operatorname{curl} \mathtt{E} = 0, \\ \partial_t \mathtt{E} - \operatorname{curl} \mathtt{B} + \dfrac{1}{\varepsilon}\sqrt{\gamma}\mathtt{Q}^\sharp = 0, \\ \partial_t \mathtt{Q}^\sharp + \varepsilon^{1+p}\omega_1 \mathtt{Q}^\sharp - \dfrac{1}{\varepsilon}(\sqrt{\gamma}\mathtt{E} - \omega_0 \mathtt{P}^\sharp) = \varepsilon\dfrac{\gamma}{\omega_0^3}\big(1 + f(\varepsilon^r|\mathtt{P}^\sharp|^2)\big)|\mathtt{P}^\sharp|^2\mathtt{P}^\sharp, \\ \partial_t \mathtt{P}^\sharp - \dfrac{1}{\varepsilon}\omega_0 \mathtt{Q}^\sharp = 0, \end{cases} \tag{12}$$

where γ, ω_0, ω_1, r, and p are constants, $0 < \varepsilon \ll 1$ is a small parameter (the ratio of the duration of an optical cycle over the duration of the pulse, see section "The Case Without Charge nor Current Density" in Appendix 1), while f is a smooth function vanishing at the origin and

$$\mathtt{P}^\sharp = \frac{\omega_0}{\sqrt{\gamma}}\mathtt{P}, \quad \mathtt{Q}^\sharp = \frac{\varepsilon}{\omega_0}\partial_t \mathtt{P}^\sharp = \frac{\varepsilon}{\sqrt{\gamma}}\partial_t \mathtt{P}.$$

[4]The constitutive laws (8) are omitted because they are propagated by the equations if they are initially satisfied.

2.1.2 The Case with Charge and Current Density

The main mechanism at stake in laser filamentation is certainly the local ionization of the medium: once a powerful self-focusing laser beam reaches high enough intensities, it ionizes the medium around itself. It leaves behind a narrow channel of plasma, hereby causing local defocusing that prevents blowup.

Taking current density into account, we come back to the set of Eqs. (10)–(11), and a physical description of the current density $\mathtt{J}$ is needed. This current density has the form

$$\mathtt{J} = \mathtt{J}_\mathrm{e} + \mathtt{J}_\mathrm{i}, \tag{13}$$

where $\mathtt{J}_\mathrm{e}$ and $\mathtt{J}_\mathrm{i}$ are, respectively, the free electron and ionization current densities.

Free Electron Current Density

Partial ionization of the material medium by the laser generates free electrons, with charge q_e ($= -1.6 \times 10^{-19}$ C). This induces a free electron current density $\mathtt{J}_\mathrm{e} = q_\mathrm{e}\rho_\mathrm{e}v_\mathrm{e}$, where ρ_e is the electron density, and v_e is the electron velocity. A rough,[5] but standard model in nonlinear optics is to take (see [19] and the references therein),

$$\mathtt{E}(t, X) \sim E_{01}(t, X)\mathrm{e}^{\mathrm{i}(\mathbf{k}_\mathrm{l}\cdot X - \omega_\mathrm{l} t)} + \mathrm{c.c.}, \tag{14}$$

where $\mathbf{k}_\mathrm{l}$ and ω_l are the laser wave number and pulsation, respectively, with $|\partial_t E_{01}| \ll |\omega_\mathrm{l} E_{01}|$ (slowly varying envelope approximation (SVEA)); the polarization current is then taken under the form

$$\mathtt{J}_\mathrm{e} \sim J_{01}(t, X)\mathrm{e}^{\mathrm{i}(\mathbf{k}_\mathrm{l}\cdot X - \omega_\mathrm{l} t)} + \mathrm{c.c.} \quad \text{with} \quad J_{01} = \mathrm{i}\frac{q_\mathrm{e}^2}{\omega_\mathrm{l} m_\mathrm{e}}\rho_\mathrm{e} E_{01}, \tag{15}$$

where m_e is the electron mass. The drawback of this model is that it assumes that the electric field and the current density field can be written at leading order as wave packets (i.e., are given under the form (14)–(15)). In particular, it does not provide any relation between the current density $\mathtt{J}$ and the electric field $\mathtt{E}$ that could be used in Maxwell's equations (10). We therefore propose here such a relation, namely,

$$\mathtt{J}_\mathrm{e} = \frac{q_\mathrm{e}^2}{\omega_\mathrm{l} m_\mathrm{e}}\mathscr{H}\left(\frac{\mathbf{k}_\mathrm{l}}{k_\mathrm{l}^2}\cdot D\right)(\rho_\mathrm{e}\mathtt{E}),$$

[5] This approximation can be deduced formally by assuming that ions are at rest and that electron motion is described by the compressible Euler system (see, for instance, [18]). Neglecting electron collisions, such a model yields $\partial_t \mathtt{J}_\mathrm{e} = (q_\mathrm{e}^2/m_\mathrm{e})\rho_\mathrm{e}\mathtt{E}$ which formally yields (15) assuming that $\mathtt{E}$ is as in (14) and that ρ_e is not oscillating at leading order. It would of course be interesting to provide a rigorous justification to these approximations.

where $k_l = |\mathbf{k}_l|$, and $\mathcal{H}$ is the regularization of the Hilbert transform given by the Fourier multiplier

$$\mathcal{H}(D_z) = \frac{\sqrt{2}\, iD_z}{(1 + D_z^2)^{1/2}}. \tag{16}$$

Quite obviously, this is consistent with the usual model (15) since this latter is recovered at leading order when the electric field is a wave packet under the form (14).

Finally, the evolution of the electron density ρ_e is given by a source term S representing external plasma sources. Taking into account photo-ionization and collisional ionization, but neglecting electron recombination (see, for instance, [18] for richer models), we have

$$S = W(I)(\rho_{nt} - \rho_e) + \frac{\sigma}{U_i}\rho_e I,$$

where the intensity is $I = |\mathrm{E}|^2$ and ρ_{nt} is the constant density of neutral species. In the regime considered here,[6] ρ_e is negligible compared to ρ_{nt} and the photo-ionization rate $W(I)$ takes the form

$$W(I) = \sigma_K I^{2K},$$

for some constant coefficient $\sigma_K > 0$ and with $K > 1$ the number of photons needed to liberate one electron. The collisional ionization cross-section σ depends on the laser frequency, and U_i is the ionization potential. Summing up, we get the following expression for the free electron current J_e and $\rho = \rho_e$,

$$\begin{cases} \mathrm{J}_e = \dfrac{q_e^2}{\omega_l m_e}\mathcal{H}\left(\dfrac{\mathbf{k}_l}{k_l^2}\cdot D\right)(\rho \mathrm{E}), \\ \partial_t \rho = \sigma_K \rho_{nt}|\mathrm{E}|^{2K} + \dfrac{\sigma}{U_i}\rho|\mathrm{E}|^2. \end{cases} \tag{17}$$

2.1.3 Ionization Current Density

It is also necessary to take into account losses due to photo-ionization. We therefore introduce an ionization current density J_i such that $\mathrm{J}_i \cdot E$ represents the energy lost by the laser to extract electrons (per time and volume unit). More precisely, $\mathrm{J}_i \cdot E$ is equal to the energy necessary to extract one electron (given by the ionization

[6]For higher intensities, electrons can tunnel out the Coulomb barrier of atoms, and $W(I)$ is modified.

potential U_i) multiplied by the number of electrons per time and volume unit (given by $\partial_t \rho$). Using the second equation of (17), this gives

$$\mathtt{J}_\mathrm{i} \cdot \mathtt{E} = U_\mathrm{i}\sigma_K \rho_\mathrm{nt} |\mathtt{E}|^{2K} + \sigma\rho|\mathtt{E}|^2.$$

We therefore take

$$\mathtt{J}_\mathrm{i} = (U_\mathrm{i}\sigma_K \rho_\mathrm{nt} |\mathtt{E}|^{2K-2} + \sigma\rho)\mathtt{E}. \tag{18}$$

We show in section "The Case with Charge and Current Density" in Appendix 1 that after nondimensionalization, the set of equations (10), (11), (13), (17), (18) (for the nonlinearities considered in Example 1) becomes

$$\begin{cases} \partial_t \mathtt{B} + \operatorname{curl} \mathtt{E} = 0, \\ \partial_t \mathtt{E} - \operatorname{curl} \mathtt{B} + \dfrac{1}{\varepsilon}\sqrt{\gamma}\mathtt{Q}^\sharp \\ \quad = -\varepsilon \mathscr{H}\left(\varepsilon \dfrac{\mathbf{k}}{k^2} \cdot D_x\right)(\rho \mathtt{E}) - \varepsilon c_0 \big(c_1 |\mathtt{E}|^{2K-2} + c_2 \rho\big)\mathtt{E}, \\ \partial_t \mathtt{Q}^\sharp + \varepsilon^{1+p}\omega_1 \mathtt{Q}^\sharp - \dfrac{1}{\varepsilon}(\sqrt{\gamma}\mathtt{E} - \omega_0 \mathtt{P}^\sharp) = \varepsilon \dfrac{\gamma}{\omega_0^3}\big(1 + f(\varepsilon^r |\mathtt{P}^\sharp|^2)\big)|\mathtt{P}^\sharp|^2 \mathtt{P}^\sharp, \\ \partial_t \mathtt{P}^\sharp - \dfrac{1}{\varepsilon}\omega_0 \mathtt{Q}^\sharp = 0, \\ \partial_t \rho = \varepsilon c_1 |\mathtt{E}|^{2K} + \varepsilon c_2 \rho |\mathtt{E}|^2, \end{cases} \tag{19}$$

with the same notations as in (12) for the constants $\gamma, \omega_0, \omega_1, r$, and p, the small parameter ε, and the function f. In addition, we have here constants $c_0, c_1, c_2 \geq 0$, and we also recall that the definition of the regularized Hilbert transform $\mathscr{H}$ is given in (16).

2.2 *Abstract Formulations*

2.2.1 The Case Without Charge nor Current Density

We show in section "The Case Without Charge nor Current Density" in Appendix 1 that the Maxwell equations can be put under the dimensionless form (12), which itself has the form

$$\partial_t \mathbf{U} + A(\partial)\mathbf{U} + \frac{1}{\varepsilon}E\mathbf{U} + \varepsilon^{1+p}A_0\mathbf{U} = \varepsilon F(\varepsilon, \mathbf{U}), \tag{20}$$

where $\mathbf{U}$ is an $\mathbb{R}^n$ ($n \geq 1$) valued function depending on the time variable t and the space variable $x \in \mathbb{R}^d$ ($d \geq 1$),

$$\mathbf{U}: (t, x) \in \mathbb{R} \times \mathbb{R}^d \to \mathbb{R}^n.$$

The operator $A(\partial)$ is defined as

$$A(\partial) = \sum_{j=1}^{d} A_j \partial_j,$$

where ∂_j is the differentiation operator with respect to the jth space coordinate. The matrix A_0 has size $n \times n$, p is a positive number, and the following assumption is made on the matrices A_j and E, and on the nonlinearity F.

Assumption 1. (i) *The matrices A_j ($j = 1, \ldots, d$) are constant coefficient $n \times n$, real-valued,* symmetric *matrices.*

(ii) *The matrix E is a constant coefficient $n \times n$, real-valued,* skew symmetric *matrix.*

(iii) *There exists a smooth mapping $f: \mathbb{R}^+ \to \mathbb{R}$ vanishing at the origin, a real number $r > 0$, a quadratic form $Q: \mathbb{C}^n \mapsto \mathbb{R}^+$, and a trilinear symmetric mapping $T: (\mathbb{C}^n)^3 \to \mathbb{C}^n$ (with $T(\mathbb{R}^n \times \mathbb{R}^n \times \mathbb{R}^n) \subset \mathbb{R}^n$) such that*

$$\forall U \in \mathbb{C}^n, \quad F(\varepsilon, U) = \big(1 + f\big(\varepsilon^r Q(U)\big)\big)\, T(U, \overline{U}, U).$$

Remark 5. There exist of course situations where the leading order of the nonlinearity is not cubic (it can be quadratic for non-centro-symmetric crystals for instance) or not of this form; since we are interested here in deriving variants of the standard cubic NLS equation, we restrict ourselves to this framework for the sake of simplicity.

Example 2. As previously said, the dimensionless version (12) of the Maxwell equations can be put under the form (20) and it satisfies Assumption 1 with $n = 12$, $\mathbf{U} = (\mathrm{B}, \mathrm{E}, \mathrm{Q}^\sharp, \mathrm{P}^\sharp)^T$. See Appendix 2 for more details.

2.2.2 The Case with Charge and Current Density

As shown in section "The Case with Charge and Current Density" in Appendix 1, the system (19) of Maxwell's equations with partial ionization can be put under the general form

$$\begin{cases} \partial_t \mathbf{U} + A(\partial)\mathbf{U} + \dfrac{1}{\varepsilon} E\mathbf{U} + \varepsilon^{1+p} A_0 \mathbf{U} \\ \quad = \varepsilon F(\varepsilon, \mathbf{U}) - \varepsilon \mathcal{H}\Big(\varepsilon \dfrac{\mathbf{k}}{k^2} \cdot D\Big)(\mathbf{W} C_1^T C_1 \mathbf{U}) - \varepsilon c\, C_1^T G(C_1 \mathbf{U}, \mathbf{W}), \\ \partial_t \mathbf{W} = \varepsilon G(C_1 \mathbf{U}, \mathbf{W}) \cdot C_1 \mathbf{U}, \end{cases} \tag{21}$$

where, as in Sect. 2.2.1, $\mathbf{U}$ is an $\mathbb{R}^n$-valued function, whereas $\mathbf{W}$ is an $\mathbb{R}$-valued function of $(t, x) \in \mathbb{R} \times \mathbb{R}^d$. The matrices A_j and E, as well as the nonlinearity F, satisfy Assumption 1. Concerning the other coefficients of the system, we assume the following.

Assumption 2. (i) *The real, constant matrix C_1 has size $m \times n$ (with $m \in \mathbb{N}$).*
(ii) *The constant c is positive.*
(iii) *There exist two real, positive constants c_1 and c_2 and an integer $K \geq 1$ such that*

$$\forall E \in \mathbb{C}^m,\ \forall w \in \mathbb{C}, \quad G(E, w) = c_1 |E|^{2K-2} E + c_2 w E.$$

Remark 6. As in Remark. 5 for the case without ionization, we can put Maxwell's equation with ionization terms (19) under the abstract form (21). Using the same notations as in Remark. 5, the matrix C_1 is the projection matrix such that $C_1 \mathbf{U} = \mathtt{E}$, and $\mathbf{w} = \rho$.

2.3 The Cauchy Problem

We are considering initial conditions that correspond to laser pulses. In the case without charge nor current density (20), they are fast oscillating *wave packets* slowly modulated by an envelope,

$$\mathbf{U}_{|t=0} = u^0(x) e^{i \frac{\mathbf{k} \cdot x}{\varepsilon}} + \text{c.c.}, \tag{22}$$

where $\mathbf{k} \in \mathbb{R}^d$ is the (spatial) *wave-number* of the oscillations. Taking charge and current density into account (21), we need to provide initial conditions for $\mathbf{W}$; since we are interested here in the situation where this quantity is created by the laser when it reaches high intensities near self-focusing, we take these initial conditions to be initially zero[7] for the sake of clarity.

$$\mathbf{W}_{|t=0} = 0. \tag{23}$$

The evolution equation (20) (as well as (21)) being of semilinear nature, it is natural to work with Banach algebra in view of a resolution by Picard iterations. Throughout this article, we assume that $u^0 \in B$, with

$$B = H^{t_0}(\mathbb{R}^d)^n, \quad (t_0 > d/2)$$

[7]One could more generally and without supplementary difficulty consider non-oscillating initial conditions for the charge density $\mathbf{W}$.

or

$$B = W(\mathbb{R}^d)^n := \{f \in \mathcal{S}'(\mathbb{R}^d)^n, |f|_B := |\hat{f}|_{L^1} < \infty\}$$

(the so-called Wiener algebra, which is better adapted than $H^{t_0}(\mathbb{R}^d)^n$ to handle short pulses, see [20, 21]). In both cases, B is stable by translations in Fourier space (this ensures that if $u^0 \in B$ in (22) then $\mathbf{U}_{|_{t=0}} \in B$) and is a Banach algebra in the sense that

$$\forall f, g \in B, \qquad f \cdot g \in B \quad \text{and} \quad |f \cdot g|_B \lesssim |f|_B |g|_B.$$

For all $k \in \mathbb{N}$, we also define

$$B^{(k)} = \{f \in B, \forall \alpha \in \mathbb{N}^d, \forall |\alpha| \leq k, \partial^\alpha f \in B\},$$

endowed with its canonical norm.

We are interested in deriving asymptotics to the solution formed by (20)–(22), with initial envelope $u^0 \in B$, and more generally (21)–(23), if we want to be able to handle ionization processes in nonlinear optics. This requires a further assumption on the nonlinearity F, namely that F acts on B and is locally Lipschitz.

Assumption 3. *In addition to* (iii) *of Assumption* 1, *the mapping F satisfies, uniformly with respect to $\varepsilon \in [0, 1)$,*

(i) *For all $f \in B$, one has $F(\varepsilon, f) \in B$ and*

$$|F(\varepsilon, f)|_B \leq C(|f|_B)|f|_B.$$

(ii) *For all $f, g \in B$, one has*

$$|d_f F(\varepsilon, \cdot)g|_B \leq C(|f|_B)|g|_B.$$

Example 3. When $B = H^{t_0}(\mathbb{R}^d)^n$ $(t_0 > d/2)$, Assumption 3 is always satisfied (by Moser's inequality); when $B = W(\mathbb{R}^d)^n$, the assumption holds for analytic nonlinearities.

3 Derivation of NLS-Type Equations

The Schrödinger approximation takes into account the *diffractive* effects that modify over large times the propagation along rays of standard geometrical optics. These diffractive effects are of linear nature and are known [15–17, 22, 23] to appear for time scales of order $O(1/\varepsilon)$ for the initial value problem formed by the linear part of (20) and (22). This is the reason why we are interested in proving the existence and describing the solutions to the (nonlinear) initial value problem (20)–(22) over such time scales.

For the sake of simplicity, the initial value problem (20)–(22) (no ionization) is first considered. Up to minor modifications, the results of Sects. 3.1–3.3 are known [20]; we reproduce them here because they are necessary steps to derive the family of NLS equations (3) and, for the sake of clarity, their proof is sketched in a few words. The general idea to derive the Schrödinger equations of Sect. 3.5 with improved dispersion was introduced in [20] but the computations are carried further here. We then derive in Sect. 3.6 a new class of models with derivative nonlinearity for which a local well-posedness result is proved. When applied to Maxwell's equations, these derivative nonlinearities yield the so-called self-steepening operators; to our knowledge, this is the first rigorous explanation of these terms.

The asymptotic description of (21)–(22) (i.e., ionization is now included) is then addressed in Sect. 3.7.

3.1 The Profile Equation

We show here that under reasonable assumptions on F, solutions to the initial value problem (20)–(22) exist for times of order $O(1/\varepsilon)$ and that they can be written under a very convenient form using a *profile* U,

$$\mathbf{U}(t,x) = U\Big(t, x, \frac{\mathbf{k}\cdot x - \omega t}{\varepsilon}\Big), \tag{24}$$

with $U(t,x,\theta)$ periodic with respect to θ and for any $\omega \in \mathbb{R}$, provided that U solves the *profile equation*

$$\begin{cases} \partial_t U + A(\partial)U + \frac{i}{\varepsilon}\mathcal{L}(\omega D_\theta, \mathbf{k}D_\theta)U + \varepsilon^{1+p}A_0 U = \varepsilon F(\varepsilon, U), \\ U|_{t=0}(x,\theta) = u^0(x)e^{i\theta} + \text{c.c.}. \end{cases} \tag{25}$$

Here, we used the notation

$$\mathcal{L}(\omega D_\theta, \mathbf{k}D_\theta) = -\omega D_\theta + A(\mathbf{k})D_\theta + \frac{E}{i}, \tag{26}$$

with $D_\theta = -i\partial_\theta$ and $A(\mathbf{k}) = \sum_{j=1}^d A_j k_j$.

Theorem 1. *Let $B = H^{t_0}(\mathbb{R}^d)^n$ or $B = W(\mathbb{R}^d)^n$ and $u^0 \in B$. Under Assumptions 1 and 3, there exists $T > 0$ such that for all $0 < \varepsilon \le 1$ there exists a unique solution $\mathbf{U} \in C([0, T/\varepsilon]; B)$ to (20)–(22). Moreover, one can write $\mathbf{U}$ under the form*

$$\mathbf{U}(t,x) = U\Big(t, x, \frac{\mathbf{k}\cdot x - \omega t}{\varepsilon}\Big),$$

where U solves the profile equation (25).

Proof. The proof is a slight adaptation of the one given in [20] in the trilinear case; consequently, we just give the main steps of the proof. Quite obviously, a solution $\mathbf{U}$ to (20)–(22) is given by (24) if (25) admits a solution $U \in C\big([0, T/\varepsilon]; H^k(\mathbb{T}; B)\big)$ $(k \geq 1)$ where

$$H^k(\mathbb{T}; B) = \Big\{ f = \sum_{n\in\mathbb{Z}} f_n e^{in\theta}, \, |f|_{H^k(\mathbb{T},B)} < \infty \Big\} \tag{27}$$

and with $|f|^2_{H^k(\mathbb{T},B)} = \sum_{n\in\mathbb{Z}} (1+n^2)^k |f_n|_B^2$. For $k \geq 1$, $H^k(\mathbb{T}, B)$ is a Banach algebra; moreover, the evolution operator $S(t)$ associated with the linear part of (25),

$$S(t) = \exp\Big(-tA(\partial) - \frac{i}{\varepsilon} t\mathcal{L}(\omega D_\theta, kD_\theta)\Big)$$

is unitary on $H^k(\mathbb{T}; B)$ (thanks to points i and ii of Assumption 1). One can therefore construct a (unique) solution to (25) by a standard iterative scheme

$$U^{l+1}(t) = S(t)U^0 + \varepsilon \int_0^t S(t-t')\big[F(\varepsilon, U^l)(t') - \varepsilon^p A_0 U^l\big]\, dt',$$

with $U^0 = U|_{t=0}$. Indeed, one has thanks to Assumption 3,

$$|U^{l+1}(t)|_{H^k(\mathbb{T};B)} \leq |U^0|_{H^k(\mathbb{T};B)} + \varepsilon \int_0^t \big[\varepsilon^p |A_0 U^l|_{H^k(\mathbb{T};B)} + C(|U^l|_{H^k(\mathbb{T};B)})|U^l|_{H^k(\mathbb{T};B)}\big];$$

thanks to the ε in front of the integral. An estimate of the same kind is valid for a difference of iterates, by point (ii) of Assumption 3. By a fixed point argument, this ensures that the sequence converges to a solution on $[0, T/\varepsilon]$ for some $T > 0$ *independent of* ε. Uniqueness then follows classically from an energy estimate on the difference of two solutions. □

3.2 The Slowly Varying Envelope Approximation

The SVEA consists in writing the profile U under the form

$$U(t, x, \theta) \sim u_{\text{env}}(t, x)e^{i\theta} + \text{c.c.}; \tag{28}$$

plugging this approximation into the profile equation (25) and keeping only the first harmonic in the Fourier expansion yields easily (writing $u = u_{\text{env}}$),

$$\partial_t u + A(\partial)u + \frac{i}{\varepsilon}\mathcal{L}(\omega, \mathbf{k})u + \varepsilon^{1+p} A_0 u = \varepsilon F^{\text{env}}(\varepsilon, u),$$

where

$$F^{\mathrm{env}}(\varepsilon, u) = \frac{1}{2\pi}\int_0^{2\pi} \mathrm{e}^{-\mathrm{i}\theta} F(\varepsilon, u\mathrm{e}^{\mathrm{i}\theta} + \text{c.c.})\,\mathrm{d}\theta. \tag{29}$$

Example 4. With $F(u) = |u|^2 u$, one gets $F^{\mathrm{env}}(u) = (u \cdot u)\overline{u} + 2|u|^2 u$.

Denoting $D = -\mathrm{i}\nabla$, we observe that

$$\begin{aligned} A(\partial) + \frac{\mathrm{i}}{\varepsilon}\mathcal{L}(\omega, \mathbf{k}) &= A(\partial) + \frac{\mathrm{i}}{\varepsilon}(-\omega\,\mathrm{Id} + A(\mathbf{k})) \\ &= \frac{\mathrm{i}}{\varepsilon}(-\omega\,\mathrm{Id} + A(\mathbf{k} + \varepsilon D)) := \frac{\mathrm{i}}{\varepsilon}\mathcal{L}(\omega, \mathbf{k} + \varepsilon D), \end{aligned}$$

where the last notation is of course consistent with (26).

As a consequence of these computations, we see that in order for (28) to hold, it is necessary that $u = u_{\mathrm{env}}$ satisfies the *envelope equation*

$$\begin{cases} \partial_t u + \dfrac{\mathrm{i}}{\varepsilon}\mathcal{L}(\omega, \mathbf{k} + \varepsilon D)u + \varepsilon^{1+p} A_0 u = \varepsilon F^{\mathrm{env}}(\varepsilon, u), \\ u|_{t=0} = u^0. \end{cases} \tag{30}$$

As implicitly assumed by omitting the fast oscillating scale in the argument of the envelope function $u_{\mathrm{env}}(t, x)$, the envelope must not contain any fast oscillation. However,

- The singular part of the linear term in (30) creates fast oscillations with frequencies $\omega - \omega_j(\mathbf{k})$, where the $\omega_j(\mathbf{k})$ stand for the eigenvalues of $\mathcal{L}(0, \mathbf{k})$.
- The nonlinearity creates other oscillations that may resonate with the linear propagator.

There is one way to avoid the catastrophic effects of these two scenarios. Choosing $\omega = \omega_j(\mathbf{k})$ for some j and assuming that, up to $O(\varepsilon)$ terms, the initial envelope u^0 is contained in the corresponding eigenspace prevents the creation of oscillations by the linear propagator. This is the *polarization condition*. The nonlinearity will, however, create harmonics of the main oscillation $\mathbf{k}\cdot x - \omega_j(\mathbf{k})t$ and it is necessary to make a non-resonance assumption. What is called *characteristic variety* in the assumption below is the set $\mathcal{C}_{\mathcal{L}} \subset \mathbb{R}^{d+1}$ defined as

$$\mathcal{C}_{\mathcal{L}} = \{(\omega', \mathbf{k}') \in \mathbb{R}^{1+d}, \det \mathcal{L}(\omega', \mathbf{k}') = 0\}.$$

Let us also recall that we assumed that the nonlinearity is under the form

$$F(\varepsilon, U) = \Big(1 + f\big(\varepsilon^r Q(U)\big)\Big)\, T(U, \overline{U}, U),$$

with $f(0) = 0$, Q a quadratic form and T a trilinear symmetric mapping. If U is a monochromatic oscillation, the nonlinearity $\varepsilon F(\varepsilon, U)$ creates third harmonic with size $O(\varepsilon)$, a fifth harmonic (if $f'(0) \neq 0$) with size $O(\varepsilon^{1+r})$, etc. The non-resonance condition stated below holds for the $(2p+3)$th harmonics, for all $p \geq 0$ such that $pr < 1$ (the contribution of higher harmonics is small enough to be controlled even if it is resonant).

Assumption 4. *The characteristic variety $\mathcal{C}_{\mathcal{L}}$ and the frequency/wave number couple $(\omega, \mathbf{k})$ satisfy:*

1. *There exist m functions $\omega_j \in C^\infty(\mathbb{R}^d \backslash \{0\})$ $(j = 1, \ldots, m)$ such that*

$$\mathcal{C}_{\mathcal{L}} \backslash \{0\} = \bigcup_{j=1}^{m} \{(\omega_j(\mathbf{k}'), \mathbf{k}'), \mathbf{k}' \in \mathbb{R}^d \backslash \{0\}\};$$

   up to a renumbering, we assume that $(\omega, \mathbf{k}) = (\omega_1(\mathbf{k}), \mathbf{k})$.
2. *There exists a constant $c_0 > 0$ such that*

$$\inf_{\mathbf{k}' \in \mathbb{R}^d} |\omega - \omega_j(\mathbf{k}')| \geq c_0, \quad j = 2, \ldots m.$$

3. *(Non-resonance assumption) One has $\pm(2p+3)(\omega, \mathbf{k}) \notin \mathcal{C}_{\mathcal{L}}$, for all $p \geq 0$ such that $pr < 1$.*

Notation 2. We denote by $\pi_j(\mathbf{k})$ $(j = 1, \ldots, m)$ the eigenprojectors of the eigenvalues $\omega_j(\mathbf{k})$ of $A(\mathbf{k}) + (1/\mathrm{i})E$; in particular, we have

$$\mathcal{L}(0, \mathbf{k}) = A(\mathbf{k}) + \frac{1}{\mathrm{i}} E = \sum_{j=1}^{m} \omega_j(\mathbf{k}) \pi_j(\mathbf{k}).$$

Example 5. For Maxwell's equations, it is shown in Appendix 2 that Assumption 4 is satisfied with $m = 7$ in dimension $d = 3$, for $\omega \neq 0$. Explicit expressions for the eigenprojectors $\pi_j(\mathbf{k})$ are also provided in Appendix 2.

Theorem 2. *Let $B = H^{t_0}(\mathbb{R}^d)^n$ or $B = W(\mathbb{R}^d)^n$ and $u^0 \in B^{(1)}$, $r \in B$. Let Assumptions 1, 3, and 4 be satisfied and assume moreover that $u^0 = \pi_1(\mathbf{k}) u^0 + \varepsilon r$. Then*

(i) *There exist $T > 0$ and, for all $\varepsilon \in (0, 1]$, a unique solution $u \in C([0, T/\varepsilon]; B^{(1)})$ to (30) with initial condition u^0.*
(ii) *There exists $\varepsilon_0 > 0$ such that for all $0 < \varepsilon < \varepsilon_0$, the solution $\mathbf{U}$ to (20) provided by Theorem 1 exists on $[0, T/\varepsilon]$ and*

$$|\mathbf{U} - \mathbf{U}_{\mathrm{SVEA}}|_{L^\infty([0,T/\varepsilon] \times \mathbb{R}^d)} \leq \varepsilon C(T, |u^0|_B)(1 + |\nabla u^0|_B + |r|_B),$$

   where $\mathbf{U}_{\mathrm{SVEA}}(t, x) = u(t, x) e^{\mathrm{i}(\mathbf{k} \cdot x - \omega t)/\varepsilon} + \mathrm{c.c.}$

Proof. Here again, the proof is a small variation of the one given in [20] for the trilinear case and $B = W(\mathbb{R}^d)^n$. We just indicate the main steps of the proof:

Step 1. Existence and bounds of the solution u to (30) is established by a fixed point argument as in Theorem 1.

Step 2. We decompose u as

$$u = u_{\text{I}} + u_{\text{II}}, \quad \text{with } u_{\text{II}} = \sum_{j=2}^{m} u_j,$$

and where $u_j = \pi_j(\mathbf{k} + \varepsilon D)u$ (see Notation 2).

Step 3. Thanks to the assumption that $\omega = \omega_1(\mathbf{k})$ one gets from the equation obtained by applying $\pi_1(\mathbf{k} + \varepsilon D)$ to (30) that $|\partial_t u_1(t)|_B$ is uniformly bounded on $[0, T/\varepsilon]$.

Step 4. Using a non-stationary phase argument (on the semigroup formulation) relying on point (ii) of Assumption 4 and the bound on $\partial_t u_1$ established in Step 3, and taking advantage of Assumption 3, we get that $(1/\varepsilon)|u_{\text{II}}(t)|_B$ remains uniformly bounded on $[0, T/\varepsilon]$.

Step 5. Using the non-resonance condition iii of Assumption 4, one can show that the third and higher harmonics created by the nonlinearity remain of order $O(\varepsilon)$. More precisely, the solution $U \in H^1(\mathbb{T}; B)$ to (25) provided by Theorem 1 can be written as

$$U(t, x, \theta) = U_{\text{app}}(t, x, \theta) + \varepsilon V(t, x, \theta),$$

where $U_{\text{app}}(t, x, \theta) = u(t, x)e^{i\theta} + \text{c.c.}$, and V remains bounded (with respect to ε) in $C([0, T/\varepsilon]; H^1(\mathbb{T}; B)^n)$.

Step 6. Since $U(t) - U_{\text{app}}(t) = \varepsilon V(t)$, it follows from the above that

$$\sup_{t\in[0,T/\varepsilon]} |U(t) - U_{\text{app}}(t)|_{H^1(\mathbb{T};B)} \leq \varepsilon C(T, |u^0|_B)(1 + |\nabla u^0|_B + |r|_B),$$

and the theorem follows therefore from the observation that

$$|\mathbf{U} - \mathbf{U}_{\text{SVEA}}|_{L^\infty([0,T/\varepsilon]\times\mathbb{R}^d)} \leq \sup_{t\in[0,T/\varepsilon]} |U(t) - U_{\text{app}}(t)|_{H^1(\mathbb{T};B)}. \qquad \square$$

3.3 *The Full Dispersion Model*

The idea is to diagonalize (30) in order to work with a scalar equation. We project therefore (30) onto the eigenspace corresponding to the oscillating term. This naturally leads to introduce

$$u_{\text{FD}} = \pi_1(\mathbf{k} + \varepsilon D)u_{\text{env}}.$$

which leads to the *full dispersion* scalar[8] equation (writing $u = u_{\mathrm{FD}}$)

$$\begin{cases} \partial_t u + \frac{\mathrm{i}}{\varepsilon}(\omega_1(\mathbf{k} + \varepsilon D) - \omega)u + \varepsilon^{1+p}\pi_1(\mathbf{k} + \varepsilon D)A_0 u \\ \quad = \varepsilon\pi_1(\mathbf{k} + \varepsilon D)F^{\mathrm{env}}(\varepsilon, u) \\ u|_{t=0}(x) = u^0(x) \end{cases} \tag{31}$$

and with $\omega_1(\cdot)$ as in Assumption 4.

The following corollary shows that the full dispersion scalar equation yields an approximation of the same precision as the envelope equation for times $t \in [0, T/\varepsilon]$.

Corollary 1 (Full Dispersion Model). *Under the assumptions of Theorem 2,*

(i) *There exist $T > 0$ and, for all $\varepsilon \in (0, 1]$, a unique solution $u \in C([0, T/\varepsilon]; B^{(1)})$ to (31) with initial condition u^0.*
(ii) *There exists $\varepsilon_0 > 0$ such that for all $0 < \varepsilon < \varepsilon_0$, the solution $\mathbf{U}$ to (20) provided by Theorem 1 exists on $[0, T/\varepsilon]$ and*

$$|\mathbf{U} - \mathbf{U}_{\mathrm{FD}}|_{L^\infty([0,T/\varepsilon]\times\mathbb{R}^d)} \le \varepsilon C(T, |u^0|_B)(1 + |\nabla u^0|_B + |r|_B),$$

where $\mathbf{U}_{\mathrm{FD}}(t, x) = u(t, x)e^{\mathrm{i}(\mathbf{k}\cdot x - \omega t)/\varepsilon} + \mathrm{c.c.}$.

Remark 7. Equation (31) does not correspond exactly to the "full dispersion" model of [20, 21], where the right-hand side is $\varepsilon\pi_1(\mathbf{k})F^{\mathrm{env}}(\varepsilon, u)$ rather than $\varepsilon\pi_1(\mathbf{k} + \varepsilon D)F^{\mathrm{env}}(\varepsilon, u)$ (but it can be found as an "intermediate model" in [24]). This change does not affect the estimate given in the corollary, but it is important to keep track of the frequency dependence of the polarization of the nonlinear term to introduce the "self-steepening" operators in Sect. 3.6. Note also that the "full dispersion" model is related to the so-called unidirectional pulse propagation equation used in nonlinear optics [19, 25].

Proof. This is actually a by-product of the proof of Theorem 2 since u_{FD} coincides with u_1 in Step 2 of the proof of Theorem 2. □

3.4 The NLS Equation

The NLS equation is easily derived from (31) using Taylor expansions

$$\frac{\mathrm{i}}{\varepsilon}(\omega_1(\mathbf{k} + \varepsilon D) - \omega) = \mathbf{c}_g \cdot \nabla - \frac{\mathrm{i}}{2}\nabla \cdot H_{\mathbf{k}}\nabla + O(\varepsilon^2),$$

$$\pi_1(\mathbf{k} + \varepsilon D) = \pi_1(\mathbf{k}) + O(\varepsilon),$$

[8]The linear propagator is a scalar operator but the equation remains vectorial because of the nonlinear term. Indeed, $\pi_1(\mathbf{k})$ is in general not of rank 1 (i.e., $\omega_1(\mathbf{k})$ is in general not of multiplicity one).

where $\mathbf{c}_g = \nabla\omega_1(\mathbf{k})$ and $H_{\mathbf{k}}$ stands for the Hessian of ω_1 at $\mathbf{k}$. Neglecting the $O(\varepsilon^2)$ terms in (31) we define the NLS approximation $u = u_{\text{NLS}}$ as the solution to

$$\begin{cases} \partial_t u + \mathbf{c}_g \cdot \nabla u - \varepsilon \frac{\mathrm{i}}{2} \nabla \cdot H_{\mathbf{k}} \nabla u + \varepsilon^{1+p} \pi_1(\mathbf{k}) A_0 u = \varepsilon \pi_1(\mathbf{k}) F^{\text{env}}(\varepsilon, u); \\ u|_{t=0}(x) = u^0(x). \end{cases} \tag{32}$$

We then get easily (see [20, 21]) the following justification of the NLS approximation.

Corollary 2 (Schrödinger Approximation). *Under the assumptions of Theorem 2, one has for all* $u^0 \in B^{(3)}$ *such that* $u_0 = \pi_1(\mathbf{k}) u_0$,

(i) *There exist* $T > 0$ *and, for all* $\varepsilon \in (0, 1]$, *a unique solution* $u \in C([0, T/\varepsilon]; B^{(3)})$ *to* (32) *with initial condition* u^0.
(ii) *There exists* $\varepsilon_0 > 0$ *and* $\mathfrak{c}_{\text{NLS}} > 0$ *such that for all* $0 < \varepsilon < \varepsilon_0$, *the solution* $\mathbf{U}$ *to* (20) *provided by Theorem* 1 *exists on* $[0, T/\varepsilon]$ *and*

$$|\mathbf{U} - \mathbf{U}_{\text{NLS}}|_{L^\infty([0,T/\varepsilon]\times\mathbb{R}^d)} \le \varepsilon C(T, |u^0|_B)(1 + |\nabla u^0|_B + \mathfrak{c}_{\text{NLS}} |u^0|_{B^{(3)}}),$$

where $\mathbf{U}_{\text{NLS}}(t, x) = u(t, x) e^{\mathrm{i}(\mathbf{k}\cdot x - \omega t)/\varepsilon} + \text{c.c.}$.

Remark 8. The component $\mathfrak{c}_{\text{NLS}} |u^0|_{B^{(3)}}$ in the error estimate of the corollary is due to the bad frequency behavior of the Schrödinger equation when the envelope of the oscillations ceases to be well localized in frequency. This is, for instance, the case for short pulses, chirped pulses [20, 21, 24], and near the focusing point. To describe such extreme situations, the standard NLS approximation does a poor job, and this is why various variants have been derived in physics (e.g., [18]).

Remark 9. We assumed here that the polarization of the initial condition is exact (i.e., $r = 0$ in Theorem 2) for the sake of simplicity; indeed, the solution remains polarized along $\pi_1(\mathbf{k})$ for all times and computations on real physical models are much easier.

Example 6. In the frequent case where $\omega_1(\cdot)$ has a radial symmetry, and writing with a slight abuse of notation $\omega_1(\mathbf{k}) = \omega_1(k)$, with $k = |\mathbf{k}|$, we can write

$$\mathbf{c}_g = \omega_1'(k)\mathbf{e}_z, \quad H_{\mathbf{k}} = \frac{\omega_1'(k)}{|\mathbf{k}|}(I - \mathbf{e}_z \otimes \mathbf{e}_z) + \mathbf{e}_z \otimes \mathbf{e}_z \omega_1''(k),$$

where we assumed without loss of generality that $(0z)$ is the direction of the wave number $\mathbf{k}$, $\mathbf{k} = k\mathbf{e}_z$. In particular, (32) reads

$$\partial_t u + \omega_1'(k)\partial_z u - \varepsilon \frac{\mathrm{i}}{2} \frac{\omega_1'(k)}{k} \Delta_\perp u - \mathrm{i}\frac{\varepsilon}{2}\omega_1''(k)\partial_z^2 u + \varepsilon^{1+p}\pi_1(\mathbf{k}) A_0 u = \varepsilon \pi_1(\mathbf{k}) F^{\text{env}}(\varepsilon, u),$$

where $\Delta_\perp = \partial_x^2 + \partial_y^2$ is the Laplace operator in transverse variables. If we write $v(t,x,z) = u(\varepsilon t, x, z - \omega_1'(k)t)$, we get

$$\partial_t v - \frac{\mathrm{i}}{2}\frac{\omega_1'(k)}{k}\Delta_\perp v - \mathrm{i}\frac{1}{2}\omega_1''(k)\partial_z^2 v + \varepsilon^p \pi_1(\mathbf{k})A_0 v = \pi_1(\mathbf{k})F^{\mathrm{env}}(\varepsilon, v).$$

3.5 The NLS Equation with Improved Dispersion Relation

We propose here to investigate further the Schrödinger equation with improved dispersion relation derived in [20]. As said in Remark 8, the NLS approximation has too bad dispersive properties to capture correctly the envelope of oscillating solutions to Maxwell's equations in extreme situations, where high frequencies are released. Indeed, the dispersion relation $\omega_1(\cdot)$ of (20) is approximated by the second order polynomial

$$\omega_{\mathrm{NLS}}(\mathbf{k}') = \omega_1(\mathbf{k}) + \mathbf{c}_g \cdot (\mathbf{k}' - \mathbf{k}) + \frac{1}{2}(\mathbf{k}' - \mathbf{k}) \cdot H_{\mathbf{k}}(\mathbf{k}' - \mathbf{k}). \tag{33}$$

For Maxwell's equations and in dimension $d = 1$, Fig. 1 shows that this dispersion relation is very poor when $\mathbf{k}'$ is not close to $\mathbf{k}$. The idea introduced in [20] is to replace the *linear* part of the Schrödinger approximation by a linear operator that differs from the linear part of the Schrödinger approximation by $O(\varepsilon^2)$ terms only, but whose dispersion relation is far better.

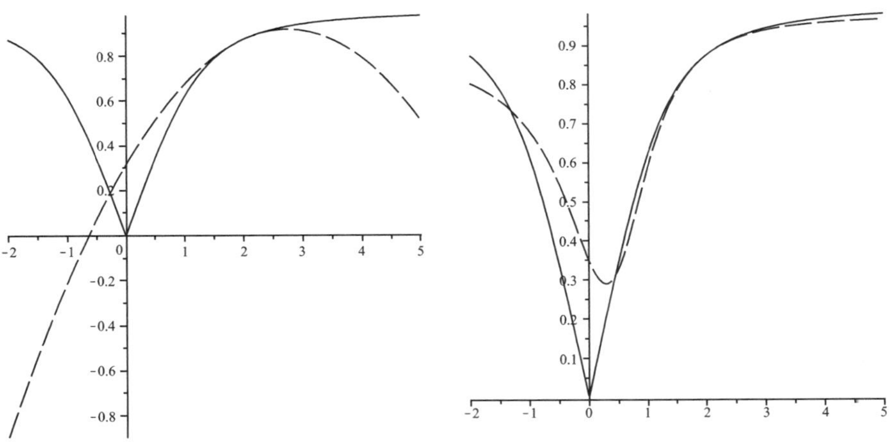

Fig. 1 The $\omega_{+,-}$ component (see Appendix 2) of the characteristic variety of Maxwell's equations (*solid*) and the dispersion relation corresponding to the Schrödinger approximation (*left*) and improved Schrödinger (*right*). Here $\mathbf{k} = 2$

More precisely, we consider an approximation under the form

$$\mathbf{U}_{\mathrm{imp}}(t,x) = u_{\mathrm{imp}}(t,x)\mathrm{e}^{\mathrm{i}(\mathbf{k}\cdot x-\omega t)/\varepsilon} + \mathrm{c.c.}, \tag{34}$$

where $u = u_{\mathrm{imp}}$ solves the *NLS equation with improved dispersion relation*

$$\begin{cases} \big(1 - \mathrm{i}\varepsilon\mathbf{b}\cdot\nabla - \varepsilon^2\nabla\cdot B\nabla\big)\partial_t u + \mathbf{c}_g\cdot\nabla u - \varepsilon\frac{\mathrm{i}}{2}\nabla\cdot\big(H_{\mathbf{k}} + 2\mathbf{c}_g\otimes\mathbf{b}\big)\nabla u \\ \quad +\varepsilon^2 C_3(\nabla)u + \varepsilon^{1+p}\pi_1(\mathbf{k})A_0 u = \varepsilon\pi_1(\mathbf{k})F^{\mathrm{env}}(\varepsilon,u) \\ u|_{t=0}(x) = u^0(x), \end{cases} \tag{35}$$

where $\mathbf{b}\in\mathbb{C}^d$, $B\in\mathcal{M}_{d\times d}(\mathbb{R})$ and $C_3(\nabla)$ is a third-order homogeneous differential operator. We assume moreover that

$$B \text{ is symmetric positive}, \quad \mathbf{b}\in\mathrm{Range}(B), \quad \text{and} \quad 4 - \mathbf{b}\cdot(B^{-1}\mathbf{b}) > 0 \tag{36}$$

(note that even though $B^{-1}\mathbf{b}$ is not unique when B is not definite, the scalar $\mathbf{b}\cdot(B^{-1}\mathbf{b})$ is uniquely defined). These assumptions ensure that the operator $(1 - \mathrm{i}\varepsilon\mathbf{b}\cdot\nabla - \varepsilon^2\nabla\cdot B\nabla)$ is invertible. This new model allows one to replace (33) by

$$\omega_{\mathrm{imp}}(\mathbf{k}') = \omega_1(\mathbf{k}) + \frac{\mathbf{c}_g\cdot(\mathbf{k}'-\mathbf{k}) + \frac{1}{2}(\mathbf{k}-\mathbf{k}')\cdot(H_{\mathbf{k}} + 2\mathbf{c}_g\otimes\mathbf{b})(\mathbf{k}'-\mathbf{k}) - C_3(\mathbf{k}'-\mathbf{k})}{1 + \mathbf{b}\cdot(\mathbf{k}'-\mathbf{k}) + (\mathbf{k}'-\mathbf{k})\cdot B(\mathbf{k}'-\mathbf{k})}.$$

A good choice of $\mathbf{b}$, B, and C_3 allows a much better approximation of $\omega_1(\cdot)$, as shown in Fig. 1 for Maxwell's equation in dimension $d = 1$.

Exactly as for Corollary 2 we get the following result, where the only difference in the error estimate with respect to Corollary 2 is the constant $\mathfrak{c}_{\mathrm{imp}}$ (which is much smaller than $\mathfrak{c}_{\mathrm{NLS}}$ for good choices of $\mathbf{b}$, B, and C). We refer to [20] for the proof and numerical validations of this model for the approximation of short pulses and chirped pulses.

Corollary 3 (Schrödinger Approximation with Improved Dispersion). *Under the assumptions of Theorem 2, one has, for all $u^0\in B^{(3)}$ such that $\pi_1(\mathbf{k})u_0 = u_0$,*

(i) *There exist $T > 0$ and, for all $\varepsilon\in(0,1]$, a unique solution $u\in C([0,T/\varepsilon];B^{(3)})$ to (35) with initial condition u^0.*
(ii) *There exist $\varepsilon_0 > 0$ and $\mathfrak{c}_{\mathrm{imp}} > 0$ such that for all $0 < \varepsilon < \varepsilon_0$, the solution $\mathbf{U}$ to (20) provided by Theorem 1 exists on $[0,T/\varepsilon]$ and*

$$|\mathbf{U} - \mathbf{U}_{\mathrm{imp}}|_{L^\infty([0,T/\varepsilon]\times\mathbb{R}^d)} \le \varepsilon C(T, |u^0|_B)(1 + |\nabla u^0|_B + \mathfrak{c}_{\mathrm{imp}}|u^0|_{B^{(3)}}),$$

where $\mathbf{U}_{\mathrm{imp}}(t,x) = u(t,x)\mathrm{e}^{\mathrm{i}(\mathbf{k}\cdot x-\omega t)/\varepsilon}$ + c.c.

Example 7. In the framework of Example 6, i.e., if $\omega_1(\mathbf{k}) = \omega_1(k)$ with $k = |\mathbf{k}|$ and $\mathbf{k} = k\mathbf{e}_z$, (35) can be written

$$
\begin{aligned}
&(1 - \mathrm{i}\varepsilon\mathbf{b}\cdot\nabla - \varepsilon^2\nabla\cdot B\nabla)\partial_t u + \omega_1'(k)\partial_z u - \varepsilon\frac{\mathrm{i}}{2}\Big(\frac{\omega_1'(k)}{k}\Delta_\perp + \omega_1''(k)\partial_z^2\Big)u \\
&-\varepsilon\mathrm{i}\omega_1'(k)\mathbf{b}\cdot\nabla\partial_z u + \varepsilon^2 C_3(\nabla)u + \varepsilon^{1+p}\pi_1(\mathbf{k})A_0 u \\
&= \varepsilon\pi_1(\mathbf{k})F^{\mathrm{env}}(\varepsilon, u).
\end{aligned}
$$

If we write $v(t,x,z) = u(\varepsilon t, x, z - \omega_1'(k)t)$ and choose $C_3(\nabla) = -\omega_1'(k)\nabla\cdot B\nabla\partial_z$, we get

$$
\begin{aligned}
&(1 - \mathrm{i}\varepsilon\mathbf{b}\cdot\nabla - \varepsilon^2\nabla\cdot B\nabla)\partial_t v - \frac{\mathrm{i}}{2}\Big(\frac{\omega_1'(k)}{k}\Delta_\perp + \omega_1''(k)\partial_z^2\Big)v + \varepsilon^p\pi_1(\mathbf{k})A_0 v \\
&= \pi_1(\mathbf{k})F^{\mathrm{env}}(\varepsilon, v).
\end{aligned} \tag{37}
$$

A similar equation has been proposed in [26, Sect. 8.5.3] in the framework of water wave equations.

3.6 The NLS Equation with Frequency Dependent Polarization

The idea here is to improve the rough approximation $\pi_1(\mathbf{k} + \varepsilon D) \sim \pi_1(\mathbf{k}) + O(\varepsilon)$ used to derive the NLS approximation (see Sect. 3.4). Indeed, when the description of the envelope of the laser pulse requires a broad band of frequencies as in the situations mentioned in Sect. 3.5, the variations of the polarization term $\pi_1(\mathbf{k} + \varepsilon D)$ in front of the nonlinearity in (31) become important and should be taken into account. We therefore make the following approximation,

$$
\pi_1(\mathbf{k} + \varepsilon D) \sim (1 - \mathrm{i}\varepsilon\mathbf{b}\cdot\nabla - \varepsilon^2\nabla\cdot B\nabla)^{-1}\big[\pi_1(\mathbf{k}) + \varepsilon\pi_1'(\mathbf{k})\cdot D - \mathrm{i}\varepsilon(\mathbf{b}\cdot\nabla)\pi_1(\mathbf{k})\big],
$$

where $\mathbf{b}$ and B are the same as in the NLS approximation with improved dispersion (35). In particular, if $\mathbf{b} = 0$ and $B = 0$ (standard NLS), then the above approximation coincides with the first order Taylor expansion

$$
\pi_1(\mathbf{k} + \varepsilon D) = \pi_1(\mathbf{k}) + \varepsilon\pi_1'(\mathbf{k})\cdot D.
$$

The general formula has the same accuracy as this Taylor expansion as $\varepsilon \to 0$. The corresponding approximation is given by

$$
\mathbf{U}_{\mathrm{pol}}(t,x) = u_{\mathrm{pol}}(t,x)\mathrm{e}^{\mathrm{i}(\mathbf{k}\cdot x - \omega t)/\varepsilon} + \mathrm{c.c.}, \tag{38}
$$

where $u = u_{\rm pol}$ solves the *NLS equation with frequency dependent polarization*

$$\begin{cases} \big(1 - {\rm i}\varepsilon\mathbf{b}\cdot\nabla - \varepsilon^2\nabla\cdot B\nabla\big)\partial_t u + \mathbf{c}_g\cdot\nabla u \\ \quad -\varepsilon\frac{\rm i}{2}\nabla\cdot\big(H_{\mathbf{k}} + 2\mathbf{c}_g\otimes\mathbf{b}\big)\nabla u + \varepsilon^2 C_3(\nabla)u + \varepsilon^{1+p}\pi_1(\mathbf{k})A_0 u \\ \quad = \varepsilon\big[\pi_1(\mathbf{k}) + \varepsilon\pi_1(\mathbf{k})\pi_1'(\mathbf{k})\cdot D - i\varepsilon(\mathbf{b}\cdot\nabla)\pi_1(\mathbf{k})\big]F^{\rm env}(\varepsilon,\pi_1(\mathbf{k})u) \\ u_{|t=0}(x) = u^0(x), \end{cases} \tag{39}$$

where $\mathbf{b}$, B, and $C_3(\nabla)$ are the same as in (31).

Contrary to all the previous models, the nonlinearity in (39) seems to be of quasilinear rather than nonlinear nature. It turns out, however, that the presence of the operator $\big(1-{\rm i}\varepsilon\mathbf{b}\cdot\nabla-\varepsilon^2\nabla\cdot B\nabla\big)$ in front of the time derivative plays a smoothing role allowing the control of one or several first order derivatives (see the discussion in point 4 in p. 22). If the first order derivatives involved in the nonlinearity are all controlled by this smoothing operator, then the nonlinearity remains semilinear in nature. As shown in the proof below, the component $-i\varepsilon(\mathbf{b}\cdot\nabla)\pi_1(\mathbf{k})F^{\rm env}(\varepsilon,\pi_1(\mathbf{k})u)$ of the nonlinearity is always semilinear in this sense. This is not the case for the component $\varepsilon\pi_1(\mathbf{k})\pi_1'(\mathbf{k})\cdot DF^{\rm env}(\varepsilon,\pi_1(\mathbf{k})u)$ that may be of quasilinear nature, in which case a symmetry assumption is needed on the nonlinearity to ensure local well-posedness. In order to state this assumption, it is convenient to introduce the norm $|\cdot|_*$ defined as

$$|u|_*^2 = \big(u, (1 - {\rm i}\varepsilon\mathbf{b}\cdot\nabla - \varepsilon^2\nabla\cdot B\nabla)u\big).$$

Assumption 5. *For all $v \in W^{1,\infty}(\mathbb{R}^d)^n$ and $u \in L^2(\mathbb{R}^d)^n$ such that $\pi_1(\mathbf{k})u = u$, one has*

$$\forall 1 \le j \le d, \quad \Re\big(\pi_1(\mathbf{k})\pi_1'(\mathbf{k})\cdot\mathbf{e}_j d_v F^{\rm env} D_j u, u\big) \le {\rm Cst}|v|_{W^{1,\infty}}|u|_*^2,$$

where $\mathbf{e}_j$ is the unit vector in the jth direction of $\mathbb{R}^d$ and $d_v F^{\rm env}$ is the derivative at v of the mapping $u \mapsto F^{\rm env}(\varepsilon, u)$.

Example 8. The computations performed in Appendix 2 show that this assumption is satisfied by the dimensionless Maxwell equations (73).

The approximation furnished by (39) is justified by the following corollary (the difference in the estimate with respect to Corollary 3 is just a better nonlinear constant, denoted $C_{\rm pol}$ to insist on this point). For the sake of simplicity, we take $B = H^{t_0}(\mathbb{R}^d)^n$ ($t_0 > d/2$) here, but adaptation to Wiener spaces is possible.

Corollary 4 (Schrödinger Approximation with Frequency Dependent Polarization). *Let the assumptions of Theorem 2 be satisfied and assume moreover that Assumption 5 holds with $B = H^{t_0}(\mathbb{R}^d)$ ($t_0 > d/2$). Then for all $u_0 \in B^{(3)}$ such that $\pi_1(\mathbf{k})u_0 = u_0$, one has*

(i) *There exist* $T > 0$ *and, for all* $\varepsilon \in (0,1]$, *a unique solution* $u \in C([0, T/\varepsilon]; B^{(3)})$ *to* (39) *with initial condition* u^0.

(ii) *There exist* $\varepsilon_0 > 0$ *and* $\mathfrak{c}_{\rm imp} > 0$ *such that for all* $0 < \varepsilon < \varepsilon_0$, *the solution* $\mathbf{U}$ *to* (20) *provided by Theorem* 1 *exists on* $[0, T/\varepsilon]$ *and*

$$|\mathbf{U} - \mathbf{U}_{\rm pol}|_{L^\infty([0,T/\varepsilon]\times\mathbb{R}^d)} \leq \varepsilon C_{\rm pol}(T, |u^0|_B)(1 + |\nabla u^0|_B + \mathfrak{c}_{\rm imp}|u^0|_{B^{(3)}}),$$

where $\mathbf{U}_{\rm pol}(t,x) = u(t,x)e^{i(\mathbf{k}\cdot x - \omega t)/\varepsilon} + {\rm c.c.}$.

Remark 10. We have introduced the variation of the polarization on the NLS equation with improved dispersion (35), but the two steps are independent (i.e., one can take $\mathbf{b} = 0$, $B = 0$ and $C_3(\nabla) = 0$ in (39)).

Remark 11. Note that in (39), we have applied $\pi_1(\mathbf{k})$ to the full nonlinearity (hence the term $\pi_1(\mathbf{k})\pi_1'(\mathbf{k}) \cdot D$ instead of $\pi_1'(\mathbf{k}) \cdot D$). This means that we keep the effect of the frequency dependent polarization on the main polarized space Range$(\pi_1(\mathbf{k}))$ only. Similarly, we have replaced $F^{\rm env}(\varepsilon, u)$ by $F^{\rm env}(\varepsilon, \pi_1(\mathbf{k})u)$. This latter substitution would not change anything to the previous NLS models since we have seen that polarized initial conditions remain polarized. Its purpose in (39) is to make Assumption 5 much easier to check.

Proof. The justification of (39) is performed as for the other models. The only difference here is that local well-posedness for a time scale of order $O(1/\varepsilon)$ must be established. We therefore show here that such a local well-posedness result holds if $u_0 \in H^{s+1}(\mathbb{R}^d)^n$ with $s > t_0 + 1$. We just prove a priori energy estimates for (39); existence, uniqueness, and stability with respect to perturbations can be deduced classically.

The natural energy associated with (39) is given for all $s \geq 0$ by

$$E^s(u) = \frac{1}{2}\big((1 - i\varepsilon\mathbf{b}\cdot\nabla - \varepsilon^2\nabla\cdot B\nabla)\Lambda^s u, \Lambda^s u\big) = \frac{1}{2}|\Lambda^s u|_*^2.$$

Under Assumptions (36) on B and $\mathbf{b}$, $E^s(u)^{1/2}$ defines a norm that controls uniformly the H^s-norm. It may also control a certain number of first order derivatives. The important point for the local well-posedness of (39) is that it always controls first order derivatives in the direction $\mathbf{b}\cdot\nabla$; more precisely, we claim that there exists $c > 0$ independent of ε such that for all u,

$$E^s(u) \geq c\big(|u|_{H^s}^2 + \varepsilon^2|\mathbf{b}\cdot\nabla u|_{H^s}^2\big). \tag{40}$$

A quick look on the Fourier side shows that it is equivalent to prove that

$$\forall X \in \mathbb{R}^d, \quad 1 + \mathbf{b}\cdot X + X\cdot BX \geq c(1 + \mathbf{b}\cdot X)^2,$$

which is a consequence of (36).

Multiplying $\Lambda^s(39)$ by $\Lambda^s\bar{u}$ and integrating by parts, we get

$$\begin{aligned}\frac{\mathrm{d}}{\mathrm{d}t}E^s(u) =& -\varepsilon^{1+p}\Re\big(\pi_1(\mathbf{k})A_0\Lambda^s u, \Lambda^s u\big) + \varepsilon\big(\pi_1(\mathbf{k})\Lambda^s F(u), \Lambda^s u\big)\\ &+\varepsilon^2\sum_{j=1}^d \Re\big(\pi_1(\mathbf{k})\pi_1(\mathbf{k})'\cdot\mathbf{e}_j\Lambda^s d_u F D_j u, \Lambda^s u\big)\\ &-\varepsilon^2\Re\big(\mathrm{i}\pi_1(\mathbf{k})\Lambda^s(\mathbf{b}\cdot\nabla)F(u), \Lambda^s u\big)\\ :=& I_1 + I_2 + I_3 + I_4.\end{aligned}$$

where we denoted $F(u) = F^{\mathrm{env}}(\varepsilon, \pi_1(\mathbf{k})u)$. It is straightforward to control I_1, and Moser's estimate gives a control of I_2,

$$I_1 \lesssim \varepsilon^{1+p}|u|_{H^s}^2, \quad I_2 \leq \varepsilon C(|u|_\infty)|u|_{H^s}^2.$$

In order to control I_3, we must split it into two parts,

$$\begin{aligned}I_3 =& \varepsilon^2\sum_{j=1}^d \Re\big(\pi_1(\mathbf{k})\pi_1(\mathbf{k})'\cdot\mathbf{e}_j d_u F D_j\Lambda^s u, \Lambda^s u\big)\\ &+\varepsilon^2\sum_{j=1}^d \Re\big(\pi_1(\mathbf{k})\pi_1(\mathbf{k})'\cdot\mathbf{e}_j[\Lambda^s, d_u F]D_j u, \Lambda^s u\big);\end{aligned}$$

the first component is controlled using Assumption 5 while the Kato–Ponce and Moser estimate give a control of the second one,

$$I_3 \leq \varepsilon^2\mathrm{Cst}|u|_{W^{1,\infty}}|u|_{H^s}^2 + \varepsilon^2 C(|u|_{W^{1,\infty}})|u|_{H^s}^2.$$

Remarking that $\Lambda^s(\mathbf{b}\cdot\nabla)F(u) = \Lambda^s d_u F(\mathbf{b}\cdot\nabla)u$ and using the tame product estimate $|fg|_{H^s} \lesssim |f|_{L^\infty}|g|_{H^s} + |f|_{H^s}|g|_{L^\infty}$ and Moser's inequality, we get

$$I_4 \leq \varepsilon C(|u|_{W^{1,\infty}})(|u|_{H^s} + \varepsilon|\mathbf{b}\cdot\nabla u|_{H^s})|u|_{H^s}.$$

Gathering all the above estimates and using (40), we obtain

$$\frac{\mathrm{d}}{\mathrm{d}t}E^s(u) \leq \varepsilon C(|u|_{W^{1,\infty}})E^s(u).$$

Since moreover $|u|_{W^{1,\infty}}^2 \lesssim E^s(u)$ for $s > d/2+1$, we deduce from Gronwall's estimate that for such s, the energy $E^s(u)$ remains bounded for times of order $O(1/\varepsilon)$. □

Example 9. In the framework of Examples 6 and 7, we can check that $v(t,x,z) = u(\varepsilon t, x, z - \omega_1'(k)t)$ solves

$$(1 - \mathrm{i}\varepsilon \mathbf{b}\cdot\nabla - \varepsilon^2\nabla\cdot B\nabla)\partial_t v - \frac{\mathrm{i}}{2}\left(\frac{\omega_1'(k)}{k}\Delta_\perp + \omega_1''(k)\partial_z^2\right)v + \varepsilon^p\pi_1(\mathbf{k})A_0 v$$
$$= \big[\pi_1(\mathbf{k}) + \varepsilon\pi_1'(\mathbf{k})\cdot D - \mathrm{i}\varepsilon(\mathbf{b}\cdot\nabla)\pi_1(\mathbf{k})\big]F^{\mathrm{env}}(\varepsilon, \pi_1(\mathbf{k})v). \tag{41}$$

3.7 *Including Ionization Processes*

3.7.1 The Profile Equation

As in Sect. 3.1, we solve the initial value problem (21)–(23) for times of order $O(1/\varepsilon)$, writing the solution under a profile form,

$$(\mathbf{U}, \mathbf{W})(t,x) = (U, W)\left(t, x, \frac{\mathbf{k}\cdot x - \omega t}{\varepsilon}\right). \tag{42}$$

Again, $U(t,x,\theta)$ and $W(t,x,\theta)$ are periodic with respect to θ, and we use any $\omega \in \mathbb{R}$. The action of the Fourier multiplier $\mathcal{H}(\varepsilon(\mathbf{k}/k^2)\cdot D)$ from (19) (and (16)) is transferred at the profile level into the operator $\mathcal{H}(D_\theta + \varepsilon(\mathbf{k}/k^2)\cdot D)$, with $k = |\mathbf{k}|$,

$$\left(\mathcal{H}\left(\varepsilon\frac{\mathbf{k}}{k^2}\cdot D\right)\mathbf{U}\right)(t,x) = \left(\mathcal{H}\left(D_\theta + \varepsilon\frac{\mathbf{k}}{k^2}\cdot D\right)U\right)\left(t, x, \frac{\mathbf{k}\cdot x - \omega t}{\varepsilon}\right),$$

where $\mathcal{H}(D_\theta + \varepsilon(\mathbf{k}/k^2)\cdot D)$ is the Fourier multiplier

$$\mathcal{H}\left(D_\theta + \varepsilon\frac{\mathbf{k}}{k^2}\cdot D\right)\sum_{n\in\mathbb{Z}} u_n(x)\mathrm{e}^{\mathrm{i}n\theta} = \sum_{n\in\mathbb{Z}}\left(\mathcal{H}\left(n + \varepsilon\frac{\mathbf{k}}{k^2}\cdot D\right)u_n\right)(x)\mathrm{e}^{\mathrm{i}n\theta},$$

which acts continuously on $H^k(\mathbb{T}, B)$, for any $k \in \mathbb{N}$. In order to get a solution to the original problem, it is sufficient that (U, W) solves

$$\begin{cases} \partial_t U + A(\partial)U + \dfrac{\mathrm{i}}{\varepsilon}\mathcal{L}(\omega D_\theta, \mathbf{k}D_\theta)U + \varepsilon^{1+p}A_0 U \\ = \varepsilon F(\varepsilon, U) - \varepsilon\mathcal{H}\left(D_\theta + \varepsilon\dfrac{\mathbf{k}}{k^2}\cdot D\right)(WC_1^T C_1 U) - \varepsilon c\, C_1^T G(C_1 U, W), \\ \partial_t W - \dfrac{\mathrm{i}}{\varepsilon}\omega D_\theta W = \varepsilon G(C_1 U, W)\cdot C_1 U, \end{cases} \tag{43}$$

with initial conditions

$$(U, W)|_{t=0}(x,\theta) = \big(u^0(x)\mathrm{e}^{\mathrm{i}\theta} + \mathrm{c.c.}, 0\big). \tag{44}$$

Theorem 3. *Let $B = A^n \times A$, with $A = H^{t_0}(\mathbb{R}^d)$ or $A = W(\mathbb{R}^d)$, and $u^0 \in A^n$. Under Assumptions* 1–3, *there exists $T > 0$ such that for all $0 < \varepsilon \leq 1$ there exists a unique solution $\mathbf{Z} = (\mathbf{U}, \mathbf{W}) \in C([0, T/\varepsilon]; B)$ to* (21)–(23). *Moreover, one can write $\mathbf{Z}$ in the form*

$$\mathbf{Z}(t,x) = Z\left(t, x, \frac{\mathbf{k}\cdot x - \omega t}{\varepsilon}\right),$$

where $Z = (U, W)$ solves the profile equation (43), *with the initial condition* (44).

Proof. Similar to the one of Theorem 1, by an iterative scheme in $H^k(\mathbb{T}; B)$, with $k \geq 1$.

3.7.2 The Slowly Varying Envelope Approximation

In this section, as in Sect. 3.2, we shall assume that $\omega = \omega_1(\mathbf{k})$ is some characteristic frequency for $\mathcal{L}(\cdot, \mathbf{k})$. Postulating the Ansatz

$$Z(t,x,\theta) \sim \big(u_{\mathrm{env}}(t,x)\mathrm{e}^{\mathrm{i}\theta} + \mathrm{c.c.}, w_{\mathrm{env}}(t,x)\big), \tag{45}$$

we obtain formally the following system for $(u_{\mathrm{env}}, w_{\mathrm{env}})$ (denoted (u, w)),

$$\begin{cases} \partial_t u + \dfrac{\mathrm{i}}{\varepsilon}\mathcal{L}(\omega, \mathbf{k} + \varepsilon D)u + \varepsilon^{1+p}A_0 u \\ \quad = \varepsilon F^{\mathrm{env}}(\varepsilon, u) - \varepsilon \mathrm{i} w C_1^T C_1 u - \varepsilon c C_1^T G^{\mathrm{env}}(C_1 u, w), \\ \partial_t w = 2\varepsilon G^{\mathrm{env}}(C_1 u, w)\cdot \overline{C_1 u}, \end{cases} \tag{46}$$

where we used that $\mathcal{H}(1 + \varepsilon(\mathbf{k}/k^2)\cdot D) = \mathrm{i} + O(\varepsilon)$. Here, F^{env} is given by (29) and G^{env} is defined in the same way, filtering oscillations,

$$G^{\mathrm{env}}(u,w) = \frac{1}{2\pi}\int_0^{2\pi} \mathrm{e}^{-\mathrm{i}\theta} G(u\mathrm{e}^{\mathrm{i}\theta} + \mathrm{c.c.}, w)\,\mathrm{d}\theta.$$

Remark 12. The equation one obtains for w from direct computations is actually

$$\begin{aligned} \partial_t w &= \frac{\varepsilon}{2\pi}\int_0^{2\pi} G(C_1 u\mathrm{e}^{\mathrm{i}\theta} + \mathrm{c.c.}, w)\cdot (C_1 u \mathrm{e}^{\mathrm{i}\theta} + \mathrm{c.c.})\,\mathrm{d}\theta \\ &= c_1\varepsilon\left(2^K|C_1 u|^{2K} + 2\sum_{k=1}^{\lfloor K/2\rfloor}\binom{K-1}{k}\binom{K-k}{k}(2|C_1u|^2)^{K-2k}|C_1 u\cdot C_1 u|^{2k}\right) \\ &\quad + 2c_2\varepsilon w|C_1 u|^2 \\ &= 2\varepsilon G^{\mathrm{env}}(C_1 u, w)\cdot\overline{C_1 u}, \end{aligned}$$

in view of Assumption 2.

The approximation (45) is justified in the following theorem.

Theorem 4. *Let $B = A^n \times A$, with $A = H^{t_0}(\mathbb{R}^d)$ or $A = W(\mathbb{R}^d)$. Let $u^0 \in A^n$, with $\nabla u^0 \in A^{nd}$, and $r \in A^n$. Let* Assumptions 1, 3, *and* 4 *be satisfied and assume moreover that $\omega \neq 0$ and that $u^0 = \pi_1(\mathbf{k})u^0 + \varepsilon r$. Then*

(i) *There exist $T > 0$ and, for all $\varepsilon \in (0,1]$, a unique solution $(u,w) \in C([0,T/\varepsilon]; B^{(1)})$ to* (46) *with initial condition $(u^0, 0)$.*
(ii) *There exists $\varepsilon_0 > 0$ such that for all $0 < \varepsilon < \varepsilon_0$, the solution $\mathbf{Z}$ to* (43) *provided by Theorem* 3 *exists on $[0, T/\varepsilon]$ and*

$$|\mathbf{Z} - \mathbf{Z}_{\mathrm{SVEA}}|_{L^\infty([0,T/\varepsilon]\times\mathbb{R}^d)} \leq \varepsilon C(T, |u^0|_A)(1 + |\nabla u^0|_A + |r|_A),$$

where $\mathbf{Z}_{\mathrm{SVEA}}(t,x) = \big(u(t,x)\mathrm{e}^{\mathrm{i}(\mathbf{k}\cdot x - \omega t)/\varepsilon} + \mathrm{c.c.}, w(t,x)\big)$.

Proof. The arguments are similar to the ones in the proof of Theorem 2.

Step 1. As in the proof of Theorems 1 and 3, we have local in time T/ε (with T independent of ε) existence of (u,w), solution to (46), together with (uniform with respect to ε) bounds.

Step 2. Decomposing $u = u_{\mathrm{I}} + u_{\mathrm{II}}$ as in Step 2 of the proof of Theorem 2, one obtains in the same way that $|\partial_t u_{\mathrm{I}}|_{L^\infty([0,T/\varepsilon],A^n)}$ is bounded with respect to ε. This is also the case for $|\partial_t w|_{L^\infty([0,T/\varepsilon],A)}$ (even of order $O(\varepsilon)$), as shows directly the third equation in (46).

Step 3. As in Step 4 of the proof of Theorem 2, we deduce that

$$\frac{1}{\varepsilon}|u_{\mathrm{II}}|_{L^\infty([0,T/\varepsilon],A)} \leq C\big(T, |u^0|_A\big)\big(1 + |\nabla u^0|_A + |r|_A\big).$$

This is obtained by integration by parts in the integral formulation giving the u_j's of u_{II}, using Step 2 and Assumption 4 to have non-stationary phase; we conclude by a Gronwall argument.

Step 4. Approximation of $\mathbf{Z}$ by $\mathbf{Z}_{\mathrm{SVEA}}$. Compared to Theorem 2, the new point is the component w; for the sake of simplicity, we therefore set $F = 0$ throughout this proof. Denote

$$Z_{\mathrm{app}}(t,x,\theta) = (U_{\mathrm{app}}, W_{\mathrm{app}})(t,x,\theta) := \big(u(t,x)\mathrm{e}^{\mathrm{i}\theta} + \mathrm{c.c.}, w(t,x)\big),$$
$$\varepsilon\tilde{Z} = Z - Z_{\mathrm{app}},$$

where $Z = (U,W)$ is the solution to the profile equation (43) provided by Theorem 3. We estimate $\tilde{Z} = (\tilde{U}, \tilde{W})$ in $H^k(\mathbb{T}; B)$ (defined in (27)), $k \geq 1$.

- For $\tilde{W}$. We have $\tilde{W}|_{t=0} = 0$ and

$$\begin{aligned}\partial_t \tilde{W} + \frac{\mathrm{i}}{\varepsilon}\omega D_\theta \tilde{W} &= G(C_1 U, W)\cdot C_1 U - 2G^{\mathrm{env}}(C_1 u, w)\cdot \overline{C_1 u}\\ &= c_1 \sum_{k\neq 0}\left(\frac{1}{2\pi}\int_0^{2\pi} \mathrm{e}^{-\mathrm{i}\theta} G(C_1 U, W)\cdot C_1 U\,\mathrm{d}\theta\right)\mathrm{e}^{\mathrm{i}k\theta}\\ &\quad + c_2 W|C_1 U|^2 - 2c_2 w|C_1 u|^2.\end{aligned}$$

The terms in c_1 and c_2 can be treated similarly. For the sake of clarity, we therefore set $c_1 = 0$ and $c_2 = 1$ in this proof, so that the right-hand side is given by

$$W|C_1 U|^2 - W_{\mathrm{app}}|C_1 U_{\mathrm{app}}|^2 + 2w\big((C_1 u)\cdot (C_1 u)\mathrm{e}^{2\mathrm{i}\theta} + \mathrm{c.c.}\big).$$

Since

$$\big|W|C_1 U|^2 - W_{\mathrm{app}}|C_1 U_{\mathrm{app}}|^2\big| \le \varepsilon C(T, |u^0|_A)|\tilde{Z}|,$$

we easily deduce that

$$\begin{aligned}&|\tilde{W}(t)|_{H^k(\mathbb{T};A)}\\ &\quad \le \varepsilon C(T, |u^0|_A)\int_0^t |\tilde{Z}(t')|_{H^k(\mathbb{T};A)}\,\mathrm{d}t' + 2\left|\int_0^t \mathrm{e}{-2\mathrm{i}t'\omega/\varepsilon} w(C_1 u)\cdot (C_1 u)\,\mathrm{d}t'\right|_A.\end{aligned}$$

Splitting $u = u_{\mathrm{I}} + u_{\mathrm{II}}$ as in Theorem 2, we have a uniform (in ε) bound for $(1/\varepsilon)|w(C_1 u)\cdot (C_1 u_{\mathrm{II}})|_{L^\infty([0,T/\varepsilon],A^n)}$ by Steps 1 and 3. The only component left to control is therefore the one involving the product $(C_1 u_{\mathrm{I}})\cdot (C_1 u_{\mathrm{I}})$, for which we write

$$\begin{aligned}\int_0^t \mathrm{e}^{-2\mathrm{i}t'\omega/\varepsilon} w(C_1 u_{\mathrm{I}})\cdot (C_1 u_{\mathrm{I}}) &= -\mathrm{i}\frac{\varepsilon}{2\omega}\int_0^t \mathrm{e}^{-2\mathrm{i}t'\omega/\varepsilon}\partial_t\big[w(C_1 u_{\mathrm{I}})\cdot (C_1 u_{\mathrm{I}})\big]\,\mathrm{d}t'\\ &\quad + \mathrm{i}\frac{\varepsilon}{2\omega}\mathrm{e}^{-2\mathrm{i}t\omega/\varepsilon}\big[w(C_1 u_{\mathrm{I}})\cdot (C_1 u_{\mathrm{I}})\big](t).\end{aligned}$$

Using the equation satisfied by w to control $\partial_t w$ and Step 2 to control $\partial_t u_{\mathrm{I}}$, one readily deduces that

$$|\tilde{W}(t)|_{H^k(\mathbb{T};A)} \quad \le \quad C(T, |u^0|_A)\left(1 + |\nabla u^0|_A + |r|_A + \varepsilon\int_0^t |\tilde{Z}(t')|_{H^k(\mathbb{T};A)}\,\mathrm{d}t'\right). \tag{47}$$

- For $\tilde{U}$, we have $\tilde{U}|_{t=0} = 0$ and

$$\partial_t \tilde{U} + \frac{\mathrm{i}}{\varepsilon}\mathcal{L}(\omega D_\theta, \mathbf{k}D_\theta + \varepsilon D)\tilde{U} + \varepsilon^{1+p}A_0\tilde{U}$$
$$= -\Big(\mathcal{H}\Big(D_\theta + \varepsilon\frac{\mathbf{k}}{k^2}\cdot D\Big) - \mathrm{i}\Big)(WC_1^T C_1 U)$$
$$-\mathrm{i}\big(W_{\mathrm{app}} + \varepsilon\tilde{W}\big)C_1^T C_1\big(U_{\mathrm{app}} + \varepsilon\tilde{U}\big) + \mathrm{i}W_{\mathrm{app}}C_1^T C_1 U_{\mathrm{app}},$$

  where, for the sake of simplicity, we also have taken $c = 0$ because the corresponding terms do not raise any difficulty. Since the Fourier multiplier $\mathcal{H}(D_\theta + \varepsilon(\mathbf{k}/k^2)\cdot D) - \mathrm{i}$ has a norm of order ε when acting on A and because the difference of the last two terms in easily bounded in terms of $\varepsilon\tilde{Z}$, we get that $|\tilde{U}(t)|_{H^k(\mathbb{T};A)}$ satisfies the same upper bound as $|\tilde{W}(t)|_{H^k(\mathbb{T};A)}$ in (47).

Gathering the upper bounds for $\tilde{W}$ and $\tilde{U}$ we therefore get

$$\sup_{t'\in[0,t]} |\tilde{Z}(t')|_{H^k(\mathbb{T};A)} \le C(T, |u^0|_A)(1 + |\nabla u^0|_A + |r|_A)$$
$$+\varepsilon C(T, |u^0|_A)\int_0^t \sup_{t''\in[0,t']} |\tilde{Z}(t'')|_{H^k(\mathbb{T};A)}\,\mathrm{d}t' .$$

By a Gronwall estimate, we finally obtain a bound on $\tilde{Z}$,

$$\sup_{t\in[0,T/\varepsilon]} |\tilde{Z}(t)|_{H^k(\mathbb{T};B)} \le C(T, |u^0|_A)(1 + |\nabla u^0|_A + |r|_A),$$

which, since the A-norm controls the L^∞-norm, immediately leads to the desired estimate on $|\mathbf{Z} - \mathbf{Z}_{\mathrm{SVEA}}|_{L^\infty([0,T/\varepsilon]\times\mathbb{R}^d)}$. □

3.7.3 The NLS Equation with Ionization

As in Sect. 3.4, a Schrödinger type equation can be derived in presence of ionization,

$$\begin{cases} \partial_t u + \mathbf{c}_g\cdot\nabla u - \varepsilon\frac{\mathrm{i}}{2}\nabla\cdot H_{\mathbf{k}}\nabla u + \varepsilon^{1+p}\pi_1(\mathbf{k})A_0 u \\ \qquad = \varepsilon\pi_1(\mathbf{k})\left(F^{\mathrm{env}}(\varepsilon, u) - \mathrm{i}wC_1^T C_1 u - c\,C_1^T G^{\mathrm{env}}(C_1 u, w)\right), \\ \partial_t w = 2\varepsilon G^{\mathrm{env}}(C_1 u, w)\cdot\overline{C_1 u}. \end{cases} \tag{48}$$

Using Theorem 4, this model can be justified with a straightforward adaptation of the case without ionization.

Corollary 5 (Schrödinger Approximation). *Under the assumptions of Theorem 4, one has for all* $u^0 \in A^{(3)}$ *such that* $u_0 = \pi_1(\mathbf{k})u_0$:

(i) *There exist* $T > 0$ *and, for all* $\varepsilon \in (0,1]$, *a unique solution* $(u,w) \in C([0,T/\varepsilon];B^{(3)})$ *to* (48) *with initial condition* $(u^0,0)$.
(ii) *There exist* $\varepsilon_0 > 0$ *and* $\mathfrak{c}_{\mathrm{NLS}} > 0$ *such that for all* $0 < \varepsilon < \varepsilon_0$, *the solution* $\mathbf{Z}$ *to* (43) *provided by Theorem 3 exists on* $[0,T/\varepsilon]$ *and*

$$|\mathbf{Z} - \mathbf{Z}_{\mathrm{NLS}}|_{L^\infty([0,T/\varepsilon]\times\mathbb{R}^d)} \le \varepsilon C(T, |u^0|_A)(1 + |\nabla u^0|_A + \mathfrak{c}_{\mathrm{NLS}}|u^0|_{A^{(3)}}),$$

where $\mathbf{Z}_{\mathrm{NLS}}(t,x) = \big(u(t,x)e^{i(\mathbf{k}\cdot x - \omega t)/\varepsilon} + \text{c.c.}, w(t,x)\big)$.

3.7.4 The Most General Model

In (48), ionization effects have been added to the standard cubic NLS equation. It is of course possible to add them to a most sophisticated model that takes into account more general nonlinearities, improved frequency dispersion, and frequency dependent polarization. We then obtain the following generalization of (48) (for which the same justification as in Corollary 5 holds),

$$\begin{cases} (1 - i\varepsilon\mathbf{b}\cdot\nabla - \varepsilon^2\nabla\cdot B\nabla)\partial_t u + \mathbf{c}_g\cdot\nabla u \\ \qquad - \varepsilon\frac{i}{2}\nabla\cdot\big(H_{\mathbf{k}} + 2\mathbf{c}_g\otimes\mathbf{b}\big)\nabla u + \varepsilon^2 C_3(\nabla)u + \varepsilon^{1+p}\pi_1(\mathbf{k})A_0 u \\ \quad = \varepsilon\big[\pi_1(\mathbf{k}) + \varepsilon\pi_1(\mathbf{k})\pi_1'(\mathbf{k})\cdot D - i\varepsilon(\mathbf{b}\cdot\nabla)\pi_1(\mathbf{k})\big] \\ \qquad \times\Big(F^{\mathrm{env}}(\varepsilon,\pi_1(\mathbf{k})u) - \big(iwC_1^T C_1 u + c\,C_1^T G^{\mathrm{env}}(C_1 u, w)\big)\Big), \\ \partial_t w = 2\varepsilon G^{\mathrm{env}}(C_1 u, w)\cdot\overline{C_1 u}. \end{cases} \tag{49}$$

As shown in section "The Case with Charge and Current Density" in Appendix 2, this system of equations takes the following form corresponding to (4) in the case of Maxwell's equations,

$$\begin{cases} i(P_2(\varepsilon\nabla)\partial_t + c_g\partial_z)u + \varepsilon(\Delta_\perp + \alpha_1\partial_z^2)u + i\varepsilon\alpha_2 u \\ \qquad + \varepsilon(1 + i\varepsilon\mathbf{c}_3\cdot\nabla)[(|u|^2 - \rho)u + ic(\alpha_4|u|^{2K-2}u + \alpha_5\rho u)] = 0, \\ \partial_t\rho = \varepsilon\alpha_4|u|^{2K} + \varepsilon\alpha_5\rho|u|^2. \end{cases} \tag{50}$$

4 Analysis of (3) and (4), and Open Problems

In this section, we analyze and formulate open problems for the NLS-type equations (3), as well as the NLS-type equations (4) taking the ionization processes into account. We first consider in Sect. 4.1 the equation (3) in the case of, respectively, no and anomalous GVD (resp. $\alpha_1 = 0$ and $\alpha_1 = 1$), and also briefly discuss (4). Then, we discuss (3) in the case of normal GVD (i.e., $\alpha_1 = -1$) in Sect. 4.2. Finally, we formulate additional open problems in Sects. 4.3, 4.4, and 4.5.

4.1 The Case of No and Anomalous GVD (Resp. $\alpha_1 = 0$ and $\alpha_1 = 1$)

Let us first consider the case where $P_2, \alpha_2, \boldsymbol{\alpha}_3$, and f take the following values,

$$P_2 = I, \quad \alpha_2 = 0, \quad \boldsymbol{\alpha}_3 = 0 \quad \text{and} \quad f = 0,$$

in which case Eq. (3) corresponds to the focusing cubic NLS (1) in dimension $d = 2$ in the case of no GVD, and in dimension $d = 3$ in the case of anomalous GVD. As recalled in the introduction, (1) is critical in dimension 2, and supercritical in dimension 3, and there exist finite-time blow-up solutions in both cases. Let us also recall that some of these blow-up dynamics are stable (see, for example, [4–8] in the critical case, and [10] in a slightly supercritical case).

Now, let us recall that the breakdown of solutions is not always observed in physical experiments on the propagation of laser beams and that instead a *filamentation* phenomenon occurs. In this section, we would like to analyze the possibility that the modified model (3) in the case of no or anomalous GVD prevents the formation of singularities. Below, we analyze the role of each parameter of (3) separately starting with the nonlinearity f.

4.1.1 The Nonlinearity

In the case where $P_2 = I, \alpha_2 = 0$, and $\boldsymbol{\alpha}_3 = 0$, (3) takes the following form,

$$i\partial_t v + \Delta v + \big(1 + f(\varepsilon^r |v|^2)\big)|v|^2 v = 0\,, \quad t > 0,\ x \in \mathbb{R}^d, \tag{51}$$

where the dimension is $d = 2$ in the case of no GVD, and $d = 3$ in the case of anomalous GVD.

As recalled in the introduction, standard modifications of the cubic nonlinearity consist either of the cubic/quintic nonlinearity, i.e., $f(s) = -s$, or a saturated nonlinearity, i.e., f is a smooth function on $\mathbb{R}^+$ vanishing at the origin and such that $\big(1 + f(s)\big)s$ is bounded on $\mathbb{R}^+$ (e.g., $f(s) = -s/(1+s)$). Let us first consider the case of a saturated nonlinearity. In that case, the fact that $\big(1 + f(s)\big)s$ is bounded on $\mathbb{R}^+$ implies the following control of the nonlinear term in (51):

$$\big\|\big(1 + f(\varepsilon^r |v|^2)\big)|v|^2 v\big\|_{L^2} \lesssim \frac{\|v\|_{L^2}}{\varepsilon^r}.$$

Thus, running a fixed point argument yields the fact that this equation is locally well-posed in $L^2(\mathbb{R}^d)$ for any integer $d \geq 1$, with a time of existence only depending on the size of $\|u_0\|_{L^2}$. Since Eq. (51) still satisfies the conservation of mass, this immediately implies global existence for any initial data in $L^2(\mathbb{R}^d)$ and for any $d \geq 1$. Therefore, modifying the cubic nonlinearity in the standard NLS equation by a saturated nonlinearity does indeed prevent finite-time breakdown of the solutions.

Next, we consider the modification by a cubic/quintic nonlinearity. In that case, (51) becomes

$$i\partial_t v + \Delta v + |v|^2 v - \varepsilon^r |v|^4 v = 0, \quad t > 0,\ x \in \mathbb{R}^d. \tag{52}$$

Note that (52) still satisfies the conservation of mass, and the conserved energy is now given by

$$E(v(t)) = \frac{1}{2}\int |\nabla v(t,x)|^2\,dx - \frac{1}{4}\int |v(t,x)|^4\,dx + \frac{\varepsilon^r}{6}\int |v(t,x)|^6\,dx = E(v_0)\,.$$

Note also that

$$\begin{aligned} E(v(t)) &+ \left(\frac{1}{2} + \frac{3}{32\varepsilon^r}\right)\|v(t)\|_{L^2}^2 \\ &= \frac{1}{2}\|v(t)\|_{H^1}^2 + \frac{\varepsilon^r}{6}\int |v(t)|^2\left(|v(t|^2 - \frac{3}{4\varepsilon^r}\right)^2 dx \\ &\geq \frac{1}{2}\|v(t)\|_{H^1}^2 \end{aligned}$$

which together with the conservation of mass and energy yields

$$\|v(t)\|_{H^1} \leq \sqrt{2E(v_0) + \left(1 + \frac{3}{16\varepsilon^r}\right)\|v_0\|_{L^2}^2}. \tag{53}$$

Let us first consider the case of dimension 2. In this case, both the cubic focusing NLS and the quintic defocusing NLS are H^1 subcritical in the sense that the equation is locally well-posed in $H^1(\mathbb{R}^2)$ with a time of existence only depending on the size of $\|u_0\|_{H^1}$ (see [1]). The proof extends to (52) which is thus locally well-posed in $H^1(\mathbb{R}^2)$ with a time of existence only depending on the size of $\|u_0\|_{H^1}$. Together with the bound (53), this immediately implies global existence for any initial data in $H^1(\mathbb{R}^2)$. The case of dimension 3 is more involved, since the quintic defocusing NLS is H^1 critical in the sense that the equation is locally well-posed in $H^1(\mathbb{R}^3)$ with a time of existence depending on the shape of u_0 (see [27]). Thus, one cannot rely solely on the bound (53) to prove global existence. However, the case of combined nonlinearities is addressed in [28], where global existence of solutions with H^1 initial data is proved, when the nonlinearity is a sum of two H^1-subcritical powers, or the sum of an H^1-subcritical power and of a defocusing H^1-critical power. This includes Eq. (52). Therefore, modifying the cubic nonlinearity in the standard NLS equation by a quintic defocusing nonlinearity does indeed prevent finite-time breakdown of the solutions both in dimensions 2 and 3. Let us mention an interesting phenomenon regarding the qualitative behavior of the solutions. Physical experiments suggest an oscillatory behavior of the solution which focuses, then defocuses, refocuses,... in an almost periodic fashion (see [18] and the references therein). Such a behavior is also observed in numerical simulations and suggested by heuristic arguments (see, for example, [3, 29]).

Open Problen 1. Establish rigorously the "oscillatory" phenomenon of the solutions to (51) observed in physical experiments (see, e.g., [18]).

Another open problem concerns the behavior of the solution as $\varepsilon \to 0$. Pick an initial data $v_0 \in H^1(\mathbb{R}^d)$, $d = 2, 3$, leading to a finite-time blow-up solution v at time $T > 0$ to the cubic NLS (1). Consider the solution v_ε of (51) with the same initial data v_0. In the case of dimension 2, it is shown in [30] that $v_\varepsilon(t) \to v(t)$ in $H^1(\mathbb{R}^2)$ as $\varepsilon \to 0$ on $[0, T)$ and,

$$\lim_{\varepsilon \to 0} \|v_\varepsilon(T)\|_{H^1(\mathbb{R}^2)} = +\infty.$$

An interesting question is the understanding of the limit of $v_\varepsilon(t)$ as $\varepsilon \to 0$ for $t > T$. Partial results have been obtained in this direction in [31]. There, it is proved that only few scenarios are possible, but one would like to establish whether all scenarios do occur or only some of them, and which scenarios are generic.

Open Problen 2. Let $v_0 \in H^1(\mathbb{R}^d)$, $d = 2, 3$ be leading to a finite-time blow-up solution v at time $T > 0$ to the cubic NLS (1). Consider the solution v_ε of (51) with the same initial data v_0. Describe the behavior of $v_\varepsilon(t)$ as $\varepsilon \to 0$ for $t > T$.

4.1.2 Taking the Ionization Process into Account

Let us discuss (4) and (5) in the case of no or anomalous GVD which take the ionization process into account and also correspond to a modification of the nonlinearity in the focusing cubic NLS (1). Ionization is certainly the most important phenomenon leading to the formation of laser filaments. Recall that (4) is given by

$$\begin{cases} \mathrm{i}(\partial_t + c_g\partial_z)u + \varepsilon(\Delta_\perp + \alpha_1\partial_z^2)u + \varepsilon(|u|^2 - \rho)u \\ \quad = -\mathrm{i}\varepsilon c(\alpha_4|u|^{2K-2}u + \alpha_5\rho u), \\ \partial_t\rho = \varepsilon\alpha_4|u|^{2K} + \varepsilon\alpha_5\rho|u|^2, \end{cases} \tag{54}$$

where $\alpha_4, \alpha_5 \geq 0$, $c > 0$, where ρ is the density of electron created by ionization, while $\mathbf{c}_g = c_g\mathbf{e}_z$ is the group velocity associated with the laser pulse, and where $d = 2$ in the case of no GVD, and $d = 3$ in the case of anomalous GVD. Also, recall that (5) is given by

$$\begin{cases} i\partial_\tau v + (\Delta_\perp + \alpha_1\partial_z^2)v + (|v|^2 - \tilde{\rho})v = -ic(\alpha_4|v|^{2K-2}v + \alpha_5\tilde{\rho}v), \\ -c_g\partial_z\tilde{\rho} = \varepsilon\alpha_4|v|^{2K} + \varepsilon\alpha_5\tilde{\rho}|v|^2, \end{cases} \tag{55}$$

where $\tilde{\rho}$ corresponds to ρ written in a frame moving at the group velocity $\mathbf{c}_g = c_g\mathbf{e}_z$ and with respect to a rescaled time $\tau = \varepsilon t$, i.e.,

$$\rho(t, X_\perp, z) = \tilde{\rho}(\varepsilon t, X_\perp, z - c_g t).$$

In view of the second equation of (55), a boundary condition for $\tilde{\rho}$ has to be prescribed at $z = z_0$ for some z_0 in order to obtain a well-posed problem. A natural choice, which ensures that $\tilde{\rho} \geq 0$, is to prescribe $\tilde{\rho}$ at $+\infty$:v

$$\lim_{z\to+\infty} (\nabla^\perp)^l\tilde{\rho}(\tau, X_\perp, z) = 0, \quad \forall\, l \geq 0. \tag{56}$$

Let us first discuss the local well-posedness theory for Eqs. (54) and (55), starting with the first one. One may obtain the existence of solutions to (54) over an interval of time with size $O(\varepsilon^{-1})$. Indeed, for an integer $N > d/2$, differentiating N times with respect to space variables the equations both for v and ρ, multiplying (54), respectively, by $\nabla^N\bar{v}$ and $\nabla^N\rho$, integrating on $\mathbb{R}^d$, taking the imaginary part and integrating by parts for the first equation, and summing both equations yields:

$$\begin{aligned} \frac{\mathrm{d}}{\mathrm{d}t}[\|v\|_{H^N}^2 + \|\rho\|_{H^N}^2] &\lesssim \varepsilon(\|v\|_{H^N}^2 + \|\rho\|_{H^N}^2)(\|v\|_{L^\infty} + \|v\|_{L^\infty}^{2K-2} + \|\rho\|_{L^\infty}) \\ &\lesssim \varepsilon(\|v\|_{H^N}^2 + \|\rho\|_{H^N}^2)(1 + \|v\|_{H^N}^2 + \|\rho\|_{H^N}^2)^{K-1} \end{aligned}$$

where we used the fact that $N > d/2$ together with simple product rules and the Sobolev embedding. Integrating this differential inequality, we obtain a time of existence with size $O(\varepsilon^{-1})$.

Next, we discuss the local well-posedness theory for Eq. (55) supplemented with the boundary condition (56). For any integer N, we introduce the space $\mathcal{H}^N$,

$$\mathcal{H}^N = \left\{ f\colon \mathbb{R}^d \to \mathbb{R}, \sum_{j=0}^{N} \|\nabla^j f\|_{L^2_{X_\perp} L^\infty_z} < +\infty \right\}.$$

For an integer $N > d$, differentiating N times with respect to space variables the equations for v, multiplying the first equation of (55) by $\nabla^N\bar{v}$, integrating on $\mathbb{R}^d$, taking the imaginary part and integrating by parts yields:

$$\frac{\mathrm{d}}{\mathrm{d}t}\|v\|_{H^N}^2 \lesssim \|v\|_{H^N}^4 + \|v\|_{H^N}^{2K} + \|v\|_{H^N}^2\|\tilde{\rho}\|_{\mathcal{H}^N}$$

where we used the fact that $N > d$ together with the Sobolev embedding in the last inequality. Next, we estimate $\|\tilde{\rho}\|_{\mathcal{H}^N}$. In view of the second equation of (55) and the boundary condition (56), we have

$$\tilde{\rho}(t,x,y,z) = \int_z^{+\infty} \alpha_4 \varepsilon |v|^{2K}(t,x,y,\sigma) \exp\left(\frac{\alpha_5 \varepsilon}{c} \int_z^{\sigma} |v|^2(t,x,y,s)\, \mathrm{d}s\right) \mathrm{d}\sigma.$$

We infer

$$|\tilde{\rho}(t,x,y,z)| \lesssim \left(\int_{-\infty}^{+\infty} |v|^{2K}(t,x,y,\sigma)\, \mathrm{d}\sigma\right) \exp(\|v\|_{H^N}^2),$$

which yields

$$\|\tilde{\rho}\|_{\mathcal{H}^0} \lesssim \|v\|_{H^N}^{2K} \exp(\|v\|_{H^N}^2),$$

where we used the fact that $N > d$ together with the Sobolev embedding. Next, differentiating the second equation of (55) N times, we obtain

$$-c\partial_z \nabla^N \tilde{\rho} = \alpha_4 \varepsilon \nabla^N(|v|^{2K}) + \alpha_5 \varepsilon \nabla^N(\tilde{\rho}|v|^2).$$

In view of the boundary condition (56), this yields

$$\begin{aligned}
&\nabla^N \tilde{\rho}(t,x,y,z) \\
&= \int_z^{+\infty} \big(\alpha_4 \varepsilon \nabla^N(|v|^{2K})(t,x,y,\sigma) + \alpha_5 \varepsilon (\nabla^N(|v|^2 \tilde{\rho}) - |v|^2 \nabla^N \tilde{\rho})(t,x,y,\sigma)\big) \\
&\quad \times \exp\left(\frac{\alpha_5 \varepsilon}{c} \int_z^{\sigma} |v|^2(t,x,y,s)\, \mathrm{d}s\right) \mathrm{d}\sigma.
\end{aligned}$$

Hence, we deduce

$$\begin{aligned}
&|\nabla^N \tilde{\rho}(t,x,y,z)| \\
&\lesssim \left(\int_{-\infty}^{+\infty} \big(|\nabla^N(|v|^{2K})(t,x,y,\sigma) + |\nabla^N(|v|^2 \tilde{\rho}) - |v|^2 \nabla^N \tilde{\rho}|(t,x,y,\sigma)\big)\, \mathrm{d}\sigma\right) \\
&\quad \times \exp(\|v\|_{H^N}^2),
\end{aligned}$$

which yields

$$\|\tilde{\rho}\|_{\mathcal{H}^N} \lesssim \big(\|\tilde{\rho}\|_{\mathcal{H}^{N-1}} \|v\|_{H^N}^2 + \|v\|_{H^N}^{2K}\big) \exp\big(\|v\|_{H^N}^2\big).$$

By induction, we obtain

$$\|\tilde{\rho}\|_{\mathcal{H}^N} \lesssim \big(1 + \|v\|_{H^N}^{2N}\big) \|v\|_{H^N}^{2K} \exp\big((N+1)\|v\|_{H^N}^2\big).$$

This yields

$$\frac{\mathrm{d}}{\mathrm{d}t}\|v\|_{H^N}^2 \lesssim \|v\|_{H^N}^4 + \|v\|_{H^N}^{2K} + \|v\|_{H^N}^2 \big(1 + \|v\|_{H^N}^{2N}\big) \|v\|_{H^N}^{2K} \exp\big((N+1)\|v\|_{H^N}^2\big).$$

Integrating this differential inequality, we obtain local existence for $(v, \tilde{\rho})$ in $H^N \times \mathcal{H}^N$ for $N > d$.

Remark 13. Note that the factor ε in the second equation of (55), while present in the model, is not needed for the well-posedness theory.

Let us now come back to the issue of global well-posedness/finite time singularity formation. For equations (54) and (55), note that the L^2 norm of u (resp. v) is dissipated. Indeed, in the case of (55), multiplying by v, integrating on $\mathbb{R}^d$, taking the imaginary part and integrating by parts yields:

$$\frac{\mathrm{d}}{\mathrm{d}t}\|v\|_{L^2}^2 = -c\left(\alpha_4\|v\|_{L^{2K}}^{2K} + \alpha_5\int |v|^2\tilde{\rho}\,\mathrm{d}x\right),$$

from which we deduce that the L^2 norm is dissipated since $\tilde{\rho} \geq 0$. On the other hand, the energy is not conserved, nor dissipated. We thus cannot carry out the analysis of Sect. 4.1.1 for (52), even in the case of dimension 2.[9] An interesting problem would then be to prove that taking the ionization process into account (i.e., replacing the cubic focusing NLS (1) by either (54) or (55)) does indeed prevent finite-time breakdown of the solutions in dimensions 2 and 3. This is formulated in the following open problem.

Open Problen 3. Prove that the solutions to (54) and (55) are global in dimensions 2 and 3.

4.1.3 The Damping

In the case where $P_2 = I$, $\boldsymbol{\alpha}_3 = 0$, and $f = 0$, (3) takes the following form,

$$\mathrm{i}\partial_t v + \Delta v + i\alpha_2 v + |v|^2 v = 0, \tag{57}$$

where $d = 2$ in the case of no GVD, and $d = 3$ in the case of anomalous GVD. The mass and energy are not conserved quantities anymore. The mass decreases,

$$\|v(t)\|_{L^2} = \mathrm{e}^{-\alpha_2 t}\|v_0\|_{L^2}, \tag{58}$$

while for the energy we have

$$\frac{\mathrm{d}}{\mathrm{d}t}\left[\frac{1}{2}\|\nabla v\|_{L^2}^2 - \frac{1}{4}\int |v|^4\,\mathrm{d}x\right] = -\alpha_2\left(\|\nabla v\|_{L^2}^2 - \int |v|^4\,\mathrm{d}x\right), \tag{59}$$

[9]Note that a simpler model of damped NLS where the ρ or $\tilde{\rho}$ term is not present has been investigated in [32]. The authors obtain global existence for a certain range of parameters by controlling a modified energy even if it is not conserved.

so that the energy is neither increasing nor decreasing. Equation (57) has been analyzed in several works (see, e.g., [3, 33–35]). In particular, from standard arguments using Strichartz estimates, there is in dimensions 2 and 3 a continuous and increasing function θ such that global existence holds for $\alpha_2 > \theta(\|u_0\|_{H^1})$ (see [3, 35]). Also, in the 3-dimensional case, there is a function θ such that stable blow-up solutions exist for $0 \leq \alpha_2 \leq \theta(u_0)$ (see [33, 35]). Finally, the same behavior has been observed in numerical simulations and suggested by heuristic arguments in the 2-dimensional case (see [34]). Thus, a small α_2 does not prevent finite-time blow-up. Since the constant α_2 obtained in (4) is usually small, it appears that the damping cannot by itself explain the physical observation according to which the breakdown of solutions does not occur.

4.1.4 Off-Axis Variation of the Group Velocity

In the case where $\alpha_2 = 0$, $\boldsymbol{\alpha}_3 = 0$, and $f = 0$, (3) takes the following form,

$$\mathrm{i}P_2(\varepsilon\nabla)\partial_t v + \Delta v + |v|^2 v = 0, \tag{60}$$

where $d = 2$ in the case of no GVD, and $d = 3$ in the case of anomalous GVD. The energy remains unchanged and is still conserved, while the mass is replaced by the following conserved quantity,

$$\big(P_2(\varepsilon\nabla)v(t), v(t)\big) = (P_2(\varepsilon\nabla)v_0, v_0). \tag{61}$$

Let us first consider the case of full off-axis dependence, i.e., the case where

$$(P_2(\varepsilon\nabla)v, v) \gtrsim \|v\|_{L^2}^2 + \varepsilon^2\|\nabla v\|_{L^2}^2. \tag{62}$$

The operator $P_2(\varepsilon\nabla)$ is a second-order self-adjoint operator which is invertible in view of (62). We denote by $P_2(\varepsilon\nabla)^{-1}$ its inverse. We rewrite (60) in the following form,

$$i\partial_t v + P_2(\varepsilon\nabla)^{-1}\Delta v + P_2(\varepsilon\nabla)^{-1}(|v|^2 v) = 0. \tag{63}$$

Using Duhamel's formula, and the semi-group $\mathrm{e}^{\mathrm{i}tP_2(\varepsilon\nabla)^{-1}\Delta}$, we obtain

$$v = \mathrm{e}^{\mathrm{i}tP_2(\varepsilon\nabla)^{-1}\Delta}v_0 + \int_0^t \mathrm{e}^{\mathrm{i}(t-s)P_2(\varepsilon\nabla)^{-1}\Delta}P_2(\varepsilon\nabla)^{-1}(|v|^2 v)(s)\,\mathrm{d}s. \tag{64}$$

As a simple consequence of (62) and the Sobolev embedding in dimensions 2 and 3, we have

$$\|P_2(\varepsilon\nabla)^{-1}(|v|^2 v)\|_{H^1(\mathbb{R}^d)} \lesssim \frac{1}{\varepsilon}\||v|^2 v\|_{L^2(\mathbb{R}^d)} \lesssim \frac{1}{\varepsilon}\|v\|_{H^1(\mathbb{R}^d)}^3, \quad d = 2, 3. \tag{65}$$

Now, a fixed point argument based on the formulation (64) together with the estimate (65) and the fact that the semi-group $e^{itP_2(\varepsilon\nabla)^{-1}\Delta}$ is unitary on H^1 implies that (60) is locally well-posed in H^1 for dimensions $d = 2, 3$, with a time of existence only depending on the size of $\|u_0\|_{H^1}$. Together with the lower bound (62) and the conserved quantity (61) which yield a uniform in time bound on the H^1 norm of the solution, this immediately implies global existence for any initial data in H^1 when $d = 2, 3$. Therefore, modifying the focusing cubic NLS equation by adding a full off-axis dependence does indeed prevent the finite-time breakdown of the solutions.

Next, let us consider the case of partial off-axis dependence, i.e., the case where

$$(P_2(\varepsilon\nabla)v, v) \gtrsim \|v\|_{L^2}^2 + \varepsilon^2 \sum_{k=1}^{j} \|v_k \cdot \nabla v\|_{L^2}^2, \tag{66}$$

for vectors v_k in $\mathbb{R}^d$, $k = 1, \ldots, j$, where $j = 1$ if $d = 2$, and $j = 1$ or $j = 2$ if $d = 3$ so that the right-hand side of (66) does not control the full H^1 norm. Notice that the well-posedness theory has yet to be investigated in this case. Indeed, consider the nonlinear term in the formulation (64),

$$e^{itP_2(\varepsilon\nabla)^{-1}\Delta} P_2(\varepsilon\nabla)^{-1}(|v|^2 v). \tag{67}$$

In the directions v_k of (66), the operator $P_2(\varepsilon\nabla)^{-1}$ gains two derivatives, while the semi-group $e^{itP_2(\varepsilon\nabla)^{-1}\Delta}$ does not disperse (and thus does not satisfy a useful Strichartz estimate). On the other hand, in the directions orthogonal to the vectors v_k, the operator $P_2(\varepsilon\nabla)^{-1}$ does not gain any derivative, while the semi-group $e^{itP_2(\varepsilon\nabla)^{-1}\Delta}$ should satisfy a useful Strichartz estimate. Thus, in order to investigate the well-posedness theory, one should try to combine the regularization provided by the operator $P_2(\varepsilon\nabla)^{-1}$ in the directions v_k of (66) with the dispersive properties of the semi-group $e^{itP_2(\varepsilon\nabla)^{-1}\Delta}$ in the direction orthogonal to the vectors v_k. Now, it would be interesting to investigate whether a suitable well-posedness theory together with the conserved quantities given by the energy and (61) yield global existence. This suggests the following open problem.

Open Problen 4. Investigate both the local and global well-posedness for Eq. (60) in the case of partial off-axis dependence. In particular, does the modification of the focusing cubic NLS (1) by the addition of a partial off-axis dependence prevent the finite-time breakdown of the solutions?

4.1.5 Self-steepening of the Pulse

In the case where $P_2 = I$, $\alpha_2 = 0$, and $f = 0$, (3) takes the following form,

$$i\partial_t v + \Delta v + (1 + i\varepsilon\boldsymbol{\alpha}_3 \cdot \nabla)|v|^2 v = 0, \tag{68}$$

where $d = 2$ in the case of no GVD, and $d = 3$ in the case of anomalous GVD. Adding the operator $(1 + i\varepsilon\boldsymbol{\alpha}_3 \cdot \nabla)$ in front of the nonlinearity of the focusing cubic NLS does certainly not prevent finite-time breakdown of the solutions. Indeed, one expects to obtain even more blow-up solutions since the formation of optical shocks is expected in this case. The term shock is used in view of the similarity with the Burgers equation—both in terms of the nonlinearity of the equation and the profiles of the solutions observed in some numerical simulations (see [36] for more details on optical shocks). The reason we included this modification in the discussion is because it may account for the physical observation of the self-steepening of the pulse: an initial pulse which is symmetric becomes asymmetric after propagating over a large distance and its profile seems to develop a shock. This phenomenon has been widely observed, and we refer the reader to [18] and the references therein. Thus, an interesting open problem would be to exhibit solutions of (68) for which the corresponding profile develops a shock. We formulate a slightly more general open problem below.

Open Problen 5. Describe the blow-up scenarios for the finite-time blow-up solutions to EQ. (68).

4.2 *The Case of Normal GVD (i.e., $\boldsymbol{\alpha}_1 = -1$)*

Let us consider the case where $P_2, \alpha_1, \alpha_2, \boldsymbol{\alpha}_3$, and f take the following value,

$$P_2 = I, \quad \alpha_1 = -1, \quad \alpha_2 = 0, \quad \boldsymbol{\alpha}_3 = 0 \quad \text{and} \quad f = 0,$$

in which case Eq. (3) corresponds to the following equation in dimension 2 or 3,

$$i\partial_t v + (\Delta_\perp - \partial_z^2)v + |v|^2 v = 0. \tag{69}$$

This equation is sometimes referred to as hyperbolic cubic NLS, or non-elliptic cubic NLS. Since Strichartz estimates only depend on the curvature of the corresponding characteristic manifold (see [37]), (69) satisfies the same Strichartz estimates as the cubic focusing NLS. Thus, the result of Ginibre and Velo [1] immediately extends to (69) which is locally well-posed in $H^1 = H^1(\mathbb{R}^d)$ with $d = 2, 3$. Therefore, for $v_0 \in H^1$, there exists $0 < T \leq +\infty$ and a unique solution $v(t) \in \mathcal{C}([0, T), H^1)$ to (69) and either $T = +\infty$, and the solution is global, or the solution blows up in finite time $T < +\infty$ and then $\lim_{t \uparrow T} \|\nabla u(t)\|_{L^2} = +\infty$.

Equation (69) admits the following conservation laws in the energy space H^1,

L^2-norm: $\|v(t)\|_{L^2}^2 = \|v_0\|_{L^2}^2$;

Energy: $E(v(t)) = \frac{1}{2}\int |\nabla_\perp v(t,x)|^2\,dx - \frac{1}{2}\int |\partial_z v|^2\,dx - \frac{1}{4}\int |v(t,x)|^4 dx$
$= E(v_0)$;

Momentum: $\mathrm{Im}(\int \nabla v(t,x)\overline{v(t,x)}\,dx) = \mathrm{Im}(\int \nabla v_0(x)\overline{v_0(x)}\,dx)$.

A large group of H^1 symmetries leaves the equation invariant: if $u(t,x)$ solves (69), then $\forall(\lambda_0, \tau_0, x_0, \gamma_0) \in \mathbb{R}_*^+ \times \mathbb{R} \times \mathbb{R}^3 \times \mathbb{R}$, so does

$$u(t,x) = \lambda_0 v(\lambda_0^2 t + \tau_0, \lambda_0 x + x_0) e^{i\gamma_0}. \tag{70}$$

Note that (69) is not invariant under the usual Galilean transform. However, it is invariant under a twisted Galilean transform. For $\beta = (\beta_1, \beta_2, \beta_3) \in \mathbb{R}^3$, we define

$$\hat{\beta} = (\beta_1, \beta_2, -\beta_3) \in \mathbb{R}^3.$$

Then (69) is invariant under the following twisted Galilean transform,

$$v_\beta(t,x) = v(t, x - \beta t) e^{i\hat{\beta}/2\cdot(x-(\beta/2)t)}. \tag{71}$$

Note also that the scaling symmetry $u(t,x) = \lambda_0 v(\lambda_0^2 t, \lambda_0 x)$ leaves the space $L^2(\mathbb{R}^2)$ invariant so that (69) is critical with respect to the conservation of mass in dimension 2, while it leaves the homogeneous Sobolev space $\dot{H}^{1/2}(\mathbb{R}^3)$ invariant so that (69) is supercritical with respect to the conservation of mass in dimension 3.

In contrast to the cubic focusing NLS, the existence or absence of finite-time blow-up solutions for (69) is widely open. While there exists a counterpart to the virial for (69), it is too weak to provide the existence of finite-time blow up dynamics (see the discussion in [3]). Also, there are no standing waves in the form $v(t,x) = Q(x)e^{i\omega t}$, $\omega \in \mathbb{R}$ in the energy space for the Eq. (69) (see [38]), while for the focusing cubic NLS, such an object exists and is fundamental in the analysis of the blow-up dynamics (see, e.g., [4–9] in dimension 2, and [10] in dimension 3). Now, numerical simulations suggest that the solutions to (69) do not break down in finite time (see, e.g., [3, 29] and the references therein). In particular, the simulations exhibit a phenomenon called pulse splitting where the pulse focuses until it reaches a certain threshold. Once this threshold is attained, the pulse splits into two pulses of less amplitude moving away from each other. This phenomenon might repeat itself (multiple splitting) and ultimately prevent finite-time blow-up, but this has not been clearly backed up by numerics so far. Thus an interesting problem would be to prove that the cubic NLS equation (69) (i.e., in the case of normal GVD) does not have finite-time blow-up solutions, whether this is due to pulse splitting, or to some other phenomenon. This is formulated in the two open problems below.

Open Problen 6. Prove that the solutions to (69) are global in dimensions 2 and 3.

Open Problen 7. Describe rigorously the phenomenon of pulse splitting for (69).

Remark 14. We only considered in this section the case where

$$P_2 = I, \quad \alpha_2 = 0, \quad \boldsymbol{\alpha}_3 = 0 \quad \text{and} \quad f = 0$$

which is (69). While it is of interest to investigate the qualitative role of the parameters $P_2, \alpha_2, \boldsymbol{\alpha}_3$ and f of (3) in the case of normal GVD (i.e., the case

$\alpha_1 = -1$), we chose not to investigate their effects since we are interested in this paper primary on the non-existence of focusing dynamics, and (69) is expected to have only global solutions.

Remark 15. In a recent preprint [39], standing waves and self-similar solutions have been exhibited for (69) in dimension 2. However, these solutions do not belong to the energy space. In particular, they are not in C^2 and do not decay along the two diagonals in $\mathbb{R}^2$.

4.3 Mixing Several Phenomena

In Sects. 4.1 and 4.2, we have investigated the effect of each of the parameters $P_2, \alpha_1, \alpha_2, \boldsymbol{\alpha}_3$, and f on the solutions to Eq. (3), and in particular whether these modifications of the cubic NLS (1) prevent the existence of finite-time blow-up solutions. We have also formulated several open problems. Now, instead of studying all these effects separately, one may investigate the case when all these phenomena are present at the same time, and ask which are the dominant ones? Consider, for instance, the case where there is partial off-axis variation of the group velocity in the direction z, and at the same time a potential self-steepening of the pulse in the same direction, i.e., $P_2 = 1 - \partial_z^2$ and $\boldsymbol{\alpha}_3 = e_z$,

$$\mathrm{i}(1 - \varepsilon^2\partial_z^2)\partial_t v + \Delta v + (1 + i\varepsilon\partial_z)(|v|^2 v) = 0. \tag{72}$$

Then, one may wonder whether the partial off-axis variation of the group velocity prevents the self-steepening from taking place. This is formulated in the open problem below.

Open Problen 8. Do the solutions to (72) exhibit the phenomenon of self-steepening?

4.4 The Vectorial Case

Recall that (3) is a particular case of the vectorial equation $(3)_{\text{vect}}$,

$$\mathrm{i}P_2(\varepsilon\nabla)\partial_t \mathbf{v} + (\Delta_\perp + \alpha_1\partial_z^2)\mathbf{v} + \mathrm{i}\alpha_2\mathbf{v} + \tfrac{1}{3}(1 + \mathrm{i}\varepsilon\boldsymbol{\alpha}_3\cdot\nabla)\big[(\mathbf{v}\cdot\mathbf{v})\overline{\mathbf{v}} + 2|\mathbf{v}|^2\mathbf{v}\big] = 0\,,$$

where $\mathbf{v}$ is now a $\mathbb{C}^2$-valued function, and where we consider for simplicity the equation corresponding to the cubic case (i.e., $f = 0$ in (3)). In fact, (3) is a particular case of $(3)_{\text{vect}}$ corresponding to initial data living on a one-dimensional subspace of $\mathbb{C}^2$ (see Remark. 2). Now, one may of course consider the previous questions formulated for (3), and investigate the same problems for the vectorial case. This is formulated in the open problem below.

Open Problen 9. Investigate the vectorial counterparts for $(3)_{\text{vect}}$ of the various open problems formulated for the scalar equation (3) in Sects. 4.1–4.3, and 4.5.

4.5 The Approximation of the Maxwell Equations over Longer Times

We have provided in Sect. 3 a rigorous justification for all the NLS-type models derived in this paper. More precisely, we have proved that there exists $T > 0$ such that

1. The exact solution $\mathbf{U}$ of (20)–(22) exists on $[0, T/\varepsilon]$.
2. The NLS-type model under consideration (e.g., (39)) admits a unique solution u_{app} with initial condition u^0 on the same time interval.
3. The approximation $\mathbf{U}_{\text{app}}(t, x) = u_{\text{app}}(t, x)e^{i(\mathbf{k}\cdot x-\omega t)/\varepsilon} + \text{c.c.}$ remains close to $\mathbf{U}$ on this time interval,[10]

$$|\mathbf{U} - \mathbf{U}_{\text{app}}|_{L^\infty([0,T/\varepsilon]\times\mathbb{R}^d)} \leq \varepsilon C(T, |u_{\text{app}}|_{L^\infty([0,T/\varepsilon];B)})(1 + |\nabla u^0|_B + |u^0|_{B^{(3)}}).$$

Such a justification is far from being sharp. In particular, for the standard cubic focusing NLS equation, one has $T < T_{\text{foc}}$, where T_{foc} is the time when focusing occurs. Indeed, the above justification process requires that u_{app} remains bounded on $[0, T/\varepsilon]$ (after rescaling, this is equivalent to require that the solution v to (3) is bounded on $[0, T]$). In the error estimate above, the constant $C(T, |u_{\text{app}}|_{L^\infty([0,T/\varepsilon];B)})$ therefore blows up as $T \to T_{\text{foc}}$.

It follows that all the NLS-type equations derived here are justified far enough from the focusing time. Now, relevant differences between the different models can only be observed close enough to focusing (they all differ by formally $O(\varepsilon^2)$ terms that become relevant only near the focusing point). So, roughly speaking, we have only proved that all *the models are justified on a time interval for which they are basically identical*.

As seen above, existence beyond the focusing time of the standard cubic NLS equation is proved or expected for many of the variants considered here. For such models, the above argument does not work, i.e., the constant $C(T, |u_{\text{app}}|_{L^\infty([0,T/\varepsilon];B)})$ of the error estimate does not blow up as $T \to T_{\text{foc}}$. This is, for instance, the case of the cubic/quintic NLS equation (52) that admits a global solution v^ε. However (see Open Problem 2), we have

$$\lim_{\varepsilon\to 0} \lim_{T\to T_{\text{foc}}} |v^\varepsilon(t, \cdot)|_{H^1} = +\infty,$$

[10]In the error estimates of Sect. 3, the constant on the right-hand side is $C(T, |u^0|_B)$ rather than $C(T, |u_{\text{app}}|_{L^\infty([0,T/\varepsilon];B)})$. Since $|u_{\text{app}}|_{L^\infty([0,T/\varepsilon];B)} = C(T, |u_0|_B)$, this is of course equivalent, but the first form is more convenient for the present discussion.

so that there is no reason to expect the error term $\varepsilon C(T, |v^\varepsilon|_{L^\infty([0,T/\varepsilon];B)})$ to be small near the focusing point and for small ε.

For the moment, the merits of the NLS-type models derived here to describe correctly the mechanisms at stake during focusing can only by assessed numerically. Hence the following interesting open problem,

Open Problen 10. Rigorously justify one of the NLS-type models derived here on a time scale $[0, T/\varepsilon]$ with $T \geq T_{\rm foc}$.

Acknowledgements David Lannes acknowledges support from the ANR-13-BS01-0003-01 DYFICOLTI, the ANR BOND.

The authors warmly thank Christof Sparber for pointing [28] out.

Appendix 1: Nondimensionalization of the Equations

The Case Without Charge nor Current Density

There are two characteristic times for the situation considered here. The first one, denoted $\bar{t}$, is the inverse of the frequency of the laser pulse; the second one, denoted $\overline{T}$, is the duration of the pulse. In the regimes considered here,[11] one has $\overline{T} \gg \bar{t}$, and the small parameter ε is defined as

$$\varepsilon = \frac{\bar{t}}{\overline{T}}.$$

The time variable is naturally nondimensionalized with $\overline{T}$, while the space variables are nondimensionalized by $L = c\overline{T}$, namely,

$$t = \overline{T}\tilde{t}, \quad x = L\tilde{x},$$

where dimensionless quantities are denoted with a tilde. We also nondimensionalize the unknowns E, B, and P with typical orders E_0, B_0, and P_0. The typical scale for the electric and magnetic fields is directly deduced from P_0,

$$\mathrm{E}_0 = \frac{1}{\epsilon_0}\mathrm{P}_0, \quad \mathrm{B}_0 = \frac{1}{c}\mathrm{E}_0.$$

In order to choose the remaining P_0, some consideration on the polarization equation is helpful. We recall that (11) is given by

$$\partial_t^2 \mathrm{P} + \Omega_1 \partial_t \mathrm{P} + \Omega_0^2 \mathrm{P} - \nabla V_{\rm NL}(\mathrm{P}) = \epsilon_0 b \mathrm{E}.$$

[11] We refer to [40] for variants of this nondimensionalization.

For all the applications we have in mind, one has $\nabla V_{\mathrm{NL}}(\mathrm{P}) = a_3|\mathrm{P}|^2\mathrm{P} + \text{h.o.t.}$, so that we nondimensionlize the nonlinear term as

$$\nabla_{\mathrm{P}} V_{\mathrm{NL}}(\mathrm{P}) = a_3 \mathrm{P}_0^3 \nabla_{\tilde{\mathrm{P}}} \tilde{V}_{\mathrm{NL}}(\tilde{\mathrm{P}}),$$

where $\tilde{V}_{\mathrm{NL}}$ is the dimensionless potential

$$\tilde{V}_{\mathrm{NL}}(\tilde{\mathrm{P}}) = \frac{1}{a_3 \mathrm{P}_0^4} V_{\mathrm{NL}}(\mathrm{P}).$$

Example 10. For the three cases considered in Example 1, we find, denoting $\tilde{a}\varepsilon^r = a_5 \mathrm{P}_0^2/a_3$, where $r > 0$ is chosen such that $\tilde{a} = O(1)$,

(i) Cubic nonlinearity:

$$\nabla_{\tilde{\mathrm{P}}} \tilde{V}_{\mathrm{NL}}(\tilde{\mathrm{P}}) = |\tilde{\mathrm{P}}|^2 \tilde{\mathrm{P}}.$$

(ii) Cubic/quintic nonlinearity:

$$\nabla_{\tilde{\mathrm{P}}} \tilde{V}_{\mathrm{NL}}(\tilde{\mathrm{P}}) = (1 - \tilde{a}\varepsilon^r |\tilde{\mathrm{P}}|^2)|\tilde{\mathrm{P}}|^2 \tilde{\mathrm{P}}.$$

(iii) Saturated nonlinearity:

$$\nabla_{\tilde{\mathrm{P}}} \tilde{V}_{\mathrm{NL}}(\tilde{\mathrm{P}}) = \frac{1 + (\tilde{a}\varepsilon^r/3)|\tilde{\mathrm{P}}|^2}{(1 + (2\tilde{a}\varepsilon^r/3)|\tilde{\mathrm{P}}|^2)^2} |\tilde{\mathrm{P}}|^2 \tilde{\mathrm{P}}.$$

All these examples can therefore be put under the form

$$\nabla_{\tilde{\mathrm{P}}} \tilde{V}_{\mathrm{NL}}(\tilde{\mathrm{P}}) = \big(1 + \tilde{f}(\tilde{a}\varepsilon^r |\tilde{\mathrm{P}}|^2)\big)|\tilde{\mathrm{P}}|^2 \tilde{\mathrm{P}},$$

where $\tilde{f} : \mathbb{R}^+ \to \mathbb{R}$ is a smooth function such that $\tilde{f}(0) = 0$.

In dimensionless form, the polarization equation becomes therefore

$$\partial_{\tilde{t}}^2 \tilde{\mathrm{P}} + \overline{T}\Omega_1 \partial_{\tilde{t}} \tilde{\mathrm{P}} + \overline{T}^2 \Omega_0^2 \tilde{\mathrm{P}} - a_3 \overline{T}^2 \mathrm{P}_0^2 \nabla_{\tilde{\mathrm{P}}} \tilde{V}_{\mathrm{NL}}(\tilde{\mathrm{P}}) = \epsilon_0 b \frac{\overline{T}^2 \mathrm{E}_0}{\mathrm{P}_0} \tilde{\mathrm{E}}.$$

We therefore choose P_0 to be the typical nonlinear scale for which the nonlinear terms in the above equations are of the same order as the second order time derivative, namely,

$$\mathrm{P}_0 = \frac{1}{\overline{T}\sqrt{a_3}}.$$

Finally, we need some information on the size of Ω_0, Ω_1 and b to give our final dimensionless form of the equations. The resonance frequency of the harmonic oscillator is typically of the same order as the frequency of the laser pulse,

$$\Omega_0 = \bar{t}^{-1}\omega_0;$$

the damping frequency is written under the form

$$\Omega_1 = \varepsilon^{\alpha+1}\bar{t}^{-1}\,\omega_1,$$

where the dimensionless damping frequency ω_1 has size $O(1)$ while the coefficient α depends on the medium in which the laser propagates. Typically, $\alpha = 1$ or $\alpha = 2$. Writing the coupling constant b as

$$b = \frac{\gamma}{\bar{t}^2},$$

the dimensionless polarization equations can therefore be written as

$$\partial_{\tilde{t}}^2\tilde{\mathrm{P}} + \varepsilon^{\alpha}\omega_1\partial_{\tilde{t}}\tilde{\mathrm{P}} + \varepsilon^{-2}\omega_0^2\tilde{\mathrm{P}} - \nabla_{\tilde{\mathrm{P}}}\tilde{V}_{\mathrm{NL}}(\tilde{\mathrm{P}}) = \varepsilon^{-2}\gamma\tilde{\mathrm{E}}.$$

Omitting the tildes for the sake of clarity, Maxwell's equations (10)–(11) read therefore (without charge nor current density),

$$\begin{cases} \partial_t\mathrm{B} + \operatorname{curl}\mathrm{E} = 0, \\ \partial_t\mathrm{E} - \operatorname{curl}\mathrm{B} = -\partial_t\mathrm{P}, \\ \partial_t^2\mathrm{P} + \varepsilon^{\alpha}\omega_1\partial_t\mathrm{P} + \varepsilon^{-2}\omega_0^2\mathrm{P} - \nabla V_{\mathrm{NL}}(\mathrm{P}) = \varepsilon^{-2}\gamma\mathrm{E}. \end{cases}$$

Introducing the auxiliary unknowns,

$$\mathrm{P}^\sharp = \frac{\omega_0}{\sqrt{\gamma}}\mathrm{P}, \quad \mathrm{Q}^\sharp = \frac{\varepsilon}{\omega_0}\partial_t\mathrm{P}^\sharp = \frac{\varepsilon}{\sqrt{\gamma}}\partial_t\mathrm{P},$$

and working with nonlinearities as those considered in Example 10, this system can be written as a first order system,

$$\begin{cases} \partial_t\mathrm{B} + \operatorname{curl}\mathrm{E} = 0, \\ \partial_t\mathrm{E} - \operatorname{curl}\mathrm{B} + \dfrac{1}{\varepsilon}\sqrt{\gamma}\mathrm{Q}^\sharp = 0, \\ \partial_t\mathrm{Q}^\sharp + \varepsilon^{\alpha}\omega_1\mathrm{Q}^\sharp - \dfrac{1}{\varepsilon}(\sqrt{\gamma}\mathrm{E} - \omega_0\mathrm{P}^\sharp) = \varepsilon\dfrac{\gamma}{\omega_0^3}\big(1 + f(\varepsilon^r|\mathrm{P}^\sharp|^2)\big)|\mathrm{P}^\sharp|^2\mathrm{P}^\sharp, \\ \partial_t\mathrm{P}^\sharp - \dfrac{1}{\varepsilon}\omega_0\mathrm{Q}^\sharp = 0, \end{cases} \tag{73}$$

where we wrote $f(x) = \tilde{f}(\tilde{a}(\gamma/\omega_0^2)x)$.
This system has the form (20) with $n = 12$, $\mathbf{U} = (\mathrm{B}, \mathrm{E}, \mathrm{Q}^\sharp, \mathrm{P}^\sharp)^T$ and

$$A(\partial) = \begin{pmatrix} 0 & \mathrm{curl} & 0 & 0 \\ -\mathrm{curl} & 0 & 0 & 0 \\ 0 & 0 & 0 & 0 \\ 0 & 0 & 0 & 0 \end{pmatrix}, \quad E = \begin{pmatrix} 0 & 0 & 0 & 0 \\ 0 & 0 & \sqrt{\gamma} I & 0 \\ 0 & -\sqrt{\gamma} I & 0 & \omega_0 I \\ 0 & 0 & -\omega_0 I & 0 \end{pmatrix}$$

while A_0 is the block diagonal matrix $A_0 = \mathrm{diag}(0, 0, \omega_1 I, 0)$, so that the first two points of Assumption 1 are satisfied. For the third one, remark that the nonlinearity is given by

$$F(\varepsilon, \mathbf{U}) = \left(0, 0, \frac{\gamma}{\omega_0^3}(1 + f(\varepsilon^r |\mathrm{P}^\sharp|^2))|\mathrm{P}^\sharp|^2 \mathrm{P}^\sharp, 0\right)^T, \tag{74}$$

which is of the same form as in Assumption 1 with $Q(U) = |\mathrm{P}^\sharp|^2$ and $T(U, \overline{U}, U) = (0, 0, (\gamma/\omega_0^3)|\mathrm{P}^\sharp|^2 \mathrm{P}^\sharp, 0)^T$. One can check that such nonlinearities also satisfy Assumption 3.

The Case with Charge and Current Density

Now, we deal with the nondimensionalization of (10), (11), (17). We proceed in the same way as in the previous paragraph, simply adding the characteristic sizes J_0 and ρ_0 of the free electron current density J_e and charge density ρ. Considerations on the polarization equation are the same.

We choose ρ_0 from the equation for the charge density,

$$\partial_{\tilde{t}} \tilde{\rho} = \sigma_K \rho_{\mathrm{nt}} \frac{\mathrm{E}_0^{2K}}{\rho_0} \overline{T} |\tilde{\mathrm{E}}|^{2K} + \frac{\sigma}{U_\mathrm{i}} \mathrm{E}_0^2 \overline{T} \tilde{\rho} |\tilde{\mathrm{E}}|^2.$$

Many configurations are possible, according to the numerical value of the various coefficients involved (see, for instance, [18] for experimental data). We choose here a configuration that leads to the richest model, where the coupling constants of the two nonlinear terms are of the same order; more precisely

$$\partial_{\tilde{t}} \tilde{\rho} = \varepsilon c_1 |\tilde{\mathrm{E}}|^{2K} + \varepsilon c_2 \tilde{\rho} |\tilde{\mathrm{E}}|^2,$$

for nonnegative constants c_1, c_2. This corresponds to the following choice for ρ_0,

$$\rho_0 = \frac{\sigma_K \rho_{\mathrm{nt}} E_0^{2K}}{\varepsilon c_1}.$$

Knowing ρ_0, we can then determine J_0 from the equation for the free electron current density,

$$\mathrm{J}_e = \frac{q_e^2}{\omega_1 m_e}\rho_0 E_0\,\tilde{\rho}\,\mathcal{H}(\varepsilon D_{\tilde{z}})\tilde{\mathrm{E}},$$

which naturally leads to

$$\mathrm{J}_0 = \frac{q_e^2}{\omega_1 m_e}\rho_0 E_0 = \varepsilon\frac{q_e^2 \bar{T}\rho_0 E_0}{m_e}.$$

Going back to Maxwell's equations, we get

$$\begin{cases} \partial_{\tilde{t}}\tilde{\mathrm{B}} + \operatorname{curl}\tilde{\mathrm{E}} = 0, \\ \partial_{\tilde{t}}\tilde{\mathrm{E}} - \operatorname{curl}\tilde{\mathrm{B}} = -\partial_{\tilde{t}}\tilde{\mathrm{P}} - \dfrac{1}{\epsilon_0}\bar{T}\dfrac{\mathrm{J}_0}{\mathrm{E}_0}\tilde{\mathrm{J}}_e - \dfrac{U_i\rho_0}{\epsilon_0 E_0}\big(\varepsilon c_1|\tilde{\mathrm{E}}|^{2K-2} + \varepsilon c_2\tilde{\rho}\big)\tilde{\mathrm{E}}. \end{cases}$$

Here again, many configurations can be found, and we choose one that leads to the richest models, namely,

$$\frac{1}{\epsilon_0}\bar{T}\frac{\mathrm{J}_0}{\mathrm{E}_0} = \varepsilon c_3, \qquad \frac{U_i\rho_0}{\epsilon_0 E_0} = c_0$$

where c_0, c_3 are dimensionless positive constants. Replacing $\tilde{\mathrm{J}}_e$ by $c_3\tilde{\mathrm{J}}_e$, $\tilde{\rho}$ by $c_3\tilde{\rho}$ and c_1 by c_1c_3, we can assume that $c_3 = 1$.

Finally, dropping the tildes, introducing the unknowns $\mathrm{Q}^\sharp$ and $\mathrm{P}^\sharp$, and using the same notations as in (73), we obtain

$$\begin{cases} \partial_t \mathrm{B} + \operatorname{curl}\mathrm{E} = 0, \\ \partial_t \mathrm{E} - \operatorname{curl}\mathrm{B} + \dfrac{1}{\varepsilon}\sqrt{\gamma}\mathrm{Q}^\sharp = -\varepsilon\rho\mathcal{H}(\varepsilon D_z)\mathrm{E} - \varepsilon c_0\big(c_1|\mathrm{E}|^{2K-2} + c_2\rho\big)\mathrm{E}, \\ \partial_t \mathrm{Q}^\sharp + \varepsilon^\alpha\omega_1\mathrm{Q}^\sharp - \dfrac{1}{\varepsilon}(\sqrt{\gamma}\mathrm{E} - \omega_0\mathrm{P}^\sharp) = \varepsilon\dfrac{\gamma}{\omega_0^3}\big(1 + f(\varepsilon^r|\mathrm{P}^\sharp|^2)\big)|\mathrm{P}^\sharp|^2\mathrm{P}^\sharp, \\ \partial_t \mathrm{P}^\sharp - \dfrac{1}{\varepsilon}\omega_0\mathrm{Q}^\sharp = 0, \\ \partial_t\rho = \varepsilon c_1|\mathrm{E}|^{2K} + \varepsilon c_2\rho|\mathrm{E}|^2. \end{cases} \tag{75}$$

Appendix 2: Explicit Computations for Maxwell's Equations

The Case Without Charge nor Current Density

We derive here the variants of the NLS equations derived in this paper in the particular case of the Maxwell equations (73). We derive all the versions of the NLS equations that do not take into account the frequency dependence of the polarization (this corresponds to (37)). This latter effect, which leads to the generalization (41) of the previous model, is examined separately. For the sake of simplicity, we only consider cubic nonlinearities here; we show that in this context, (41) reduces to the vectorial family of NLS equation $(3)_{\mathrm{vect}}$.

Without Frequency Dependent Polarization

In order to check that Assumption 4 is satisfied by the dimensionless Maxwell equations (73), we need to compute the eigenvalues and eigenvectors of the matrix

$$\mathcal{L}(0,\mathbf{k}) = A(\mathbf{k}) + \frac{1}{\mathrm{i}}E,$$

where $A(\mathbf{k})$ and E are given in section "The Case with Charge and Current Density" in Appendix 1, which is equivalent to find all nontrivial solutions to the equation

$$\mathcal{L}(\omega,\mathbf{k})u = 0, \quad \text{with } \mathcal{L}(\omega,\mathbf{k}) = -\omega I + A(\mathbf{k}) + \frac{1}{\mathrm{i}}E,$$

and $\omega \in \mathbb{R}$, $u = (\mathtt{b}, \mathtt{e}, \mathtt{q}^\sharp, \mathtt{p}^\sharp) \in \mathbb{C}^{12}$.

From the last two equations of (73), we can always write $\mathtt{p}^\sharp$ and $\mathtt{q}^\sharp$ in terms of $\mathtt{e}$,

$$\mathtt{p}^\sharp = -\frac{\omega_0\sqrt{\gamma}}{\omega^2-\omega_0^2}\mathtt{e}, \quad \mathtt{q}^\sharp = \mathrm{i}\frac{\omega\sqrt{\gamma}}{\omega^2-\omega_0^2}\mathtt{e}. \tag{76}$$

Let us now consider several cases:

- If $\mathtt{e} \parallel \mathbf{k}$, then by the first equation, either $\omega = 0$ and $\mathtt{b} \parallel \mathbf{k}$ or $\mathtt{b} = 0$ and $\omega^2 = \omega_0^2 + \gamma$.
- If $\mathtt{e} \not\parallel \mathbf{k}$, then $\omega \neq 0$ and combining the first two equations, we get

$$\left[\omega^2\frac{\omega^2-(\omega_0^2+\gamma)}{\omega^2-\omega_0^2} - k^2\right]\mathtt{e} = -(\mathbf{k}\cdot\mathtt{e})\mathbf{k},$$

  where we denoted as usual by k the modulus of $\mathbf{k}$, $k = |\mathbf{k}|$; this in turns implies (since $\mathbf{k}$ is not parallel to $\mathtt{e}$) that $\mathbf{k} \perp \mathtt{e}$ and

$$\omega^4 - \omega^2\big((\omega_0^2 + \gamma) + k^2\big) + \omega_0^2 k^2 = 0;$$

denoting $\Delta = \big((\omega_0^2 + \gamma) + k^2\big)^2 - 4\omega_0^2 k^2 > 0$, this equation admits four real solutions

$$\omega_{\pm,\pm}(k) = \pm\frac{1}{\sqrt{2}}\sqrt{\omega_0^2 + \gamma + k^2 \pm \sqrt{\Delta}}.$$

There are $m = 7$ different branches of $\mathcal{C}_{\mathcal{L}}$: three constant sheets given by

$$\omega = 0 \quad \text{and} \quad \omega = \pm\sqrt{\gamma + \omega_0^2}$$

(the first one being of multiplicity two, the two other ones of multiplicity one), and four curved sheets (each one being of multiplicity two) given by $\omega = \omega_{\pm,\pm}$. We denote by $\pi_{\pm,\pm}$ the associated eigenprojectors (which are therefore of rank 2). Remarking that $\pi_{\pm,\pm}(\mathbf{k})u = u$ if and only if

$$\mathtt{e} \cdot \mathbf{k} = 0, \qquad \mathtt{b} = -\frac{1}{\omega}\mathbf{k} \wedge \mathtt{e},$$

and $\mathtt{p}^\sharp$ and $\mathtt{q}^\sharp$ are given in terms of $\mathtt{e}$ by (76), one readily finds that $\pi_{\pm,\pm}(\mathbf{k})$ can be written as a block matrix

$$\pi_{\pm,\pm}(\mathbf{k}) = \frac{1}{N(k)^2}(P_{ij})_{1\le i,j\le 4} \quad \text{with } N(k)^2 = \frac{k^2}{\omega(k)^2} + 1 + \gamma\frac{\omega(k)^2 + \omega_0^2}{(\omega(k)^2 - \omega_0^2)^2}$$

(for the sake of clarity, we write $\omega(k) = \omega_{\pm,\pm}(k)$), and where the 3×3 blocks P_{1j} are given by

$$P_{11} = \frac{k^2}{\omega(k)}\Pi_{\mathbf{k}\perp}, \qquad P_{12} = -\frac{1}{\omega(k)}\mathbf{k}\wedge,$$

$$P_{13} = \mathrm{i}\frac{\sqrt{\gamma}}{\omega(k)^2 - \omega_0^2}\mathbf{k}\wedge, \qquad P_{14} = \frac{\omega_0}{\omega(k)}\frac{\sqrt{\gamma}}{\omega(k)^2 - \omega_0^2}\mathbf{k}\wedge,$$

$$P_{22} = \Pi_{\mathbf{k}\perp}, \qquad P_{23} = -\mathrm{i}\frac{\omega(k)\sqrt{\gamma}}{\omega(k)^2 - \omega_0^2}\Pi_{\mathbf{k}\perp},$$

$$P_{24} = -\frac{\omega_0\sqrt{\gamma}}{\omega(k)^2 - \omega_0^2}\Pi_{\mathbf{k}\perp}, \qquad P_{33} = \frac{\omega(k)^2\gamma}{(\omega(k)^2 - \omega_0^2)^2}\Pi_{\mathbf{k}\perp},$$

$$P_{34} = -\mathrm{i}\frac{\omega(k)\omega_0\gamma}{(\omega(k)^2 - \omega_0^2)^2}\Pi_{\mathbf{k}\perp}, \qquad P_{44} = \frac{\omega_0^2\gamma}{(\omega(k)^2 - \omega_0^2)^2}\Pi_{\mathbf{k}\perp},$$

where $\Pi_{\mathbf{k}\perp}$ is the orthogonal projector onto the orthogonal plane to $\mathbf{k}$; the remaining blocks stem from the symmetry relations $P_{ij} = P_{ji}^*$.

It follows from these computations that in the case of a cubic nonlinearity[12] (i.e., $f = 0$ in (74)), $F^{\mathrm{env}}(\varepsilon, u)$ is given for all u such that $u = \pi_{\pm,\pm}(\mathbf{k})u$ by

$$F^{\mathrm{env}}(\varepsilon, u) = \left(0, 0, -\frac{\gamma^{5/2}}{(\omega^2-\omega_0^2)^3}[(\mathsf{e}\cdot\mathsf{e})\bar{\mathsf{e}} + 2|\mathsf{e}|^2\mathsf{e}], 0\right)^T;$$

similarly,

$$\pi_{\pm,\pm}(\mathbf{k})F^{\mathrm{env}}(\varepsilon, u) = \left(*, \frac{\mathrm{i}}{N(k)^2}\frac{\gamma^3\omega}{(\omega^2-\omega_0^2)^4}[(\mathsf{e}\cdot\mathsf{e})\bar{\mathsf{e}} + 2|\mathsf{e}|^2\mathsf{e}], *, *\right)^T.$$

and

$$\pi_{\pm,\pm}A_0 u = \left(*, \omega_1\frac{\gamma\omega^2}{(\omega^2-\omega_0^2)^2}\mathsf{e}, *, *\right)^T.$$

We can now write explicitly the general NLS equation (37) derived in Example 7 in the particular case of the Maxwell equations (73). Choosing $\omega_1(\cdot) = \omega_{\pm,\pm}(\cdot)$ and therefore $\pi_1(\cdot) = \pi_{\pm,\pm}(\cdot) = \omega(\cdot)$, and remarking that the solution v to (37) remains polarized along $\pi_{\pm,\pm}(\mathbf{k})$ if it is initially polarized, it is enough to give an equation on its electric field component e (the components b, $\mathsf{p}^\sharp$, and $\mathsf{q}^\sharp$ being recovered as indicated above). Moreover, we can assume the $\mathbf{k} = k\mathbf{e}_z$, so that the electric field e only has transverse components $\mathfrak{e} \in \mathbb{C}^2$ (i.e., $\mathsf{e} = (\mathfrak{e}^T, 0)^T$) and (37) finally reduces to

$$\begin{aligned}(1 - \mathrm{i}\varepsilon\mathbf{b}\cdot\nabla - \varepsilon^2\nabla\cdot B\nabla)\partial_t\mathfrak{e} - \frac{\mathrm{i}}{2}\left(\frac{\omega'(k)}{k}\Delta_\perp + \omega''(k)\partial_z^2\right)\mathfrak{e}\\ +\varepsilon^p\omega_1\frac{\gamma\omega(k)^2}{(\omega(k)^2-\omega_0^2)^2}\mathfrak{e} = \frac{\mathrm{i}}{N(k)^2}\frac{\gamma^3\omega(k)}{(\omega(k)^2-\omega_0^2)^4}[(\mathfrak{e}\cdot\mathfrak{e})\bar{\mathfrak{e}} + 2|\mathfrak{e}|^2\mathfrak{e}].\end{aligned} \tag{77}$$

Proceeding to the following rescaling of the space variables

$$(x, y) = \left(\frac{\omega'(k)}{2k}\right)^{1/2}(x', y') \quad \text{and} \quad z = |\tfrac{1}{2}\omega''(k)|^{1/2}z'$$

(the latter one only if $\omega''(k) \neq 0$), and rescaling $\mathfrak{e}$ as

$$\mathfrak{e} = \left(\frac{1}{N(k)^2}\frac{\gamma^3\omega(k)}{3(\omega(k)^2-\omega_0^2)^4}\right)^{1/2}\mathfrak{e}',$$

[12]The treatment of other kinds of nonlinearities is absolutely similar.

the above equation becomes

$$iP_2(\varepsilon\nabla)\partial_t\mathfrak{e} + \big(\Delta_\perp + \alpha_1\partial_z^2\big)\mathfrak{e} + i\alpha_2\mathfrak{e} + \tfrac{1}{3}[(\mathfrak{e}\cdot\mathfrak{e})\bar{\mathfrak{e}} + 2|\mathfrak{e}|^2\mathfrak{e}] = 0, \tag{78}$$

where $\alpha_1 = \mathrm{sgn}(\omega''_{\pm,\pm}(k)) \in \{0, \pm 1\}$, $\alpha_2 = \varepsilon^p\omega_1\gamma\omega(k)^2/(\omega(k)^2 - \omega_0^2)^2$ and

$$P_2(\varepsilon\nabla) = 1 - i\varepsilon M\mathbf{b}\cdot\nabla - \varepsilon^2\nabla\cdot MBM\nabla,$$

where $M = \mathrm{diag}\Big(\big(\omega'(k)/(2k)\big)^{-1/2}, \big(\omega'(k)/(2k)\big)^{-1/2}, |\tfrac{1}{2}\omega''(k)|^{-1/2}\Big)$. This equation is exactly in the form $(3)_{\mathrm{vect}}$ with $\boldsymbol{\alpha}_3 = 0$ (i.e., without the terms modeling the frequency dependence of the polarization).

With Frequency Dependent Polarization

Let us now describe the modifications that have to be made in order to take into account a frequency dependent polarization, as in (41). Since initial data polarized along $\pi_{\pm,\pm}(\mathbf{k})$ conserve this polarization during the evolution in time, the only difference between (41) and (37) is the presence of the terms $\varepsilon\pi_{\pm,\pm}(\mathbf{k})\pi'_{\pm,\pm}(\mathbf{k})\cdot DF^{\mathrm{env}}(\varepsilon, u)$ and $-\mathrm{i}\varepsilon(\mathbf{b}\cdot\nabla)\pi_{\pm,\pm}(\mathbf{k})F^{\mathrm{env}}(\varepsilon, u)$. Only the first one requires a nontrivial computation. Thanks to the expression for $\pi_{\pm,\pm}(\mathbf{k})$ given above, we can write for all $F = (0, 0, f, 0)$,

$$\pi_{\pm,\pm}(\mathbf{k})F = (ig_1(k)\mathbf{k}\wedge f, -\mathrm{i}g_2(k)\Pi_{\mathbf{k}\perp}f, g_3(k)\Pi_{\mathbf{k}\perp}f, \mathrm{i}g_4(k)\Pi_{\mathbf{k}\perp}f)^T,$$

with

$$g_1(k) = \frac{1}{N(k)^2}\frac{\sqrt{\gamma}}{\omega(k)^2 - \omega_0^2}, \qquad g_2(k) = \omega(k)g_1(k),$$

$$g_3(k) = \frac{\omega(k)\sqrt{\gamma}}{\omega(k)^2 - \omega_0^2}g_2(k), \qquad g_4(k) = \frac{\omega_0\sqrt{\gamma}}{\omega(k)^2 - \omega_0^2}g_2(k).$$

One has therefore, with $h = \frac{\mathbf{k}}{k}\cdot Df$ and $l = d_{\mathbf{k}}(\mathbf{k}' \mapsto \Pi_{\mathbf{k}'\perp})\cdot Df$,

$$\begin{aligned}\pi'_{\pm,\pm}(\mathbf{k})\cdot DF = {} & (\mathrm{i}g'_1(k)\mathbf{k}\wedge h, -\mathrm{i}g'_2(k)\Pi_{\mathbf{k}\perp}h, g'_3(k)\Pi_{\mathbf{k}\perp}h, ig'_4(k)\Pi_{\mathbf{k}\perp}h)^T \\ & +(\mathrm{i}g_1(k)D\wedge f, -\mathrm{i}g_2(k)l, g_3(k)l, \mathrm{i}g_4(k)l)^T.\end{aligned}$$

If $f\cdot\mathbf{k} = 0$, we can use the identity $\Pi_{\mathbf{k}\perp}d_{\mathbf{k}}(\mathbf{k}' \mapsto \Pi_{\mathbf{k}'\perp})\cdot D\Pi_{\mathbf{k}\perp} = 0$ to deduce that

$$\pi_{\pm,\pm}(\mathbf{k})\pi'_{\pm,\pm}(\mathbf{k})\cdot DF = \left(*, -\frac{m(k)}{N(k)^2}\frac{\mathbf{k}}{k}\cdot\nabla f, *, *\right)^T,$$

where the $*$ can be easily computed, and where

$$m(k) = \frac{k}{\omega(k)}(g_1(k) + kg_1'(k)) + g_2'(k) + \frac{\sqrt{\gamma}}{\omega(k)^2 - \omega_0^2}(\omega(k)g_3'(k) - \omega_0 g_4'(k)).$$

The nonlinearity in the right-hand side of (77) must therefore be replaced by

$$\frac{\mathrm{i}}{N(k)^2}\frac{\gamma^3\omega(k)}{(\omega(k)^2-\omega_0^2)^4}(1 - i\varepsilon\mathbf{b}\cdot\nabla + i\alpha_3\partial_z)[(\mathfrak{e}\cdot\mathfrak{e})\bar{\mathfrak{e}} + 2|\mathfrak{e}|^2\mathfrak{e}],$$

with $\alpha_3 = -(\omega^2 - \omega_0^2)/(\sqrt{\gamma}\omega(k))$. The reduction to $(3)_{\text{vect}}$ then follows the same steps as for (78).

The Case with Charge and Current Density

According to (48) and more generally (49), ionizations effects are taken into account by adding

$$I := -\varepsilon\pi_1(\mathbf{k})\big(\mathrm{i}wC_1^T C_1 u + c\, C_1^T G^{\text{env}}(C_1 u, w)\big)$$

to the right-hand side of the NLS equation, and by considering the following equation for w,

$$\partial_t w = 2\varepsilon G^{\text{env}}(C_1 u, w)\cdot\overline{C_1 u}.$$

In the particular case of the Maxwell equations (19), one has $w = \rho$ and

$$I = -\varepsilon\pi_1(\mathbf{k})\big(0, \mathrm{i}\rho\mathrm{e} + c_0\big(c_1|\mathrm{e}|^{2K-2} + c_2\rho\big)\mathrm{e}, 0, 0\big)^T;$$

consequently, one must add to the right-hand side of (77) the ionization term

$$-\mathrm{i}\rho\mathfrak{e} - c_0\big(c_1|\mathfrak{e}|^{2K-2} + c_2\rho\big)\mathfrak{e},$$

so that after rescaling as for (78), the equation takes the form (54) (when written in a fixed frame rather than the frame moving at the group velocity).

References

1. J. Ginibre, G. Velo, J. Funct. Anal. **32**(1), 1 (1979)
2. V.E. Zakharov, A.B. Shabat, Sov. Phys. JETP **34**(1), 62 (1972)

3. C. Sulem, P.L. Sulem, *The Nonlinear Schrödinger Equation: Self-Focusing and Wave Collapse. Applied Mathematical Sciences*, vol. 139 (Springer, New York, 1999)
4. F. Merle, P. Raphaël, Ann. Math. **161**(1), 157 (2005)
5. F. Merle, P. Raphaël, Geom. Funct. Anal. **13**(3), 591 (2003)
6. F. Merle, P. Raphaël, Invent. Math. **156**(3), 565 (2004)
7. F. Merle, P. Raphaël, J. Am. Math. Soc. **19**(1), 37 (2006)
8. F. Merle, P. Raphaël, Commun. Math. Phys. **253**(3), 675 (2005)
9. F. Merle, P. Raphaël, J. Hyperbol. Differ. Eq. **2**(4), 919 (2005)
10. F. Merle, P. Raphaël, J. Szeftel, Geom. Funct. Anal. **20**(4), 1028 (2010)
11. J.E. Rothenberg, Opt. Lett. **17**(19), 1340 (1992)
12. H.A. Lorentz, *Theory of Electrons* (Teubner, New York, 1909). Reprint (Dover, Englewood Cliffs, NJ, 1952)
13. N. Bloembergen, *Nonlinear Optics. Frontiers in Physics*, vol. 21 (W. A. Benjamin, Reading, MA, 1977)
14. A. Owyoung, The origins of the nonlinear refractive indices of liquids and gases. Ph.D. thesis, California Institute of Technology (1971)
15. P. Donnat, J.L. Joly, G. Métivier, J. Rauch, *Équations aux dérivées partielles 1995–1996* (École Polytechnique, Palaiseau, 1996). Exp. No. XVII
16. J.L. Joly, G. Métivier, J. Rauch, Indiana U. Math. J. **47**(4), 1167 (1998)
17. D. Lannes, Asymptot. Anal. **18**(1–2), 111 (1998)
18. L. Bergé, S. Skupin, Discrete Contin. Dyn. Syst. **23**(4), 1099 (2009)
19. L. Bergé, S. Skupin, R. Nuter, J. Kasparian, J.P. Wolf, Rep. Prog. Phys. **70**(10), 1633 (2007)
20. M. Colin, D. Lannes, SIAM J. Math. Anal. **41**(2), 708 (2009)
21. D. Lannes, Proc. R. Soc. Edinb. A Math. **141**(2), 253 (2011)
22. L.A. Kalyakin, Math. USSR Sb. **60**(2), 457 (1988)
23. G. Schneider, Nonlinear Differ. Equ. Appl. **5**(1), 69 (1998)
24. T. Colin, G. Gallice, K. Laurioux, SIAM J. Math. Anal. **36**(5), 1664 (2005)
25. M. Kolesik, J.V. Moloney, Phys. Rev. E **70**(3), 036604 (2004)
26. D. Lannes, *The Water Waves Problem: Mathematical Analysis and Asymptotics. Mathematical Surveys and Monographs*, vol. 188 (American Mathematical Society, Providence, RI, 2013)
27. T. Cazenave, F.B. Weissler, Nonlinear Anal. **14**(10), 807 (1990)
28. T. Tao, M. Visan, X. Zhang, Commun. Partial Differ. Equ. **32**(8), 1281 (2007)
29. G. Fibich, G. Papanicolaou, SIAM J. Appl. Math. **60**(1), 183 (1999)
30. F. Merle, C. R. Acad. Sci. Paris Sér. I Math. **304**(16), 479 (1987)
31. F. Merle, Commun. Math. Phys. **149**(2), 377 (1992)
32. P. Antonelli, R. Carles, C. Sparber, Int. Math. Res. Not. **2015**(3), 740 (2015)
33. M. Tsutsumi, SIAM J. Math. Anal. **15**(2), 357 (1984)
34. G. Fibich, SIAM J. Appl. Math. **61**(5), 1680 (2001)
35. M. Ohta, G. Todorova, Discrete Contin. Dyn. Syst. **23**(4), 1313 (2009)
36. D. Anderson, M. Lisak, Phys. Rev. A **27**(3), 1393 (1983)
37. R.S. Strichartz, Duke Math. J. **44**(3), 705 (1977)
38. J.M. Ghidaglia, J.C. Saut, J. Nonlinear Sci. **6**(2), 139 (1996)
39. P. Kevrekidis, A.R. Nahmod, C. Zeng, Nonlinearity **24**(5), 1523 (2011)
40. P. Donnat, Quelques contributions mathématiques en optique non linéaire. Ph.D. thesis, École Polytechnique, Paris (1994)

Blowing Up Solutions to the Zakharov System for Langmuir Waves

Yuri Cher, Magdalena Czubak, and Catherine Sulem

Abstract Langmuir waves take place in a quasi-neutral plasma and are modeled by the Zakharov system. The phenomenon of collapse, described by blowing up solutions, plays a central role in their dynamics. We present in this article a review of the main mathematical properties of blowing up solutions. They include conditions for blowup in finite or infinite time, description of self-similar singular solutions and lower bounds for the rate of blowup of certain norms associated with the solutions.

1 Introduction

Langmuir waves take place in a non-magnetized or weakly magnetized plasma and are described by the Zakharov system [1]

$$i\partial_t \mathbf{E} - \alpha \nabla \times (\nabla \times \mathbf{E}) + \nabla(\nabla \cdot \mathbf{E}) = n\mathbf{E}, \tag{1}$$

$$\partial_{tt} n - \Delta n = \Delta |\mathbf{E}|^2. \tag{2}$$

Equation (2) originates from the hydrodynamic system

$$n_t + \nabla \cdot \mathbf{v} = 0, \tag{3}$$

$$\mathbf{v}_t + \nabla n = -\nabla |\mathbf{E}|^2, \tag{4}$$

governing ion sound waves. $\mathbf{E}(\mathbf{x}, t)$ is the complex envelope of the electric field oscillations. $n(\mathbf{x}, t)$ denotes the fluctuations of density of ions and $\mathbf{v}(\mathbf{x}, t)$ their velocity, with $\mathbf{x} \in \mathbb{R}^d$, in dimension $d = 2$ or 3. The parameter α in (1) is defined as the square ratio of the light speed and the electron Fermi velocity and

Y. Cher • C. Sulem (✉)
Department of Mathematics, University of Toronto, Toronto, ON, M5S 2E4 Canada
e-mail: ycher@math.toronto.edu; sulem@math.toronto.edu

M. Czubak
Department of Mathematical Sciences, Binghamton University, Binghamton, NY 13902-6000, USA
e-mail: czubak@math.binghamton.edu

A.D. Bandrauk et al. (eds.), *Laser Filamentation*, CRM Series in Mathematical Physics, DOI 10.1007/978-3-319-23084-9_3

is usually large. A simplified system of equations is obtained in the electrostatic limit ($\alpha \to \infty$) expanding the electric field in the form $\mathbf{E} = \nabla\psi + (1/\alpha)\mathbf{E}_1 + \cdots$. Substituting this expansion in (1) and taking the divergence of the equation gives a system describing the interaction of the electrostatic potential with the plasma density [1, 2]

$$\Delta(\mathrm{i}\partial_t\psi + \Delta\psi) = \nabla\cdot(n\nabla\psi), \tag{5}$$

$$\partial_{tt}n - \Delta n = \Delta(|\nabla\psi|^2). \tag{6}$$

A further simplification leads to

$$\mathrm{i}\psi_t + \Delta\psi = n\psi, \tag{7}$$

$$n_{tt} - \Delta n = \Delta|\psi|^2. \tag{8}$$

with

$$n_t + \nabla\cdot\mathbf{v} = 0, \tag{9}$$

$$\mathbf{v}_t + \nabla n = -\nabla|\psi|^2, \tag{10}$$

usually called the scalar Zakharov system. Introducing the hydrodynamic potential U such that $\mathbf{v} = -\nabla U$, (4) becomes

$$\partial_t U = n + |\mathbf{E}|^2. \tag{11}$$

Heuristic derivations of the Zakharov system can be found in [3, 4]. Viewing the plasma as a two interpenetrating fluids (electrons and ions), the Zakharov system (7)–(9) can be obtained using a multiple-scale modulation analysis [5]. A rigorous derivation of the scalar model is given in [6] using techniques of geometric optics and semi-classical calculus.

Invariance properties of the system by simple transformations lead to several conserved quantities. In particular, if $(\mathbf{E}, n)$ is a smooth solution of (7)–(10), the wave energy $N = |\mathbf{E}|^2_{L^2}$ and the Hamiltonian

$$H = \alpha|\nabla\times\mathbf{E}|^2_{L^2} + |\nabla\cdot\mathbf{E}|^2_{L^2} + \tfrac{1}{2}|n|^2_{L^2} + \tfrac{1}{2}|\nabla U|^2_{L^2} + \int n|\mathbf{E}|^2\,\mathrm{d}\mathbf{x} \tag{12}$$

are conserved. Other invariants are the linear and angular momenta

$$\mathbf{P} = \int\left(\frac{\mathrm{i}}{2}\sum_j(E_j\nabla E_j^* - E_j^*\nabla E_j) + n\mathbf{v}\right)\mathrm{d}\mathbf{x} \tag{13}$$

and

$$\mathbf{M} = \int(\mathrm{i}\mathbf{E}\times\mathbf{E}^* + \mathbf{x}\times\mathbf{P})\,\mathrm{d}\mathbf{x}. \tag{14}$$

Modulational instability leads to the formation of regions where the density of the plasma is very low. In these regions referred to as cavities, high-frequency oscillations of the electric field are trapped. Their nonlinear evolution gives rise to the collapse of the cavities and a strong amplification of the amplitude of the oscillations of the electric field. Heuristic arguments and numerical simulations show that, for large enough initial conditions, solutions blowup in a finite time both in two and three dimensions (see [5] for a review).

In this article, we present an overview of mathematical results and open questions concerning blowing up solutions for the scalar Zakharov model. We also discuss the extension of some of the features of blowup to the Vectorial Zakharov system for which very few rigorous are known apart from local well-posedness and global well-posedness under the assumption of small enough initial conditions.

2 The Scalar Zakharov System

We consider the scalar Zakharov system (7) and (8) with initial conditions

$$\psi(\mathbf{x},0) = \psi_0(\mathbf{x}), \quad n(\mathbf{x},0) = n_0(\mathbf{x}), \quad n_t(\mathbf{x},0) = n_1(\mathbf{x}). \tag{15}$$

The conserved quantities are:

- the wave energy

$$N = |\psi|^2_{L^2}, \tag{16}$$

- the linear momentum

$$\mathbf{P} = \int \left(\frac{\mathrm{i}}{2}(\psi \nabla \psi^* - \psi^* \nabla \psi) + n\mathbf{v} \right) \mathrm{d}\mathbf{x}, \tag{17}$$

- the angular momentum

$$\mathbf{M} = \int \mathbf{x} \times \mathbf{P} \, \mathrm{d}\mathbf{x}, \tag{18}$$

and
- the Hamiltonian

$$H = \int \left(|\nabla \psi|^2 + n|\psi|^2 + \frac{1}{2}|\mathbf{v}|^2 + \frac{1}{2}n^2 \right) \mathrm{d}\mathbf{x}. \tag{19}$$

There is a large literature devoted to the local and global well-posedness of the initial value problem. Earlier works concern smooth solutions, in particular solutions with finite energy (Hamiltonian) [7–13]. More recently, there has been an interest in solutions with lower regularity assumptions and in particular in solutions with infinite energy [14–19]. Associated to the long time existence theory are the important questions of scattering theory, existence of wave operators [20], and precise decay of solutions for large time. In particular, in three dimensions, it is proved in [21] that, if the initial conditions are small and localized, then $\sup_x|\psi(t)| \leq C|t|^{-7/6-}$, $\sup_x|n(t)| \leq C|t|^{-1}$, and the solution (ψ, n) scatters to a solution to the associated linear problem as $|t| \to \infty$. Here the notation $7/6-$ means $7/6-\varepsilon$, for any $\varepsilon > 0$.

We denote by $H^k(\mathbb{R}^d)$ the Sobolev space of functions f such that f and its derivatives of order p, $|p| \leq k$, are bounded in the L^2-space. It is also convenient to define the product space

$$H_k = H^k(\mathbb{R}^d) \times H^{k-1}(\mathbb{R}^d) \times H^{k-2}(\mathbb{R}^d). \tag{20}$$

The energy space corresponds to H_1.

2.1 Blowup in Finite or Infinite Time

A central tool in the theory of blowup for the Nonlinear Schrödinger (NLS) equation

$$\mathrm{i}\partial_t\psi + \Delta\psi + |\psi|^{2\sigma}\psi = 0, \quad \psi(\mathbf{x}, t) = \psi_0(\mathbf{x}) \tag{21}$$

is the variance identity

$$\frac{\mathrm{d}^2}{\mathrm{d}t^2}\int |\mathbf{x}|^2|\psi|^2\,\mathrm{d}\mathbf{x} = 8H_{\mathrm{NLS}} - 4\frac{\mathrm{d}\sigma - 2}{\sigma + 1}\int |\psi|^{2\sigma+2}\,\mathrm{d}\mathbf{x},$$

where $H_{\mathrm{NLS}} = \int(|\nabla\psi|^2 - (\sigma+1)^{-1}|\psi|^{2\sigma+2})\,\mathrm{d}\mathbf{x}$ is the NLS Hamiltonian. Under the assumption that the initial condition ψ_0 is in $H^1(\mathbb{R}^d)$ has finite variance and $H_{\mathrm{NLS}}(\psi_0) < 0$, the solution of (21) blows up in a finite time if $\sigma d \geq 2$. For the Zakharov system, the usual variance $\int|\mathbf{x}|^2|\psi|^2\,\mathrm{d}\mathbf{x}$ can be replaced by the quantity

$$\mathcal{V}(t) = \frac{1}{4}\int |\mathbf{x}|^2|\psi|^2\,\mathrm{d}\mathbf{x} + \int_0^t \int (\mathbf{x}\cdot\mathbf{v})n\,\mathrm{d}\mathbf{x}\,\mathrm{d}t \tag{22}$$

which is well defined for functions in the space

$$\Sigma' = \left\{(\psi, n, \mathbf{v}) \in H_1, \int \left(|\mathbf{x}|^2|\psi|^2 + |\mathbf{x}|(|n|^2 + |\mathbf{v}|^2)\right)\mathrm{d}\mathbf{x} < \infty\right\}. \tag{23}$$

The function $\mathcal{V}(t)$ and satisfies

$$\frac{\mathrm{d}^2\mathcal{V}}{\mathrm{d}t^2}(t) = dH - (d-2)|\nabla\psi|_{L^2}^2 - (d-1)|\mathbf{v}|_{L^2}^2, \tag{24}$$

where H is the Hamiltonian defined in (19). In dimension $d \geq 2$, one has $(\mathrm{d}^2/\mathrm{d}t^2)\mathcal{V} < 0$ if the initial conditions are such that $H < 0$. However, one cannot conclude on existence of blowup solutions because, unlike the NLS case, $\mathcal{V}$ does not have a fixed sign. In particular, in dimension 2, Merle [22] proved that it tends to $-\infty$ as the singularity is approached. Nevertheless, one can get partial results, under the assumption of radial symmetry. Indeed in this case, there is a useful result referred to as the Strauss Lemma [23] which gives an upper bound of the sup norm of a function in terms of its H^1-norm far from the origin. Namely, if f is a radially symmetric function in $H^1(\mathbb{R}^d)$ with $d \geq 2$, then, for any $R > 0$,

$$|f|^2_{L^\infty(|\mathbf{x}|>R)} \leq CR^{-d+1}|\nabla f|_{L^2(|\mathbf{x}|>R)}|f|_{L^2(|\mathbf{x}|>R)}. \tag{25}$$

The radial assumption has been useful in other contexts such as in existence and scattering theory, where it allows a larger range of parameters for linear estimates of Strichartz type [24].

The method consists in modifying the quadratic weight $|\mathbf{x}|^2$ in $\mathcal{V}$ defined in (22) by a smooth function $p(\mathbf{x})$ that behaves like $|\mathbf{x}|^2$ near the origin and like $|\mathbf{x}|$ at infinity, and by considering the time derivative $y(t) = -(\mathrm{d}/\mathrm{d}t)\mathcal{U}$ of the modified variance

$$\mathcal{U}(t) = \frac{1}{2}\int p(x)|\psi|^2\,\mathrm{d}\mathbf{x} + \int_0^t \int (\nabla p \cdot \mathbf{v})n\,\mathrm{d}\mathbf{x}\,\mathrm{d}t, \tag{26}$$

as in the case of solutions of the NLS equation with infinite variance. However, the modification of the weight induces additional terms in the time evolution of the function $y(t)$ that need to be estimated. For this purpose, one uses a sequence of rescaled weights $p_m(|\mathbf{x}|) = m^2p(\mathbf{x}/m)$, and proves that the additional contributions are controlled for m sufficiently large. More precisely, one proves that

$$y_m(t) = -\Im \int \big((\nabla p_m \cdot \nabla\psi)\psi^* - (\nabla p_m \cdot \mathbf{v})n\big)\,\mathrm{d}\mathbf{x} \tag{27}$$

satisfies, for m sufficiently large

$$y_m(t) \geq \frac{d}{2}|H|t. \tag{28}$$

On the other hand, the function $y_m(t)$ is controlled by the norm of the solution in the energy space, namely

$$|y_m(t)| \leq C(|\psi_0|^2_{L^2} + |\nabla\psi|^2_{L^2} + |\mathbf{v}|^2_{L^2} + |n|^2_{L^2}). \tag{29}$$

We have the following result proved by Merle in [25]:

Theorem 1. *Consider the Zakharov system* (7), (9), (10) *in dimension* $d = 2$ *or* $d = 3$ *with initial conditions in the space* Σ'. *Assume that there exists a* *smooth solution* $(\psi, n, \mathbf{v})$ *during an interval of time* $[0, t_0]$. *In particular, its mass and Hamiltonian are conserved and its variance* $\mathcal{U}$ *is well defined. Assume in addition that the solution is radially symmetric and its Hamiltonian* $H < 0$. *Then, either* $|\psi|_{H^1} + |n|_{L^2} + |\mathbf{v}|_{L^2} \to \infty$ *as* $t \to t^*$ *with* t^* *finite, or* $(\psi, n, \mathbf{v})$ *exists for all time and* $|\psi|_{H^1} + |n|_{L^2} + |\mathbf{v}|_{L^2} \to \infty$ *as* $t \to \infty$.

Remark 1. An open question is the extension of this analysis to solutions that are not radially symmetric, and furthermore, to solutions to the full vector Zakharov system. Based on numerical observations, it is believed that blowup does indeed occur in a finite time for general initial conditions with negative Hamiltonian.

2.2 Self-similar Blowing Up Solutions

2.2.1 Dimension $d = 2$

Unlike for the NLS equation, there is no conformal mapping for the two-dimensional Zakharov system. Nevertheless one can construct exact self-similar blowing up solutions that have the form in the form [26]

$$\psi(x,t) = \frac{1}{a(t^*-t)} P\left(\frac{|x|}{a(t^*-t)}\right) \times \exp\left(\mathrm{i}\left(\theta + \frac{1}{a^2(t^*-t)} - \frac{|x|^2}{4(t^*-t)}\right)\right), \tag{30}$$

$$n(x,t) = \frac{1}{a^2(t^*-t)^2} N\left(\frac{|x|}{a(t^*-t)}\right), \tag{31}$$

where (P, N) are real functions satisfying the system of ODEs

$$\Delta P - P - NP = 0, \tag{32}$$

$$a^2(\eta^2 N_{\eta\eta} + 6\eta N_\eta + 6N) - \Delta N = \Delta P^2, \tag{33}$$

with η being the rescaled independent variable and $a > 0$ a free parameter. Glangetas and Merle have rigorously studied the system (32) and (33) in [11, 12]. We summarize below the most important properties. When $a = 0$, $N = -P^2$ and (32) becomes

$$\Delta P - P + P^3 = 0. \tag{34}$$

It is known that (34) has an infinite number of radially symmetric solutions that decay exponentially at infinity, only one of them, denoted R and called the NLS the ground state, (also known as the Townes soliton) is strictly positive and monotone decreasing (see, for example, [27, 28]). It plays a central role in the study of NLS equations.

If the coefficient a in (33) is sufficiently small, there exists a solution (P_a, N_a) in $H^1 \times L^2$ with $P_a > 0$. This solution is in fact C^∞ and its derivatives of order k satisfy the decay properties

$$|P^{(k)}(\eta)| \leq c_k \mathrm{e}^{-\delta\eta}, \quad |N^{(k)}(\eta)| \leq \frac{c_k}{|\eta|^{k+3}}$$

for large η. When the parameter a is small, the solution (P_a, N_a) is constructed by a continuation method from the solution $(R, -R^2)$ corresponding to $a = 0$.

Furthermore, for any value c strictly larger than the L^2-norm of the NLS ground state R, there exists a_c such that for any $a < a_c$, there is a unique solution (P_a, N_a) in $H^1 \times L^2$ with $P_a > 0$ and $|P_a|_{L^2} < c$.

Numerical simulations show that for a large class of radially symmetric initial conditions having a strictly negative Hamiltonian, the solutions display a self-similar collapse as $t \to t^*$ as described by (30) and (31) [29, 30]. The coefficient a in the equation for the limiting profiles (P, N) depends on the initial conditions. When considering a sequence of initial conditions with an initial L^2-norm of ψ_0 decreasing to $|R|^2_{L^2}$, where R is the NLS ground state, it was observed that the computed value of the coefficient a tends to zero. In this limit, the self-similar profile becomes (strongly) subsonic and tends to the NLS ground state R. This limit is delicate, since solutions of the scalar Zakharov equation with critical norm $|\psi_0|^2_{L^2} = |R|^2_{L^2}$ remain smooth for all time [11, 12]. Indeed, unlike the NLS equation, there are no minimal blowing up solutions to the 2D Zakharov system. For initial conditions in the energy space such that $|\psi_0|_{L^2} \leq |R|_{L^2}$ solutions remain in the energy space for all times. The case $|\psi_0|_{L^2} < |R|_{L^2}$ is straightforward and follows the NLS analysis [8, 31]. When $|\psi_0|_{L^2} = |R|_{L^2}$, the global well-posedness property is very specific to the Zakharov system.

Finally, when numerical simulations are performed with anisotropic initial conditions with negative Hamiltonian, it was observed that the solutions become isotropic near collapse with the same limiting profiles as those obtained with isotropic initial conditions [30].

2.2.2 Dimension $d = 3$

In three dimensions, there are no known explicit blowing up solutions. Self-similar solutions exist only asymptotically close to collapse and have the universal form [26, 32]

$$\psi(x,t) \sim \frac{1}{(t^* - t)} P\left(\frac{|x|}{\sqrt{3}(t^* - t)^{2/3}}\right) \exp\left(\mathrm{i}(t^* - t)^{-1/3}\right), \tag{35}$$

$$n(x,t) \sim \frac{1}{3(t^* - t)^{4/3}} N\left(\frac{|x|}{\sqrt{3}(t^* - t)^{2/3}}\right), \tag{36}$$

where $P(\eta)$ and $N(\eta)$ are radially symmetric scalar functions satisfying the coupled system of ODEs

$$\Delta P - P - NP = 0, \tag{37}$$

$$\tfrac{2}{9}(2\eta^2 N_{\eta\eta} + 13\eta N_\eta + 14N) = \Delta P^2. \tag{38}$$

This type of blowup is referred to as supersonic collapse, because, when substituting the expressions (35) and (36) into the Zakharov system, the pressure term Δn is of lower order than $\partial_{tt} n$.

Note that, unlike the 2D case, there is no free parameter in the system. As discussed in [26] and proved more recently in [33], there exists an infinite number of solutions (P_k, N_k) to (37) and (38) such that, for all k,

$$0 < P_k(\eta) < P_{k+1}(\eta) \quad \text{and} \quad N_k(\eta) < 0. \tag{39}$$

The profiles P_k decay exponentially $|P_k(\eta)| \leq C_k e^{-\delta\eta}$ for η large, while the $|N_k| \leq C_k/(1+\eta^2)$ decay algebraically.

The values of P and N at the origin satisfy $N_k(0) = 9P_k(0)^2/\big(14 - 9P_k^2(0)\big)$. We have also $P'(0) = N'(0) = 0$ due to the radial symmetry. The pair (P_k, N_k) is thus characterized by the value $P_k(0)$. It is proved in [33] that there exists a sequence $\alpha_k = \frac{1}{3}\sqrt{2k(4k+3)} > 0$ such that the values $P_k(0)$ are ordered as $\alpha_k < P_k(0) < \alpha_{k+1}$. It is of interest to see how the values α_k arise in the analysis. They appear when one writes the Taylor series expansion of P and N near the origin:

$$P(\eta) = \sum_{i=0}^{\infty} a_i \eta^{2i}; \quad N(\eta) = \sum_{i=0}^{\infty} b_i \eta^{2i}. \tag{40}$$

The series have only even powers because P and N are radially symmetric. From the substitution of the Taylor series into the system (37) and (38), one gets two relations between the coefficients a_i, b_i. The values α_i appear when solving the equation for the coefficient a_i:

$$(\alpha_i^2 - a_0^2)a_i = F(a_0, \ldots, a_{i-1}, b_0, \ldots, b_{i-1}) \tag{41}$$

when solving for the coefficients. In order to have well-defined coefficients and an analytic solution, $P(0)$ which identifies to a_0 should be different from the α_i. In [33], it is proved that there is at least one solution P_k with initial value $P_k(0) \in (\alpha_k, \alpha_{k+1})$ which is strictly positive and decays to 0 at infinity. Numerically, we found (at least for the first few that we computed) that there is only one. Figure 1 shows the first four pairs of solutions computed numerically by a shooting method (with the shooting parameter being $P_k(0)$. The values $P_k(0)$ (for $k = 1, \ldots, 4$) are:

$$P_1(0) \approx 1.38, \quad P_2(0) \approx 2.43, \quad P_3(0) \approx 3.42, \quad P_4(0) \approx 4.40. \tag{42}$$

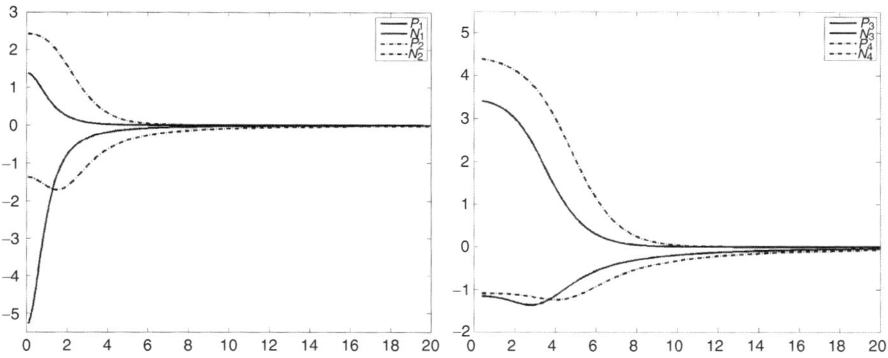

Fig. 1 Solutions (P_k, N_k) of (37) and (38) for $k = 1, \dots, 4$. *Left*: *solid line*—(P_1, N_1), *dashed line*—(P_2, N_2) corresponding to initial values $P_1(0)$ and $P_2(0)$ in (42), respectively. *Right*: *solid line*—(P_3, N_3), *dashed line*—(P_4, N_4) corresponding to initial values $P_3(0)$ and $P_4(0)$, respectively

Like in the 2D case, the dynamical stability of the asymptotically self-similar solutions to the 3D Zakharov system for both radially symmetric and anisotropic initial conditions was studied numerically in [30]. It was observed that for a large class of data, blowup solutions asymptotically display a self-similar collapse described by the above solutions. The profiles identify to the first mode (P_1, N_1) solution of (37) and (38) that has the lowest value at the origin, and for which N_1 is monotone increasing.

Remark 2. There is no rigorous proof of dynamic stability of the (2D) self-similar or (3D) asymptotically self-similar solutions even for well-prepared initial conditions (with or without radial symmetry) chosen close to the profiles (P, N) solutions of the ODE systems (32) and (33) or (37) and (38).

2.3 *Lower Bounds for Rate of Blowup*

2.3.1 Scale Invariance, Criticality, and Local Well-Posedness

An important aspect in the analysis of dispersive equations is the notion of criticality. It is closely related to the invariance properties of the equation. For example, the NLS equation (21) is invariant under the scaling transformation $\psi(\mathbf{x}, t) \rightarrow \psi_\lambda(\mathbf{x}, t) = \lambda^{1/\sigma} \psi(\lambda \mathbf{x}, \lambda^2 t)$. It is said to be $\dot{H}^s$-critical if the (homogeneous) H^s-norm is unchanged under the above scaling transformation. The corresponding critical Sobolev exponent for NLS is thus $s_c = d/2 - 1/\sigma$. The notion of criticality is not straightforward for the Zakharov system because the Schrödinger equation and the wave equation have different scale invariances. In [14], criticality is defined by considering the scaling

$$\psi \to \psi_\lambda = \lambda^{3/2}\psi(\lambda\mathbf{x}, \lambda^2 t), \quad n \to n_\lambda = \lambda^2 n(\lambda\mathbf{x}, \lambda^2 t) \tag{43}$$

that would leave the Zakharov system invariant in the absence of the term Δn. This is indeed the relevant scaling to study blowing up solutions of the three-dimensional Zakharov system as we have seen in the previous section.

In relation to the initial value problem, the Sobolev space with critical exponent often corresponds to the space with minimal regularity in which the problem is locally well-posed. For the Zakharov system, the critical values for the initial value problem in $H^k \times H^l \times H^{l-1}$ are $k = d/2 - \frac{3}{2}$ and $l = d/2 - 2$. Note that $k - l = \frac{1}{2}$, while one would have $k - l = 1$ in the classical setting of the energy space H_1. We now summarize the well-posedness results from the works of Ginibre et al. [14], Colliander et al. [17], Bejenaru et al. [18], and Bejenaru and Herr [19]. For ill-posedness results in dimension one, see [34].

Theorem 2. *In dimension $d = 1$, the Zakharov system is locally well-posed in $H^k \times H^l \times H^{l-1}$, provided that $-\frac{1}{2} < k - l \leq 1, 2k \geq l + \frac{1}{2} \geq 0$. Furthermore, global well-posedness holds in the largest space in which local well-posedness holds, that is $L^2 \times H^{-1/2} \times H^{-3/2}$.*

In dimension 2, it is locally well-posed in the critical space $L^2 \times H^{-1/2} \times H^{-3/2}$, and in dimension 3, it is locally well-posed in $H^\varepsilon \times H^{-1/2+\varepsilon} \times H^{-3/2+\varepsilon}$ which is also, up to arbitrarily small ε the critical space.

Finally, in dimension $d \geq 4$, the whole range of subcritical values $k > d/2 - \frac{3}{2}$ and $l > d/2 - 2$ is covered by the theorem as long as $l \leq k \leq l + 1$ and $2k - l - 1 > d/2 - 2$.

2.3.2 Finite Energy Solutions: The Two-Dimensional Case

The next theorem is due to Merle [22]. It concerns solutions of the 2D scalar Zakharov system with initial conditions (ψ_0, n_0, n_1) in the energy space H_1, thus having a finite Hamiltonian. Assume that there exists a finite time t^* such that

$$|\nabla\psi(t)|_{L^2} + |n(t)|_{L^2} + |\mathbf{v}(t)|_{L^2} \to \infty \quad \text{as } t \to t^*.$$

The question is to determine at what rate these norms become infinite as t approaches t^*.

Theorem 3. *Assume that the solution (ψ, n) to the 2D scalar Zakharov system blows up in the energy space H_1 at a finite time t^*. Then there exist constants $c_1 > 0$ and $c_2 > 0$ depending only on $|\psi_0|_{L^2}$ such that for t close to t^*,*

$$|\nabla\psi(t)|_{L^2} \geq \frac{c_1}{t^* - t}, \tag{44}$$

$$|n(t)|_{L^2} \geq \frac{c_2}{t^* - t}. \tag{45}$$

More precisely, the constants c_1 and c_2 scale like $(|\psi_0|_{L^2}^2 - |R|_{L^2}^2)^{-1/2}$ where R is the NLS ground state.

Remark 3. This rate is optimal in the sense that the self-similar solutions (30) and (31) satisfy

$$|\nabla\psi(t)|_{L^2} = \frac{1}{a(t^* - t)}|\nabla P|_{L^2}, \quad |n(t)|_{L^2} = \frac{1}{a(t^* - t)}|N|_{L^2}. \tag{46}$$

and thus blow up exactly at the rate stated in theorem. Notice also that the theorem provides the blowup rate for the two quantities $|\nabla\psi(t)|_{L^2}$ and $|n(t)|_{L^2}$ separately but does not give information on $|\mathbf{v}(t)|_{L^2}$.

The derivation of this result is based on scaling properties and conservation of the Hamiltonian. One defines the rescaled functions $\tilde{\psi}(\mathbf{x}, s)$, $\tilde{n}(\mathbf{x}, s)$, $\tilde{\mathbf{v}}(\mathbf{x}, s)$ (where t is seen as a parameter)

$$\tilde{\psi}(\mathbf{x}, s) = \frac{1}{\lambda(t)}\psi\left(\frac{\mathbf{x}}{\lambda(t)}, t + \frac{s}{\lambda(t)}\right), \tag{47}$$

$$\tilde{n}(\mathbf{x}, s) = \frac{1}{\lambda^2(t)}n\left(\frac{\mathbf{x}}{\lambda(t)}, t + \frac{s}{\lambda(t)}\right), \tag{48}$$

$$\tilde{\mathbf{v}}(\mathbf{x}, s) = \frac{1}{\lambda^2(t)}\mathbf{v}\left(\frac{\mathbf{x}}{\lambda(t)}, t + \frac{s}{\lambda(t)}\right). \tag{49}$$

where the scaling factor

$$\lambda(t) = \int \left(|\nabla\psi|^2 + \tfrac{1}{2}n^2 + \tfrac{1}{2}|\mathbf{v}|^2\right) d\mathbf{x}, \tag{50}$$

is associated with the energy norm. Notice that the scaling of the time variable corresponds to the wave equation rather than to the Schrödinger equation. At $s = 0$,

$$\int \left(|\nabla\tilde{\psi}(0)|^2 + \tfrac{1}{2}\tilde{n}(0)^2 + \tfrac{1}{2}|\tilde{\mathbf{v}}(0)|^2\right) d\mathbf{x} = 1. \tag{51}$$

Under the hypothesis of the theorem, $\lambda(t) \to \infty$ as t approaches t^*.

The analysis consists in establishing bounds for the individual quantities $|\nabla\tilde{\psi}(0)|$, $\tilde{n}(0)$, $\tilde{\mathbf{v}}(0)$ and estimates of $\tilde{\psi}(s)$, $\tilde{n}(s)$ and $\tilde{\mathbf{v}}(s)$ as $t \to t^*$. It uses delicate compactness arguments allowing the identification of limiting quantities as t goes to t^*. This approach, initiated in [22] and now known as profile decomposition, has led to many breakthroughs in various fields of dispersive PDEs.

2.3.3 Infinite Energy Solutions

We present in this section another approach for the derivation of a lower bound for the rate of blowing up solutions. It is more general, but less precise than the one presented in the previous section. It applies to the problem in dimensions two or three, and to initial conditions that may or may not have a finite Hamiltonian. The result below was established in 3D in [35]. The 2D result follows the same line of proof.

Theorem 4. *Let the initial data* $\big(\psi(0), n(0), n_t(0)\big)$ *be in* $\mathcal{H}_\ell := H^{\ell+1/2}(\mathbb{R}^d) \times H^{\ell}(\mathbb{R}^d) \times H^{\ell-1}(\mathbb{R}^d)$, *with the condition* $0 \le \ell \le \frac{1}{2}$ *if* $d = 2$ *and* $0 \le \ell \le 1$ *if* $d = 3$. *Assume that the solution* (ψ, n, n_t) *blows up in a finite time* t^*, *that is, as* t *approaches* t^*, $\|\psi(t)\|_{H^{\ell+1/2}} + \|n(t)\|_{H^\ell} + \|n_t(t)\|_{H^{\ell-1}} \to \infty$. *Then, the rate of blowup of the Sobolev norms satisfies the lower bound estimate*

$$\|\psi(t)\|_{H^{\ell+1/2}} + \|n(t)\|_{H^\ell} + \|n_t(t)\|_{H^{\ell-1}} > C(t^* - t)^{-\theta_\ell} \tag{52}$$

with $\theta_\ell = \frac{1}{4}(4 - d + 2\ell)^-$, $d = 2$ *or* $d = 3$.

In the above formula, the notation a^- means $(a - \varepsilon)$ for arbitrarily small $\varepsilon > 0$.

Remark 4. Unlike the method of the previous section, this approach provides a lower bound for the sums of the norms of ψ, n, n_t but not for the norms separately.

Remark 5. In 3D, the lower bound (52) is probably not optimal. Indeed, the homogeneous $\dot{H}^{\ell+1/2}$-norm of ψ and the homogeneous $\dot{H}^{\ell}$-norm of n in the expression of the asymptotic solution (35) and (36) both blowup at a faster rate, namely $\frac{1}{3}(1+2\ell)$. Note that these norms blowup at the same rate in 3D, showing that the space $H^{\ell+1/2} \times H^\ell$ is appropriate for the analysis.

Remark 6. In 2D, the norms $\|\psi\|_{\dot{H}^k}$, with $k = \ell + 1$ and $\|n\|_{H^\ell}$ of the exact self-similar solutions (30) and (31) blow up at the same rate $(t^* - t)^{-(\ell+1)}$. Merle's work [22] gives the optimal rate when $k = 1$. Theorem 4 predicts rates of blowup in the space $\mathcal{H}_\ell$. It gives almost the optimal rate of blowup for ψ when $\ell = 0$, but it is off then by $\frac{1}{2}$ for n.

Remark 7. In 3D, a particular result about blowup of a space-time norm $L^{q,r}_{x,t}$ of n is given in [33] under the assumption that blowup occurs in the energy space H_1.

Assume that the solution $u := (\psi, n, n_t)$ exists during a finite time $|t| \le T$ in the space $\mathcal{H}_\ell$. There are two elements in the proof of the theorem above:

1. A local well-posedness estimate for u in the form of the one obtained by Ginibre–Tsutsumi–Velo [14],

$$\|u\|_{X_T} \le C\|u_0\|_{\mathcal{H}_\ell} + CT^\theta \|u\|^2_{X_T}, \quad \theta > 0, \tag{53}$$

   where $\|\cdot\|_{X_T}$ is a space-time norm that will be defined later. For the purpose of local well-posedness, it is not important to determine exactly the power θ, the only requirement being that it is away from 0. On the other hand, in the blowup analysis, a key element is to maximize the power θ, because it leads to a better estimate for the lower bound of the blowup rate. We find an expression for θ that depends on ℓ, the order of the norm under consideration and increases with ℓ.

2. A classical contradiction argument introduced in [36] for semilinear heat equations and used in [37] for NLS equations that reverses the local well-posedness estimate into a blowup rate estimate.

Step 1 *(Local Well-Posedness Estimate).* Rewrite the wave equation (8) as two reduced wave equations for

$$w^{\pm} = n \pm \mathrm{i}\omega^{-1}\partial_t n,$$

where $\omega = (-\Delta)^{1/2}$. The Zakharov system then becomes

$$\mathrm{i}\partial_t\psi + \Delta\psi = (w^+ + w^-)\psi, \tag{54}$$

$$(\mathrm{i}\partial_t \mp \omega)w^{\pm} = \pm\omega(|\psi|^2). \tag{55}$$

$(\psi, w^{\pm})$ solve (54) and (55) with initial data $(\psi_0, w_0^{\pm}) = (\psi_0, n_0 \pm i\omega^{-1}n_1)$ if and only if (ψ, n) solve (7) and (8) with initial data (ψ_0, n_0, n_1).

In the analysis, one slightly modifies the above system by replacing the operator $\omega = (-\Delta)^{1/2}$ by $\omega_1 = (1-\Delta)^{1/2}$ to avoid divergence at low wavenumbers. This leads to an additional term in the wave equation of the form $\langle\nabla\rangle^{-1}\Re w^{\pm}$, which is linear with a gain in derivatives, thus it is easily controlled (see [35]).

The solution of (54) and (55) is written in its Duhamel formulation. Since the solution is considered in a fixed interval $[-T, T]$, we introduce in addition a cut-off C^∞ function $\varphi(t) = 1$ for $|t| \leq 1$, $\varphi(t) = 0$ for $|t| \geq 2$, $0 \leq \varphi(t) \leq 1$, and define $\varphi_T(t) = \varphi(t/T)$, $(T \leq 1)$. The initial value problem (54) and (55) on time interval $[-T, T]$ is equivalent to the system of integral equations

$$\psi(t) = \varphi_1(t)U(t)\psi_0 - \mathrm{i}\varphi_T(t)\int_0^t U(t-s)\varphi_{2T}^2(w^+ + w^-)\psi(s)\,\mathrm{d}s, \tag{56}$$

$$w^{\pm}(t) = \varphi_1(t)W(t)w_0 \pm \mathrm{i}\varphi_T(t)\int_0^t W(t-s)\varphi_{2T}^2\omega(|\psi|^2)\,\mathrm{d}s, \tag{57}$$

where $U(t) = \mathrm{e}^{\mathrm{i}t\Delta}$, $W(t) = \mathrm{e}^{\mp\mathrm{i}t\sqrt{-\Delta}}$ are the free Schrödinger and free reduced wave operators, respectively.

The space X_T in (53), in which the analysis performed, is a product of weighted Sobolev spaces, with space-time weights being the Fourier multipliers associated with the linear Schrödinger and linear reduced wave equation [38]. Namely, $X_T = X_S^{\ell+1/2,b} \times X_{W_{\pm}}^{l,b}$, with the norms given by

$$\|\psi\|_{X_S^{\ell+1/212.b}} = \|\langle\xi\rangle^{\ell+1/2}\langle\tau+|\xi|^2\rangle^b\hat{\psi}(\tau,\xi)\|_{L^2_{\tau,\xi}},$$
$$\|w\|_{X_{W_\pm}^{\ell,b}} = \|\langle\xi\rangle^{\ell}\langle\tau\pm|\xi|\rangle^b\hat{w}(\tau,\xi)\|_{L^2_{\tau,\xi}},$$

where $\langle\xi\rangle = (1+|\xi|^2)^{1/2}$ and $b > \frac{1}{2}$.

There are two distinct elements in the estimates, the linear estimates and the nonlinear ones.

Lemma 1 (Linear Estimates [14, Lemma 2.1]). *Consider the general linear equation*

$$\mathrm{i}u_t - \Phi(-\mathrm{i}\nabla)u = F \quad \textit{on } [0,T]\times\mathbb{R}^d, \qquad u(0)=u_0\in H^s,$$

where Φ is a real valued function. Then for $\frac{1}{2}-\varepsilon < b \le 1-\varepsilon$, $(\varepsilon > 0)$

$$\|\varphi_T u\|_{X^{s,b}} \lesssim \|u_0\|_{H^s} + T^\varepsilon\|F\|_{X^{s,b-1+\varepsilon}}, \tag{58}$$

where the norm in $X^{s,b}$ is associated with the linear operator, namely $\|\cdot\|_{X^{s,b}} = \|\langle\xi\rangle^s\langle\tau+\Phi(\xi)\rangle^b\hat{\cdot}(\tau,\xi)\|_{L^2_{\tau,\xi}}$.

An application of (58) to the solution $(\psi, w^\pm)$ gives

$$\|\psi\|_{X_S^{\ell+1/2,b}} \lesssim \|\psi_0\|_{H^{\ell+112}} + T^\varepsilon\|\varphi_{2T}^2(w^+ + w^-)\psi\|_{X_S^{\ell+1/2,b-1+\varepsilon}}, \tag{59}$$

and

$$\|w^\pm\|_{X_{W_\pm}^{\ell,b}} \lesssim \|w_0\|_{H^\ell} + T^\varepsilon\|\varphi_{2T}^2\omega|\psi|^2\|_{X_W^{\ell,b-1+\varepsilon}}. \tag{60}$$

We now explain how one gets an estimate for the nonlinear terms and produce higher powers of T. The goal is to establish

$$\|\varphi_{2T}^2 w^+\psi\|_{X_S^{\ell+1/2,b-1+\varepsilon}} \lesssim T^\theta\|\varphi_{2T}w^+\|_{X_{W_+}^{\ell,b}}\|\varphi_{2T}\psi\|_{X_S^{\ell+1/2,b}}, \tag{61}$$

$$\|\varphi_{2T}^2\omega|\psi|^2\|_{X_{W_+}^{\ell,b-1+\varepsilon}} \lesssim T^\theta\|\varphi_{2T}\psi\|^2_{X_S^{\ell+1/2,b}}, \tag{62}$$

with equivalent estimates for w^-. A classical argument is to consider the nonlinear terms on the Fourier side and use duality. This reduces (61) and (62) to showing the following inequalities

$$|N_1| \lesssim T^\theta\|v\|_2\|v_1\|_2\|v_2\|_2 \tag{63}$$

$$|N_2| \lesssim T^\theta\|v\|_2\|v_1\|_2\|v_2\|_2, \tag{64}$$

where

$$N_1 = \int \frac{\hat{v}(\xi_1 - \xi_2, \tau_1 - \tau_2)\hat{v}_1(\xi_1, \tau_1)\hat{v}_2(\xi_2, \tau_2)\, \langle \xi_1 \rangle^k}{\langle \tau + |\xi| \rangle^b \langle \tau_1 + |\xi_1|^2 \rangle^c \langle \tau_2 + |\xi_2|^2 \rangle^b \langle \xi_2 \rangle^k \langle \xi \rangle^l}\, \mathrm{d}\xi_1\, \mathrm{d}\xi_2\, \mathrm{d}\tau_1\, \mathrm{d}\tau_2, \tag{65}$$

$$N_2 = \int \frac{\hat{v}(\xi_1 - \xi_2, \tau_1 - \tau_2)\hat{v}_1(\xi_1, \tau_1)\hat{v}_2(\xi_2, \tau_2)\, |\xi|\, \langle \xi \rangle^\ell}{\langle \tau + |\xi| \rangle^c \langle \tau_1 + |\xi_1|^2 \rangle^b \langle \tau_2 + |\xi_2|^2 \rangle^b \langle \xi_1 \rangle^k \langle \xi_2 \rangle^k}\, \mathrm{d}\xi_1\, \mathrm{d}\xi_2\, \mathrm{d}\tau_1\, \mathrm{d}\tau_2. \tag{66}$$

Ginibre–Tsutsumi–Velo [14] showed the above estimates by a repeated application of an inequality obtained from Strichartz estimates and Hölder inequality in time (see [14, Lemmas 3.1–3.4]). Their analysis did not require an optimal power of θ, but needed it to be just large enough, so the final power of T was positive. They find

$$\left(b + 1 - \left(\frac{n}{2} + 1\right)b_0 - \epsilon\right)\frac{1}{2b}.$$

For the rate of blowup analysis, we seek the optimal power of θ, and obtain estimates (63) and (64) with

$$\theta = b + 1 - \left(\frac{n}{2} + 1 - \ell\right)b_0 - \epsilon. \tag{67}$$

To remove the time cutoff from the right-hand side of (61) and (62), we recall

Lemma 2 ([14]).

$$\|\varphi_T u\|_{X^{s,b}} \leq CT^{-b+1/q}\|u\|_{X^{s,b}},$$

where $s \in \mathbb{R}$, $b \geq 0$, $q \geq 2$ *and* $bq > 1$.

Applying this twice (since the nonlinearity is quadratic) with $q = 2$ and combining estimates (59) and (60) with (61) and (62) gives the final estimate

$$\begin{aligned} \|\psi\|_{X_S^{\ell+1/2,b}} + \|n\|_{X_{W_+}^{\ell,b}} + \|n_t\|_{X_{W_+}^{\ell-1,b}} &\leq C\big(\|\psi_0\|_{H^{\ell+1/2}} + \|n_0\|_{H^\ell} + \|n_1\|_{H^{\ell-1}}\big) \\ &\quad + CT^{\theta_\ell}\big(\|\psi\|_{X_S^{\ell+1/2,b}} + \|n\|_{X_{W_+}^{\ell,b}} + \|n_t\|_{X_{W_+}^{\ell-1,b}}\big)^2, \end{aligned} \tag{68}$$

with a power of θ_ℓ as stated in Theorem 4.

Step 2 (Contradiction Argument and Lower Bound).
Let us explain the contradiction argument for a general evolution PDE with a quadratic nonlinearity, and an initial data u_0 belonging to some space H. Suppose an a priori estimate of the form

$$\|u\|_{X_T} \leq C\|u_0\|_H + CT^\theta\|u\|_{X_T}^2, \quad \theta > 0 \tag{69}$$

holds, where $\|.\|_{X_T}$ is some appropriate space-time norm. (This is the a priori estimate (68) with $u = (\psi, n, n_t)$ and $H = \mathcal{H}_\ell$.) Let

$$\mathcal{X}(T, M) = \{u : u(0) = u_0, \|u\|_{X_T} \leq M\}.$$

When performing an iteration argument in $\mathcal{X}(T, M)$, we would like to show

$$C\|u_0\|_H + CT^\theta M^2 \leq M \tag{70}$$

to keep the iterates in $\mathcal{X}(T, M)$. Local well-posedness follows if (70) holds (with, for example, $M = 2C\|u_0\|_H$, and T small enough so that $2CT^\theta M < 1$). The relation between the spaces X_T and H is that X_T must be imbedded in $C([-T, T], H)$ meaning that if u belongs to X_T, it must be a continuous function of $t \in [-T, T]$ with values in H.

Let t^* be the maximal time of existence of solutions, that is

$$t^* = \sup\{T : \|u\|_{X_T} < \infty\}.$$

The blowup hypothesis implies that t^* is finite. Returning to (70), let $0 < t < t^*$ and consider $u(t)$ as an initial condition. The following statement must hold:

If there exists some $M > 0$ such that $C\|u(t)\|_H + C(T-t)^\theta M^2 \leq M$, then $T < t^*$. Or equivalently:

If $T \geq t^*$, in particular $T = t^*$, then for all $M > 0$

$$C\|u(t)\|_H + C(t^* - t)^\theta M^2 > M.$$

We now choose $M = 2C\|u(t)\|_H$, then $\frac{1}{2}M + C(t^* - t)^\theta M^2 > M$ or

$$C(t^* - t)^\theta M^2 > \frac{M}{2} \tag{71}$$

or equivalently

$$\|u(t)\|_H > c(t^* - t)^{-\theta}. \tag{72}$$

Hence, since we cannot continue the time of existence past time t^*, we have a lower bound for the blowup rate of the norm H as given by (72).

Note that the conclusion is about the rate of blowup of the norm H even though the iteration is performed in another norm. One just needs the other norm to embed into $C([-T, T], H)$.

3 The Vectorial Zakharov System

There is no rigorous analysis of blowing up solutions of the full vectorial Zakharov system (1)–(3) although wave collapse is expected when the initial conditions are large enough on the basis of numerical simulations and heuristic arguments (see [5] for review). Here we extend the results of Sect. 2.3.3.

Recalling that for a vector valued function $\mathbf{E}$

$$\Delta \mathbf{E} = \nabla(\nabla \cdot \mathbf{E}) - \nabla \times (\nabla \times \mathbf{E}),$$

we can write (1)–(3) as

$$i\partial_t \mathbf{E} + \alpha \Delta \mathbf{E} + (1-\alpha)\nabla(\nabla \cdot \mathbf{E}) = n\mathbf{E}, \tag{73}$$

$$\partial_{tt} n - \Delta n = \Delta |\mathbf{E}|^2. \tag{74}$$

The symbol of the Laplacian is $|\xi|^2$. This together with the one time derivative determines the $\langle \tau + |\xi|^2 \rangle$ weight in the $X^{s,b}$ norm for the NLS equation of the scalar Zakharov system. To determine the weight that should appear in the $X^{s,b}$ norm for the NLS equation of the vectorial Zakharov system, one needs to determine the symbol of the spatial linear operator appearing in the left-hand side of (73). A simple calculation leads to the matrices M_2 in two dimensions and M_3 in three dimensions given by

$$M_2 = (1-\alpha)\begin{pmatrix} \xi_1^2 & \xi_1\xi_2 \\ \xi_1\xi_2 & \xi_2^2 \end{pmatrix} + \alpha|\xi|^2 I_{2\times 2},$$

and

$$M_3 = (1-\alpha)\begin{pmatrix} \xi_1^2 & \xi_1\xi_2 & \xi_1\xi_3 \\ \xi_1\xi_2 & \xi_2^2 & \xi_2\xi_3 \\ \xi_1\xi_3 & \xi_2\xi_3 & \xi_3^2 \end{pmatrix} + \alpha|\xi|^2 I_{3\times 3}.$$

where $I_{d\times d}$ is the $d \times d$ unit matrix. It was observed by Tzvetkov [15] that the symbol of the operator given by matrix M_d is actually equivalent to the symbol of the Laplacian.

Lemma 3 ([15, Proposition 1]). *Let $d = 2, 3$. Then there exists a constant C such that*

$$|\xi|^2 I_{d\times d} \le M_d \le C|\xi|^2 I_{d\times d}.$$

Using this lemma, one can obtain a local well-posedness result for (73) and (74) [15] analogous to the scalar case, and a lower bound for the rate of blowup of Sobolev norms.

Theorem 5. *Let $d = 2, 3$, and the initial data $\big(E(0), n(0), n_t(0)\big)$ be in $\mathcal{H}_\ell := (H^{\ell+1/2}(\mathbb{R}^d))^d \times H^\ell(\mathbb{R}^d) \times H^{\ell-1}(\mathbb{R}^d)$, $0 \le \ell \le d/2 - \frac{1}{2}$. Assume that the solution (E, n, n_t) blows up in a finite time $t^* < \infty$. Then*

$$\|\mathbf{E}(t)\|_{H^{\ell+1/2}} + \|n(t)\|_{H^\ell} + \|n_t(t)\|_{H^{\ell-1}} > C(t^* - t)^{-\theta_\ell} \tag{75}$$

with $\theta_\ell = \frac{1}{4}(4 - d + 2\ell)^-$, in dimension $d = 2$ or $d = 3$.

Finally, as for the NLS equation, it is of interest to consider the influence of additional dispersive terms and their effect on blowing up solutions. In [39], Haas and Schukla consider the system

$$\mathrm{i}\partial_t \mathbf{E} - \alpha \nabla \times (\nabla \times \mathbf{E}) + \nabla(\nabla \cdot \mathbf{E}) = n\mathbf{E} + \Gamma \nabla \Delta(\nabla \cdot \mathbf{E}) \tag{76}$$

$$\partial_{tt} n - \Delta n = \Delta |\mathbf{E}|^2 - \Gamma \Delta^2 n. \tag{77}$$

which takes into account quantum corrections. The coefficient $\Gamma > 0$ is assumed to be very small. In [40], it is shown rigorously that quantum terms arrest collapse in two and three dimensions, for arbitrarily small values of the parameter Γ.

Acknowledgements MC is partially supported by grant #246255 from the Simons Foundation. CS is partially supported by NSERC through grant number 46179–13 and Simons Foundation Fellowship #265059.

References

1. V.E. Zakharov, Sov. Phys. JETP **35**(5), 908 (1972)
2. V.E. Zakharov, A.F. Mastryukov, V.S. Synakh, Sov. J. Plasma Phys. **1**, 339 (1975)
3. P.A. Robinson, Rev. Mod. Phys. **69**, 507 (1997)
4. L. Bergé, Phys. Rep. **303**(5–6), 259 (1998)
5. C. Sulem, P.L. Sulem, *The Nonlinear Schrödinger Equation. Self-focusing and Wave Collapse*. Applied Mathematical Sciences, vol. 139 (Springer, New York, 1999)
6. B. Texier, Arch. Ration. Mech. Anal. **184**(1), 121 (2007)
7. C. Sulem, P.L. Sulem, C. R. Acad. Sci. Paris Sér. A-B **289**(3), A173 (1979)
8. H. Added, S. Added, C. R. Acad. Sci. Paris Sér. I Math. **299**(12), 551 (1984)
9. S.H. Schochet, M.I. Weinstein, Commun. Math. Phys. **106**(4), 569 (1986)
10. T. Ozawa, Y. Tsutsumi, Publ. Res. Inst. Math. Sci. **28**(3), 329 (1992)
11. L. Glangetas, F. Merle, Commun. Math. Phys. **160**(1), 173 (1994)
12. L. Glangetas, F. Merle, Commun. Math. Phys. **160**(2), 349 (1994)
13. J. Bourgain, J. Colliander, Int. Math. Res. Not. **1996**(11), 515 (1996)
14. J. Ginibre, Y. Tsutsumi, G. Velo, J. Funct. Anal. **151**(2), 384 (1997)
15. N. Tzvetkov, Differ. Integr. Equ. **13**(4–6), 423 (2000)
16. H. Pecher, Int. Math. Res. Not. **2001**(19), 1027 (2001)
17. J. Colliander, J. Holmer, N. Tzirakis, Trans. Am. Math. Soc. **360**(9), 4619 (2008)
18. I. Bejenaru, S. Herr, J. Holmer, D. Tataru, Nonlinearity **22**(5), 1063 (2009)
19. I. Bejenaru, S. Herr, J. Funct. Anal. **261**(2), 478 (2011)
20. J. Ginibre, G. Velo, Hokkaido Math. J. **35**(4), 865 (2006)

21. Z. Hani, F. Pusateri, J. Shatah, Commun. Math. Phys. **322**(3), 731 (2013)
22. F. Merle, Commun. Pure Appl. Math. **49**(8), 765 (1996)
23. W.A. Strauss, Commun. Math. Phys. **55**(2), 149 (1977)
24. Z. Guo, K. Nakanishi, Int. Math. Res. Not. IMRN **2014**(9), 2327 (2014)
25. F. Merle, Commun. Math. Phys. **175**(2), 433 (1996)
26. V.E. Zakharov, L.N. Shur, Sov. Phys. JETP **54**(6), 1064 (1981)
27. H. Berestycki, P.L. Lions, Arch. Ration. Mech. Anal. **82**(4), 313 (1983)
28. H. Berestycki, P.L. Lions, Arch. Ration. Mech. Anal. **82**(4), 347 (1983)
29. L. Bergé, Luc, G. Pelletier, D. Pesme, Phys. Rev. A **42**(8), 4962 (1990)
30. M. Landman, G.C. Papanicolaou, C. Sulem, P.L. Sulem, X.P. Wang, Phys. Rev. A **46**(12), 7869 (1992)
31. M.I. Weinstein, Commun. Math. Phys. **87**(4), 567 (1982/1983)
32. O.B. Budneva, V.E. Zakharov, V.S. Synakh, Sov. J. Plasma Phys. **1**, 335 (1975)
33. V. Masselin, Adv. Differ. Equ. **6**(10), 1153 (2001)
34. J. Holmer, Electron. J. Differ. Equ. **24**, (2007)
35. J. Colliander, M. Czubak, C. Sulem, J. Hyperbolic Differ. Equ. **10**(3), 523 (2013)
36. F.B. Weissler, Isr. J. Math. **38**(1–2), 29 (1981)
37. T. Cazenave, F.B. Weissler, Nonlinear Anal. **14**(10), 807 (1990)
38. J. Bourgain, Geom. Funct. Anal. **3**(2), 107 (1993)
39. F. Haas, P.K. Shukla, Phys. Rev. E **79**(6), 066402 (2009)
40. G. Simpson, C. Sulem, P.L. Sulem, Phys. Rev. E **80**(5), 056405 (2009)

THz Waveforms and Polarization from Laser Induced Plasmas by Few-Cycle Pulses

Peng Liu, Ruxin Li, and Zhizhan Xu

Abstract Terahertz (THz) emission from air plasmas induced by intense few-cycle laser fields is investigated to achieve controllable waveforms and polarizations. Linearly and circularly carrier envelope phase (CEP) stabilized few-cycle laser pulses are produced using a home-built three-stage optical parametric amplifier (OPA) system and a hollow fiber compressor. Focusing the linearly polarized few-cycle pulses into air, THz waveform varies by changing the detected filament length and the CEP of driving pulses. Simulation using the photocurrent model including the propagation effects reveals the phase evolution inside the filament. The ellipticity and conversion efficiency of THz radiation are found dependent on the pulse duration of *circularly* polarized few-cycle pulses. Both the asymmetry and ellipticity of the driving pulses significantly affect the polarization of the generated THz waves. The direction of the elliptically polarized THz radiation rotates by varying the CEP of few-cycle pulses. Such waveform and polarization controllable THz emission is of great importance due to its potential application in THz sensing and coherent control of quantum systems. The evolution of THz waveform provides a sensitive probe to the variation of the CEP of propagating intense few-cycle pulses. The number and positions of the inversions of THz polarity are dependent on the initial CEP, which is constantly near 0.5π under varied driving pulse energies as two inversions become one. This provides a method of measuring the initial CEP at an accuracy that is only limited by the stability of the driving few-cycle pulses.

1 Introduction

Laser induced plasma generating terahertz (THz) waves attract much attention as it provides a pulsed, broad-band emission that can be applied for nonlinear spectroscopy, imaging and remote sensing [1–3]. As an intense femtosecond laser pulse undergoes filamentation in ambient air, radially polarized THz radiation in forward direction is generated due to the transition-Cherenkov emission [4]. Intense

P. Liu (✉) • R. Li • Z. Xu
State Key Laboratory of High Field Laser Physics, Shanghai Institute of Optics and Fine Mechanics, Chinese Academy of Sciences, Shanghai 201800, China
e-mail: peng@siom.ac.cn

A.D. Bandrauk et al. (eds.), *Laser Filamentation*, CRM Series in Mathematical Physics, DOI 10.1007/978-3-319-23084-9_4

THz radiation field ($>$ MV/cm) can be produced from the plasma pumped by a two-color ($\omega/2\omega$) laser field, where four-wave mixing model and photocurrent model are used to explain the underlying physics [5–10]. By adjusting the phase difference of the two-color intense laser fields, enhanced THz emission is generated due to the directional electron motions driven by the asymmetric laser field. And also, polarization of the THz emission has been found to be controllable through the manipulation of the electron trajectories by the two-color laser fields [11, 12]. Similarly, few-cycle pulses can also generate the enhanced THz radiation controlled by the carrier-envelope phase (CEP) of such ultra-short laser pulses [9, 13]. An ultra-short laser pulse that contains only a few cycles of carrier wave can be written as $E = E_0(t)\cos(\omega t + \phi)$, where $E_0(t)$ is the pulse envelope, ω the frequency of the carrier wave, and ϕ the CEP which indicates the phase of the carrier oscillations at the instant of maximum amplitude of the pulse envelope. For few-cycle pulses, the CEP becomes a significant parameter in modifying the electric field of laser pulses and outcomes of the interaction with a medium [14–18]. Intense CEP stabilized few-cycle pulses have been successfully applied in the generation of isolated attosecond pulses and the steering of molecular dissociative ionization dynamics [19–21]. The THz emission from plasma by few-cycle pulses has been interpreted as a result of asymmetric transient currents by ionization induced quasi-dc current produced by the few-cycle laser pulse, as confirmed in subsequent theoretical investigations [22]. By measuring the amplitude and polarity of generated THz emission, the CEP of few-cycle laser pulses can be inferred [22].

The mechanism of THz emission from the laser induced plasma is still under intensive studies [23–26]. The spatiotemporal dynamics of THz emission from the air plasma generated by the two-color laser field has been considered and the scheme of tailoring the THz emission spectrum was proposed by adjusting the tunneling ionization events [24, 27]. Recently the contribution of plasma dynamics of the residual photocurrent has also been applied to explain the wavelength detuning effects of the THz generation out of two-color pumping laser fields [26]. In parallel with the progress, the variation of the THz waveform and polarization along the air plasma driven by linearly and circularly polarized few-cycle pulses has been investigated and the result is reviewed in this chapter. As it shows, the few-cycle pulses provides a unique tool to reveal the spatiotemporal dynamics of THz generation and the propagating intense laser pulses in the filament, and also to control the waveform and polarization of the generated THz radiation. This chapter is organized as follows: first the technology of generating CEP stabilized few-cycle laser pulses is discussed, then the THz wave evolution in air plasma and the polarization of THz emission controlled by the few-cycle pulses are presented consecutively, and finally the discussion on the initial CEP and measurement is given.

2 Generation of CEP Stabilized Few-Cycle Pulses

The CEP of few-cycle pulses provides a special controlling knob for steering the electronic dynamics of atom and molecule in the ultra-short intense laser field, because the laser field varies significantly following the CEP. In laboratory the CEP has to be stabilized in order to obtain valid experimental observations since the phase of laser pulses can be easily scrambled by any forms of disturbance. There are usually two means of stabilizing the CEP: one is to measure and compensate the phase fluctuation by servo loops mounted in the Ti:saphire amplifier system [14]; the other is to take advantages of the optical parametric amplification (OPA) process, through which the differential frequency generation (DFG) of the pumping pulses and the signal pulses (white light continuum in most cases) can cancel out the phase variation in the pumping pulse so that the idler pulses is passively phase stabilized [28].

One of the advantages of the OPA output pumped by the Ti:saphire amplifier system is the generation of longer wavelength intense laser field. Because the ponderomotive potential U_p is proportional to wavelength $U_p \propto I\lambda^2$, the intense laser field process such as high harmonic generation (HHG) and above threshold ionization (ATI) demands the longer wavelength of the laser field for the generation of higher energy photons and electrons. Compared with the active CEP stabilization technique, the passive stabilization technique is relatively insensitive to the environmental disturbance. Moreover, as the phase drift is naturally eliminated in the DFG process, excellent long-term CEP stability can be expected.

There are two kinds of schemes for the CEP stabilized pulse generation using the DFG technique: (1) the CEP stabilized seed is generated in the DFG process within a single pulse, and then amplified in the following OPA or optical parametric chirped-pulse amplifier (OPCPA) stages. 1.2 mJ, 17 fs at 1.5 μm [29] and 740 μJ, 15.6 fs at 2.1 μm [30] CEP stabilized pulses were generated by using this method; (2) the CEP stabilized idler pulse is generated in an OPA process directly, if the pump and the signal pulses are from the same laser source with the same carrier phase offset. The 1.5 mJ, 19.8 fs at 1.5 μm [31] and 0.4 mJ, 11.5 fs at 1.8 μm [32, 33] CEP stabilized pulses were obtained. We adopt the second scheme to generate CEP stabilized few-cycle pulses by an OPA system followed by a hollow fiber compressor. Firstly, CEP stabilized femtosecond multi-cycle pulses are produced at center wavelength of 1.8 μm from a three-stage near-IR OPA pumped by a commercial Ti:sapphire laser amplifier at 1 kHz repetition rate. The 1.8 μm pulses are then spectrally broadened using an argon-filled hollow fiber and subsequently compressed to 8.6 fs (less than 1.5 cycles) with energy of 0.7 mJ with a pair of thin fused silica wedges compensating the dispersion. Circularly polarized few-cycle pulses are produced by inserting a broad-band waveplate in front of the hollow fiber compressor. By optimizing the pressure of the filled gases, the pulse duration of the circularly polarized pulses is compressed to 8.4 fs. Nonlinear spectral interferometry (f–$2f$) measurement is employed to monitor the CEP stability.

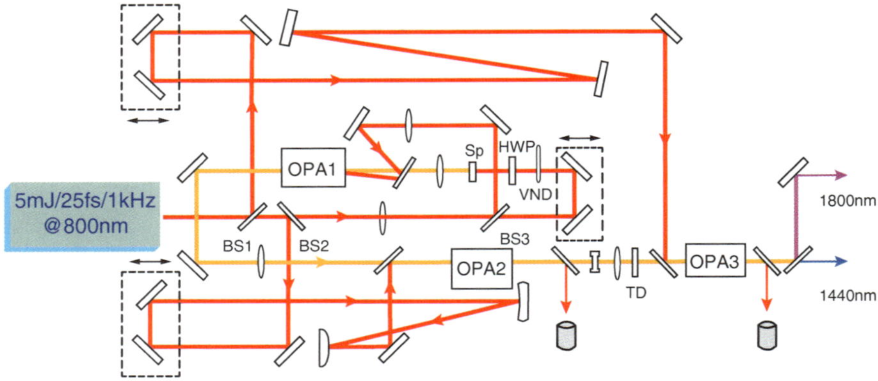

Fig. 1 Setup of the three-stage OPA system: *VND* variable neutral density filter, *HWP* half-wavelength waveplate, *Sp* sapphire plate, *TD* time delay crystal

2.1 Optical Parametric Amplifier

Pumped by a commercial Ti:sapphire laser amplifier (Coherent Elite-HP-USX) which provides 25 fs pulses with pulse energy up to 5.2 mJ at 1 kHz repetition rate, a home-built three-stage OPA system generates CEP stabilized femtosecond laser pulses at the center wavelength of 1.8 μm. The design of the OPA system is the same as the previously reported [34] as shown in Fig. 1. The pump pulses are divided into four parts using three beam splitters. The first part of the laser pulses of less than 20 μJ is focused into a 2 mm-thick sapphire plate to generate a single-filament white light continuum (WLC), providing the seed pulses for the following OPA stages.

The second fraction of the pump laser pulses with ~80 μJ pulse energy is used to pump the first near-collinear OPA stage (OPA1) consisting of a 2.5 mm-thick BBO crystal cut for type II phase matching ($\theta = 27.2°$, $\phi = 30°$). The intersection angle between the pump and seed beams is ~1°. The WLC is amplified to ~3 μJ at the OPA1 with the center wavelength of 1.44 μm. Although the type I BBO crystal in this spectral region has a broader phase matching bandwidth [35], we found a lower efficiency and strong parasitic self-diffraction so it is not used. The amplified pulses at 1.44 μm from OPA1 are collimated and injected into the second collinear OPA stage (OPA2) consisting of a 2 mm-thick BBO crystal cut for type II phase matching ($\theta = 27.2°$, $\phi = 30°$). About 0.7 mJ laser pulse with a diameter of ~4 mm is used to pump OPA2. In order to get high efficiency and prevent wave front tilt [36], both the signal and the pump beams are well collimated and the two beams are collinearly injected into the BBO crystal. The seed pulses are amplified to ~50 μJ in this stage.

The amplified signal pulses from OPA2 are enlarged and collimated to ~9 mm in diameter with a Galilean telescope. A 1.5 mm-thick a-cut YVO4 crystal is employed as a time delay crystal to separate the signal and the idler pulses in time. Then the laser beam is injected into the third collinear OPA stage (OPA3) consisting of a 2 mm-thick BBO crystal (type II, $\theta = 27.2°$, $\phi = 30°$). The remaining ~3.4 mJ pump laser with diameter of ~9 mm is used to pump OPA3. Same as OPA2, the pump and the signal beams are collimated and then collinearly injected into the BBO crystal. This collinear setup is quite necessary to avoid angular dispersion in the generated idler beam. The signal pulses are further amplified to ~1.0 mJ at 1.44 μm, corresponding to the idler pulses with energy of ~0.9 mJ at 1.8 μm. The pulse duration of the output signal pulses and the idler pulses are measured when the highest conversion efficiency is obtained by a home-built single-shot autocorrelator. The idler pulse duration is ~40 fs at 1.8 μm without any further pulse compression.

2.2 *Pulse Compression*

The 1.8 μm idler beam from the OPA system is coupled into a 1-m long hollow fiber filled with argon (400 μm inner diameter, 0.5 mm-thick fused silica window) using an $f = 750$ mm plano-convex lens. The output beam is collimated with a silver-coated concave mirror ($R = 2000$ mm). The spectrum may be broadened by keep increasing the gas pressure, but too high gas pressure may induce energy flow from the fundamental to higher-order eigenmodes of the waveguide, which deteriorates the output beam profile and even cause filamentation in front of the fiber lowering the coupling efficiency. For a balance between broad spectrum and single mode purity of the output beam, 500 mbar of argon gas is filled in the hollow fiber. As a result, the supercontinuum in the range of 1200–2100 nm is obtained which supports the Fourier transform-limited pulse duration of about 7.8 fs. A pair of fused silica wedge is used to compensate the pulse dispersion since it has negative group delay dispersion (GDD) in this spectral range [37]. After the compression, the energy of the output pulses reaches > 500 μJ per pulse.

When generating circularly polarized pulses, the polarization of 1.8 μm idler pulses out of the OPA system is changed from linear to circular by passing through a quarter-wave plate (Thorlabs AQWP05M-1600) in front of the compressor. The beam is then focused into the same hollow fiber chamber for compression using an $f = 900$ mm plano-convex lens. In the compressor, different noble gases and gas pressures are filled in the hollow fiber for the optimized condition of the spectrum broadening and output beam quality. Pumped by 0.9 mJ linearly polarized laser pulses, the optimum gas pressure for argon, krypton, and xenon is found to be 500 mbar, 250 mbar, and 100 mbar, respectively, to achieve a spectrum of 1300–2100 nm. Because the circularly polarized laser fields reduce both the Kerr nonlinearity and ionization rate of the gas, the pressure to drive the self-phase modulation (SPM) is expected to be higher for equivalent laser intensity. The broadened spectrum using different gases indicates that, for the circularly polarized laser pulses the spectrum broadened by xenon is smoother and broader than argon and krypton. The output pulse energy is 0.55 mJ, achieving a high transmittance (>60 %) through hollow fiber which should be attributed to not only the circular polarization but also the good pulse profile of the input beam and the proper gas pressure in the fiber. The polarization of the laser pulses is measured by a Glan

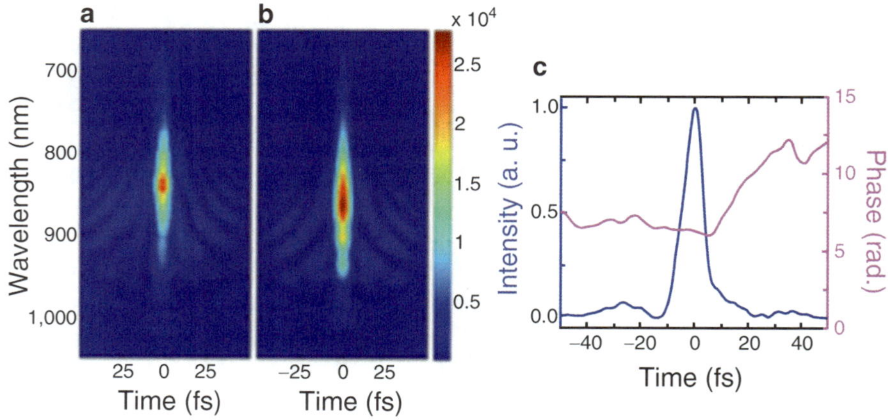

Fig. 2 (**a**) Measured and (**b**) retrieved SHG-FROG traces, and (**c**) the retrieved temporal profile (*blue*) and phase (*purple*)

prism and a power meter. Rotating the Glan prism, the transmitted intensity changes less than 3 % that indicates the pulses having a near circularly polarization.

After the pulse compression, the pulse duration is measured by a home-built second harmonic generation frequency resolved optical grating (SHG-FROG) [38] and a home-built autocorrelator. No transmissive optics is used in the SHG-FROG which is suitable to measure the pulse duration down to a few cycles. A 20 μm-thick type I BBO crystal ($\theta = 20.2°$) is used in the SHG-FROG which ensures broad SHG bandwidth. In the routine experiments, the duration of linearly polarized pulses is about 10 fs while the optimized duration can reach to 8.6 fs. For the circularly polarized pulses, the best pulse compression is achieved by inserting a phase compensator of fused silica with a thickness of 2 mm. The measured and retrieved SHG-FROG traces are shown in Fig. 2a, b. The retrieval error is down to 0.007 and the pulse duration is 8.6 fs as shown in Fig. 2c.

2.3 CEP Stability

The CEP stability is characterized with a home-built collinear f–$2f$ interferometer whose layout is similar to [39]. The only difference is a 5 mm-thick c-cut LiNbO3 crystal inserted between the WLC and SHG crystal. Because the group delay is almost the same for 900 and 1800 nm lights in glass, the fringe period cannot be observed if the time delay crystal is not used. Frequency overlap between the spectral broadened fundamental frequency and the second harmonic frequency is achieved in the 750–820 nm spectral range. Figure 3 shows the CEP fringes and phase measurement for the compressed 1.8 μm idler pulses over a 27 min observation time, whose fluctuation is 248 mrad.

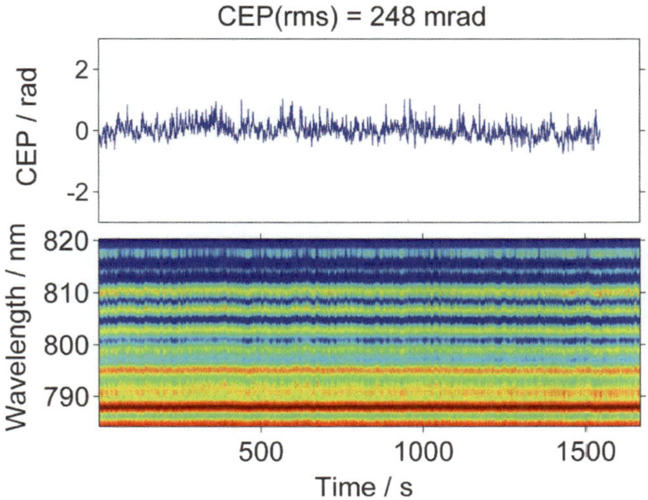

Fig. 3 *Upper*: CEP phase shift; *lower*: the measured CEP fringes

3 Variation of THz Waves in Air Plasma by Few-Cycle Pulses

Recent investigations show that waveform-controlled THz radiation is of great importance due to its potential application in THz imaging and coherent control of molecular dynamics [40, 41]. Can we control the waves of the THz emission form the femtosecond laser induced filament all-optically? Understanding the laser phase effect and propagation effects on the THz radiation from the few-cycle laser pulse produced filament is crucial for the THz generation based CEP metrology [13] and the generation of intense and waveform-controlled THz radiation.

We demonstrate a novel scheme to measure the spatially resolved THz radiation in the filament produced by few-cycle laser pulses. Variation of THz waveform and even its polarity inversion are found along the propagating plasma. It is revealed that the Gouy phase shift of the converging driving laser pulses and nonlinear effects in air and plasma both affect the pulse front of the few-cycle pulses. THz waveform can be controlled by varying the filament length and the initial CEP of the driving laser pulses.

3.1 Variation of THz *Waveform in Air Plasma*

In experiments the output beam of the Ti:sapphire laser amplifier is split into two: a major part of the beam, 4.6 mJ, is used to pump the OPA system while the left part as a probe pulse for sampling the generated THz temporal waveforms. The 40 fs-long CEP-stabilized laser pulses at 1.8 μm are compressed to be averaged 10 fs

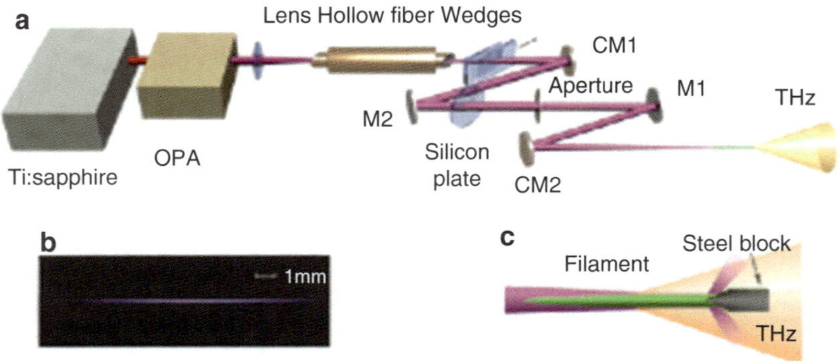

Fig. 4 (**a**) Experimental layout; (**b**) image of the filament formed by the few-cycle pulses; and (**c**) the filament length control [43]

routinely with the maximum pulse energy of 450 μJ, which is focused into the air by a spherical mirror ($f = 150$ mm) to produce a stable luminescence filament of ∼12 mm-long, as shown in Fig. 4b. The generated THz radiation is measured by using the standard balanced diode geometry EO sampling technique [42]. We then insert a sharp stainless steel blade (∼0.2 mm ×4 mm ×20 mm) into the plasma column to stop the filament, and the blade is moved by a motor stage to vary the length of filament in the laser propagation direction (z coordinate). As shown in Fig. 4c, the blade is positioned to allow the laser beam hit onto its sharp edge, so that the detection of THz radiation from the leftover filament is less influenced. A sharp piece of quartz wedge is also inserted into the filament at some positions (both in the front and the end of filament) same as the blade and no difference is observed, indicating that no secondary emission from the laser ablation influences the detection of THz emission in the forward direction. In so doing, the integrated THz signal as a function of filament length is measured. In analysis we calibrate experimentally the deviation of the collection efficiency of the THz detection optics varied by the filament length to the detector.

Figure 5a shows the measured THz waveforms from the filaments of 3 mm (from the visible starting position of the filament to the blade) and 10 mm, respectively. One can note that the recorded THz radiation from the 10 mm-long filament reverses its polarity comparing to that from the 3 mm-long filament, indicating a phase variation of the driving laser field in the filament. It is reasonable that the measured THz emission from the 3 mm-long filament is much weaker than that from the 10 mm-long one, since the measured signal comes from the integration of the unblocked filament. By moving the blocking blade continuously, the THz waveforms as a function of filament length are recorded and plotted in Fig. 5b. One can see that the THz emission gradually changes its amplitude and polarity when the length of filament increases, and at certain length (∼5 mm-long filament) the THz emission signal goes to zero and then reverses the polarity thereafter.

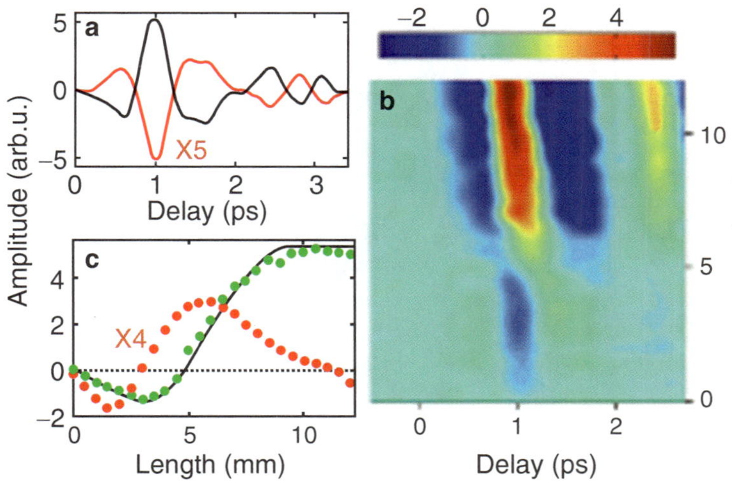

Fig. 5 (**a**) THz waveforms measured from the filaments of 3 mm (*red solid line*) and 10 mm (*black dashed line*), respectively; (**b**) THz waveforms measured by moving the block continuously along the filament; (**c**) the measured THz amplitudes (*green solid squares*), the calculated results (*black solid line*), and the retrieved THz signal amplitude (*red solid circles*) as a function of the position in the filament [43]

For analyzing the amplitude modulation of THz radiation, the measured amplitude values at the time of maximum THz signal, ∼1 ps in Fig. 5b, are plotted in Fig. 5c. And through the disintegration, the spatially resolved THz emission in the filament is retrieved and shown in Fig. 5c. One can see that the THz emission first increases in one (negative) polarity and then decreases to zero at ∼3 mm, after which the amplitude increases in the other (positive) polarity and decreases a little at the tail of the filament.

3.2 Simulation of THz *Emission in Air Plasma by Few-Cycle Pulses*

The polarization of THz emission was measured to be nearly linear so that we can exclude that the generated THz radiation originates from the transition-Cherenkov emission. We consider the THz emission originate from the transient photocurrent driven by the propagating intense few-cycle laser fields in the plasma [7, 8]. The propagation of intense few-cycle laser fields in a dispersive medium can be described by using a propagation equation in an axial-symmetric coordinates [44],

$$\partial_z \tilde{E}(\mathbf{r}, z, \omega) = \left[\frac{\mathrm{i}}{2k(\omega)}\nabla_\perp^2 + \mathrm{i}k(\omega)\right]\tilde{E}(\mathbf{r}, z, \omega) + \frac{\mathrm{i}\omega^2}{2c^2\varepsilon_0 k(\omega)}\tilde{P}_{\mathrm{NL}} - \frac{\omega}{2c^2\varepsilon_0 k(\omega)}\tilde{J}_{\mathrm{ioni}} , \quad (1)$$

where $\tilde{E}(\mathbf{r}, z, \omega)$ is the frequency domain laser field, the nonlinear polarization $\tilde{P}_{\mathrm{NL}}$ accounts for the Kerr effect, and the polarization $\tilde{J}_{\mathrm{ioni}}$ is caused by photoelectrons from the tunneling ionization of N_2 and O_2 in air. The collective motion of the tunneling ionized electrons results in a directional nonlinear photocurrent surge described by [24]

$$\partial_t J_{\mathrm{e}}(\mathbf{r}, z, \omega) + v_{\mathrm{e}} J_{\mathrm{e}}(\mathbf{r}, z, \omega) = \frac{e}{m}\rho_{\mathrm{e}}(\mathbf{r}, z, \omega)E(\mathbf{r}, z, \omega) , \quad (2)$$

where v_{e}, e, m, and ρ_{e} denote the electron–ion collision rate, electron charge, mass, and electron density, respectively. The transient current at each propagation step of the calculation is treated as radiation source of far field THz emission [45]

$$E_{\mathrm{THz}}(\mathbf{r}', t) = -\frac{1}{4\pi\varepsilon_0}\int \frac{1}{c^2 R}\rho_{\mathrm{e}}(\mathbf{r}, z, \omega)\,\mathrm{d}^3\mathbf{r} , \quad (3)$$

where R is the distance between the THz point source and the detection plane. Equation (3) describes the THz radiation towards all directions. The forward far field THz emission near the propagation axis from the unblocked air plasma has been added up to simulate the measured results. The initial few-cycle laser field is of a Gaussian beam with the waist size of $w_0 = 260\,\mu\mathrm{m}$ $(1/e^2)$ at center wavelength of $1.8\,\mu\mathrm{m}$, duration of $\tau = 10\,\mathrm{fs}$ in full width at half maximum (FWHM) and pulse energy of $300\,\mu\mathrm{J}$. By optimizing the initial CEP of the few-cycle laser field to be $\phi_0 = 0.33\pi$, the THz amplitude as a function of filament length is found consistent with the measured results, as shown in black solid line in Fig. 5c.

3.3 Variation of CEP and Phase of Few-Cycle Pulses in Filament

Since the THz emission is shown to be sensitively determined by the intense few-cycle laser field, we then investigate the CEP variation in the air plasma. As indicated previously [46], the CEP is the difference of the pulse phase, i.e., carrier phase, and the pulse front. The carrier phase shift in a dispersive medium can be written as

$$\phi_{\omega_0}(\mathbf{r}, z) = n(\mathbf{r}, z)\frac{\omega_0}{c} + \phi_{\mathrm{G}}(z, z_{\mathrm{eff}}) + \phi_{\text{off-a}}(\mathbf{r}, z) \quad (4)$$

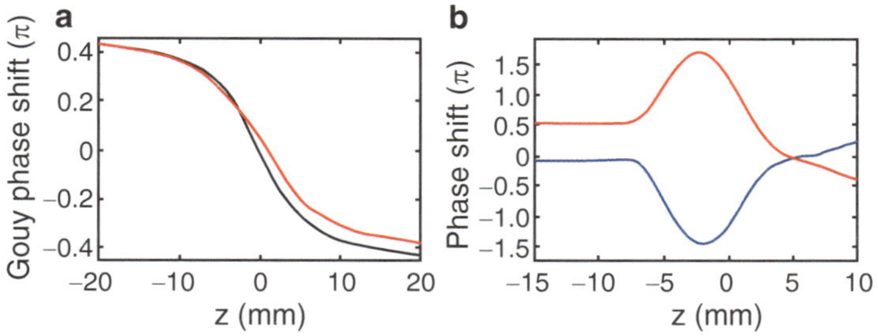

Fig. 6 (**a**) The modified Gouy phase shift (*red dashed line*) and the Gouy phase shift (*black solid line*) of linear focus condition; (**b**) the pulse front shift (*blue solid line*) and the CEP shift (*red dotted solid line*) of the laser field in air plasma [43]

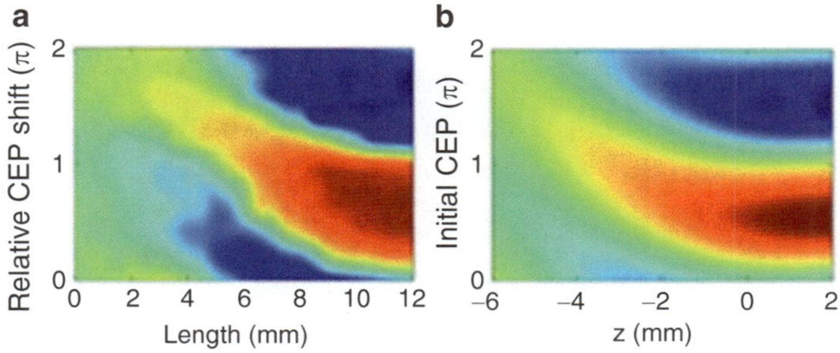

Fig. 7 (**a**) Measured and (**b**) simulated THz amplitude as functions of relative CEP shift and filament length [43]

where the first term is the phase shift at the carrier frequency induced by spatially dependent refractive index $n(\mathbf{r}, z)$, the second term is the Gouy phase shift with the effective focal length z_{eff}, and the third term denotes the off-axial phase shift due to diffraction effects. For simplicity we only consider the axial phase shift that includes the first two terms. We define the second term as the modified Gouy phase shift that is characterized by z_{eff}, since the Rayleigh length is no longer valid in filament.

The calculated modified Gouy phase shift decreases with a relatively slower slope than the Gouy phase shift under the linear focus condition, as shown in Fig. 6a. The CEP of laser fields is also determined by the propagating pulse front, which is defined as [46, (17)] and can be retrieved from the calculation results. The axial pulse front is shown in Fig. 6b. As one can see the pulse front experiences a jump around the center of the plasma region ($z = -7$–4 mm), as a result of the plasma-induced dispersion. The total axial CEP shift is $\sim 1.4\pi$ with a hump structure in the plasma region, which is originated from the modified Gouy phase shift and plasma effects.

By increasing the initial CEP gradually in a step size of 0.2π, we obtained a series of two-dimensional THz waveform maps similar to that shown in Fig. 5b. The amplitudes of THz emission along the filament from experiments are retrieved and plotted as a function of ϕ_0 in Fig. 7a. We found that the THz waveforms change back to be the same as those shown in Fig. 5b after a change of initial CEP by 2π. This indicates that in air plasma the propagation effect on the CEP variation is independent of the initial CEP of driving laser fields, which is also verified by our numerical simulations. The phases of few-cycle pulses experience a fixed shift of larger than π, which is consistent with the CEP shift through the air plasma. This observation helps to validate the CEP metrology based on THz generation [13]. However, due to the variation of CEP through the plasma, an offset of CEP must be determined and taken into account for measuring the CEP of laser pulses. As shown in Fig. 7b, the simulated results using the photocurrent model and propagation equation agree well with experimental results. Since the polarity of THz radiation can be controlled by varying the initial CEP of driving laser pulses, one can maximize the output THz radiation intensity from the filament by optimizing the initial CEP to minimize the intensity cancellation due to opposite polarities.

4 Polarization of THz Wave by Circularly Polarized Few-Cycle Pulses

Polarization control of THz waves presents a new opportunity for its potential applications. It has been shown that the electron trajectory in plasma is controlled by adjusting the phase and polarization of the two color laser fields, and the polarization direction of the consequent linearly polarized THz emission can be tuned continuously [11, 12]. The control of linearly polarized THz emission enables ultrafast manipulation of the polarization state with application to imaging, dichroism, and polarimetry at THz frequencies [47–49]. However, THz sources with controlled and adjustable elliptical polarization states are needed for a wide range of new experiments, such as the control of excitons in semiconductor nanostructures [50] and of molecular rotational wave packets [51]. Controllable elliptically polarized THz is also potentially important to study macromolecular chiral structures such as proteins and DNA because of strong absorption in the THz regime due to collective vibrational modes [52].

Recently, X. Lu et al. reported the generation of elliptically polarized THz emission by applying an external rotating electric field on the air plasma column [53]. The helical electric field imposes an external force that accelerates the photoelectrons produced from the ionization of intense laser fields to orthogonal directions. Such a "downstream" means can effectively control the orthogonal accelerations and consequently the transient photocurrents for the elliptically polarized THz emission. However, this scheme is not valid for the remote generation of polarization-controlled THz waves. Fazel et al. reported the generation of elliptically polarized THz emission from multi-cycle circularly polarized laser induced air plasma [54]. But the polarization cannot be controlled, and the THz conversion is low because of the weak asymmetry of the multi-cycle laser field. Therefore, the all-optical control of elliptically or circularly polarized broadband THz emission remains a challenge so far, and the insightful physical mechanism is required for further polarization control and manipulation. Here we present the experimental demonstration and theoretical explanation of the all-optical generation and control of elliptically polarized terahertz waves.

4.1 Generation of Elliptically Polarized THz Radiation

Focusing the circularly polarized few-cycle laser pulses into ambient air by a concave mirror ($f = 150$ mm), an 8 mm-long filament is produced with the driving pulse energy of 350 μJ. The driving laser pulses are blocked by a 1 mm-thick polyethylene plate, and THz waves pass through a metal wire polarizer to analyze the polarization state. The transmitted THz radiation of a selected polarization direction is measured by a THz power meter (Gentec THZ1.5B-MT-USB) mounted with an 8.8 THz low pass filter (Tydex LPF 8.8-35) for the output power. The balanced diode geometry EO sampling method [42] can replace the power meter for the waveform measurement using a 1 mm ZnTe crystal.

Figure 8 shows the measured THz intensity (I) as a function of the rotating angle (θ) of the polarizer. The experimental data obtained using 12 fs (blue square) and 15 fs (red dot) driving laser pulses can be very well fit with the expression of an elliptically polarized THz field $I = a^2 \cos^2 \theta + b^2 \sin^2 \theta$, where a and b are major and minor axes of ellipse, respectively. It is calculated that $b/a = 2.48$ for 12 fs and 2.08 for 15 fs. Measurement results indicate that the generated THz radiation is elliptically polarized. And when the driving laser pulse duration increases from 12 to 15 fs, the THz polarization turns to be more circular and the THz intensity decreases. We also carried out the EO sampling measurement of THz time domain spectrum, and the result is consistent with the result of power measurement.

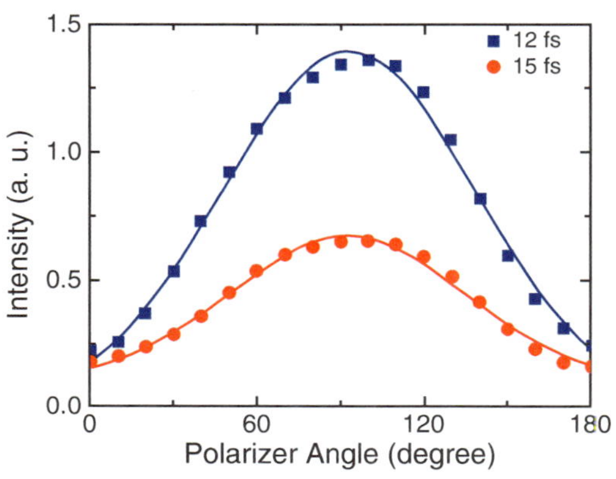

Fig. 8 Polarization analysis of THz radiation, measured for 12 fs (*blue square*) and 15 fs (*red dot*) driving laser pulses [55]

4.2 Ellipticity and Conversion Efficiency on Pulse Duration

In order to understand the generation mechanism of the elliptically polarized THz emission driven by the circularly polarized few-cycle laser pulses, we simulate the THz generation using the transient photocurrent model. The circularly polarized few-cycle laser field is written as:

$$E(z, r, t) = E_{\mathrm{env}}(z, r, t)[\exp(-\mathrm{i}\omega t - \mathrm{i}\phi)\mathbf{x} + \exp(-\mathrm{i}\omega t - \mathrm{i}\phi - \mathrm{i}\pi/2)\mathbf{y}] \quad (5)$$

in which $E_{\mathrm{env}}(z, r, t)$ represents the Gaussian envelope in space and temporal domain. In the simplified model we only consider the on-axis laser field since the minor distortion of off-axis laser field does not affect the far field THz emission significantly. The far field THz emission can then be calculated by (2) and (3). In the calculation, the retard time of THz emission and few-cycle laser pulses is considered by the term $(n_{\mathrm{opt}} - n_{\mathrm{THz}})$[53]. Finally the time domain THz waveform can be obtained by the inverse Fourier transform of the calculated far field THz emission. In the model, electron trajectory, retard delay of THz propagation regarding the driving pulse, and the Gouy phase variation are all included.

In Fig. 9a–c we plot the calculated results of the THz electric field driven by the circularly polarized laser pulses of different durations (6, 12, and 20 fs in FWHM). As one can see, the polarization of THz emission changes from elliptical to circular as the duration of laser pulses increases, meanwhile the amplitude of THz

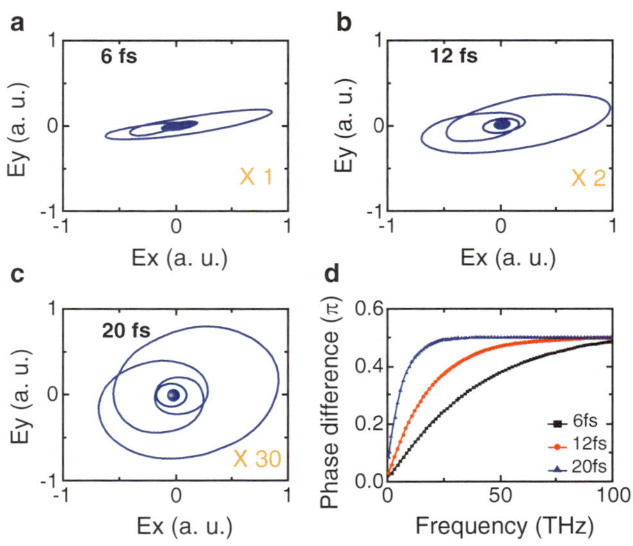

Fig. 9 Calculated THz electric field driven by circularly polarized laser pulses with the durations of (**a**) 6 fs, (**b**) 12 fs, and (**c**) 20 fs; (**d**) the phase difference between x and y components of THz emission driven by laser pulses of different duration [55]

emission decreases significantly. For the circularly polarized few-cycle laser pulses, the transient vector of the laser field has asymmetric amplitudes towards all the directions in the transverse plane of the propagation direction. The amplitude peak is at the center of the pulse envelope and decreases to zero at the edge. Such a spatially and temporally varying intense laser field produces a broadband transient current towards the rotating directions in transverse plane which induces THz emission in the rotating directions. And according to the simulation, the phase of the THz waves partially inherits from the driving laser. As shown in Fig. 9d, the phase difference between x and y components of THz field is about $\pi/2$ in high frequency (the same as that of the driving laser pulses) and decreases as the frequency getting lower. As the result of the THz emission in the rotating directions and phase difference between x and y components, elliptically polarized THz emission is generated with handedness inheriting from the driving laser. Figure 9d also shows that, as the laser pulse duration increases, the phase difference in low frequency is closer to $\pi/2$ which results in more circular polarization. The calculated result for the 12 fs driving laser pulses shown in Fig. 9b has the similar ellipticity with which we observed experimentally. For THz waves produced by multi-cycle circularly polarized laser pulses, the polarization has been demonstrated to be nearly linear [12, 56]. Simulation also shows that, when the pulse duration becomes longer, the asymmetry of the laser field decreases. As a result, the net transient photocurrent decreases significantly, leading to a much weaker (1/30) THz signal for the 20 fs driving pulses than the 6 fs driving pulses.

4.3 Polarization Dependence on CEP

Our experiment also indicates that the polarization direction of generated THz waves can be controlled by adjusting the CEP of the driving laser pulses. Measurement results of THz emission versus the angle of polarizer and the CEP of driving laser pulses are shown in Fig. 10a. It is indicated the polarization plane of generated THz waves rotates following the CEP variation while the ellipticity and intensity of the THz waves remains the same. Simulation results are shown in Fig. 10b, because the direction of the maximum laser field vector of the circularly polarized few-cycle laser pulses rotates with the variation of CEP, the polarization plane of generated THz waves rotates following the CEP variation with a constant amplitude field, and the major axis of the THz ellipse is rotating right-handed/left-handed as we increase/decrease the CEP. On the other hand, the CEP of few-cycle laser pulses can be determined by measuring the polarization of generated THz waves [22], which provides a convenient method for CEP measurement without using sophisticated devices.

The retardation of THz waves and driving laser pulses has been shown to have a remarkable effect on the elliptical THz emission driven by linearly polarized multi-cycle femtosecond laser pulses [53]. In our experiment, because of the limited pulse energy (<400 μJ), the laser-induced air filament is too short to observe the effect of

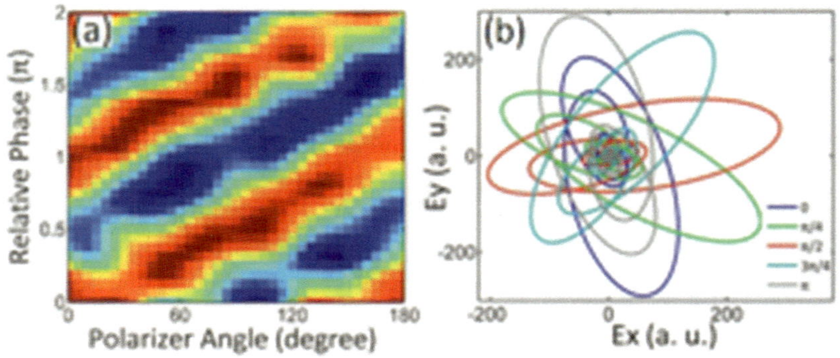

Fig. 10 The experimental (**a**) and simulation (**b**) results of THz polarization rotating with the CEP of driving laser pulses [55]

retardation between THz waves and driving laser pulses. We block the THz waves by applying an aperture through which the filament passes. The polarization of THz radiation from half of the filament is almost the same as that of the whole filament. We compare the THz emission induced by laser pulses of different energy and find that the laser energy has little effect on the polarization of generated THz waves under the experimental condition. This is due to the fact that the ionization rate for the circularly polarized laser field is much lower than the linearly polarized case.

5 Initial CEP and Its Determination Through THz Waveform Variation

For fully characterizing the phase of few-cycle laser pulses, several schemes of measuring the actual CEP of few-cycle pulses have been proposed and demonstrated. The stereo-ATI method measures the ATI electrons in opposite directions along the laser polarization and relates the asymmetry of cut-off electrons to the CEP value [57]. The recent advance indicates a capability of single-shot measurement with an accuracy of ~100 mrad [58]. The detection of THz emission from air plasma driven by intense few-cycle laser fields has shown an alternative method of probing the phase of the few-cycle pulses, based on the CEP dependent asymmetry of tunneling ionization [13]. The high-harmonic emission at individual half-cycles of a few-cycle pulse has also been shown to be able to measure the CEP of the driving laser field in situ [59]. These methods all aim to measure the CEP of the localized laser field where intense few-cycle pulses cross with the detection medium by assuming the variation of CEP is negligible during the interaction.

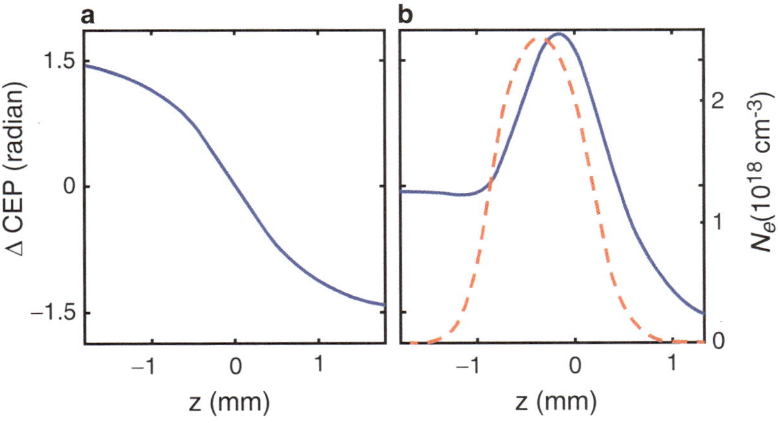

Fig. 11 (**a**) The CEP shift (*blue solid line*) around the focus of a few-cycle pulse in vacuum. (**b**) The CEP shift (*blue solid line*) in the interaction region between an intense few-cycle laser field and N2 molecules, and the electron density in the air plasma (*red dashed line*) [60]

5.1 Initial CEP

For the intense few-cycle laser fields, Gouy phase shift and diffraction dominate the CEP variation even under the condition of linear focusing [46]. For example, if one considers few-cycle laser pulses of 8 fs in FWHM and beam diameter of 8 mm, focused by a mirror of $f = 200$ mm into vacuum, the CEP shift over a distance of 1 mm at the center of focus is 1.43 radian [46], as shown in Fig. 11a. Interacting with a nonlinear medium, the variation of CEP becomes more complex because of the dispersion of the medium and the plasma effect. Taking the N_2 molecules (with the peak pressure of 100 torr, thickness of 1 mm) ionized by the intense few-cycle laser fields (with the pulse energy of 40 μJ and assuming the diameter of filament is 140 μm), the CEP shift over the interaction length is calculated by solving the propagation equation [44] and plotted in Fig. 11b. One can see that the CEP variation over the interaction length of 1 mm is 1.34 radian, and the value becomes larger as the focusing length decreases and the pulse energy increases.

An alternative option of characterizing the CEP of few-cycle laser pulses is to define the initial CEP, which is the actual CEP value of few-cycle pulses before focusing. The initial CEP can be experimentally determined, as will be shown in this work, and its value is free of the uncertainties caused by the focusing and interaction with nonlinear medium. With the knowledge of the initial CEP, the actual CEP of the local laser field interacting with a medium can be estimated in applications.

It is known that the THz waveform reverses its polarity in air plasma because of the variation of the phase and the pulse front (namely the CEP) of the driving few-cycle fields. One can in principle determine the initial CEP of driving few-cycle pulses by taking into account the variation of CEP in the air plasma. However, the value of CEP shift changes as the laser pulse energy varies, so it cannot be predetermined in experiments and therefore hinders the accurate measurement of the initial CEP. We find that the positions where THz waveform reverses are related to the initial CEP of the intense few-cycle fields, from which a method of measuring the initial CEP of few-cycle pulses is given.

5.2 Determination of the Initial CEP

We focus on the investigation of inversion number and positions of the THz waveform from different length of air plasma. THz emissions from air plasma can be described by the integration of far field emission from the transient photocurrents driven by an intense few-cycle laser field propagating in the plasma as indicated in (1). The integrated THz amplitude as a function of filament length and initial CEP at different input energy is shown in Fig. 12. It can be seen that the THz amplitude, the number of the THz waveform inversion, and the inversion positions are all related to the initial CEP. As shown in Fig. 12a, the white color represents that the THz amplitude is zero. At a low input energy of 113 μJ, one may note that the variation of THz amplitude tilts a little when the initial CEP is near $\phi_0 = 0.5\pi$. For the initial CEP apart from $\phi_0 = 0.5\pi$, the THz amplitude increases monotonically along the filament. The positions of THz inversion are extracted for all the initial

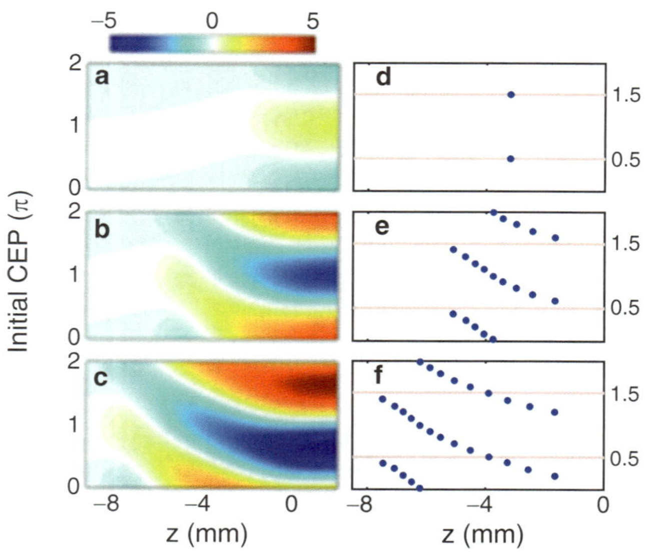

Fig. 12 (**a**)–(**c**) The calculated THz amplitude as a function of plasma length and the initial CEP of few-cycle pulses with the input energy of 109 μJ, 250 μJ, 400 μJ, respectively. (**d**)–(**f**) The inversion positions of THz emission in a step-size of 0.1π, for the energy of 109 μJ, 250 μJ, 400 μJ, respectively [60]

CEP values in Fig. 10d, which shows inversion of THz waveform only at 0.5π and 1.5π. It is known that because of the shift of pulse front and phase in air plasma, the variation of CEP results in varied THz waveforms [43]. The integrated far field THz emission shows a reversed polarity when the CEP of laser pulses changes across the value of 0.5π.

Increasing the energy of laser pulses to 250 μJ, the starting position of THz generation moves to the upstream of the air plasma because of Kerr effect, as shown in Fig. 12b, e. Inversions of THz emission begin to appear for some initial CEP of few-cycle pulses, because of the increased CEP variation in air plasma. It shows that in the plasma the THz emission crosses with zero only once for most of the initial CEP. When the energy of laser pulses increases to 400 μJ, one can see that for some initial CEP of laser pulses, $\phi_0 = 0.2\pi$–0.4π in Fig. 12f, there are two inversions appear in the plasma, and the inversion positions of THz waveform gradually shift to the beginning of the filament. Finally the first inversion disappears at $\phi_0 = 0.5\pi$. We also find that when the energy of laser pulses increases, the initial CEP of $\phi_0 = 0.5\pi$, where the first inversion disappears into the beginning of filament, does not change.

The observation that no inversion appears in front of the plasma for the initial CEP of 0.5π indicates a smooth variation of CEP in the beginning of plasma. In the middle and end part of the filament, the significant increase and decrease of CEP results in the reversed polarity of THz emission, and consequently the integrated THz emission shows an inversion of polarity. For the intense laser fields of $\phi_0 \geq 0.5\pi$, the THz polarity is positive at the beginning of the plasma, the superposed THz radiation keep increasing as the CEP increases in the plasma, since there is no polarity change. It eventually decreases and reverses its polarity at the downstream of the plasma when the increment of CEP is larger than $-\pi$. As a result, an inversion of THz waveform appears in the downstream of plasma. For the initial CEP of $\phi_0 < 0.5\pi$, the THz polarity is negative at the beginning of the plasma, where the CEP can increase to be larger than 0.5π so that the instantaneous THz emission changes its polarity. As a result, the superposed THz emission becomes zero and reverses its polarity as a function of plasma length. As the energy of laser pulses increases, the variation of CEP in plasma becomes larger than π, by which the second polarity inversion of THz emission may appear if the first inversion shows at the beginning of the plasma.

Such energy-independent CEP values, $\phi_0 = 0.5\pi$ or 1.5π, at which the inversion of THz waveforms disappears into the front of air plasma, provide a flag that one can use to determine the initial CEP of the driving few-cycle laser pulses.

5.3 *Experimental Observation*

In our experiment, the CEP stabilized infrared (IR) few-cycle laser pulses (center wavelength of 1.8 μm, 11 fs in FWHM and pulse energy of 0.46 mJ) are focused into the ambient air by a spherical mirror ($f = 150$ mm), and a stable luminescence filament of ~12 mm is formed by the input pulse energy of maximum 450 μJ

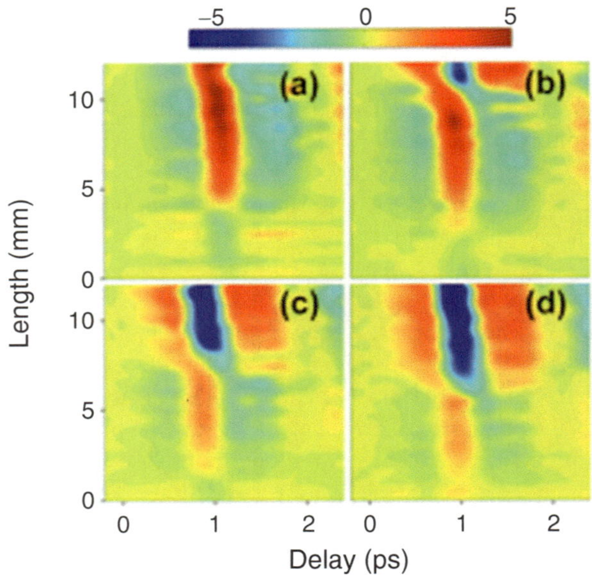

Fig. 13 THz waveform as a function of plasma length by few-cycle laser fields with the initial CEP of (**a**)–(**d**): $\delta\phi_0 = 0.0\pi, 0.2\pi, 0.4\pi, 0.6\pi$, respectively [60]

adjusted by an iris diaphragm. The generated THz waveforms are measured using the balanced diode geometry of EO sampling method. Inserting a sharp stainless steel blade into the plasma, the THz emission from different lengths of the filament is detected. Figure 13a shows the THz waveforms as a function of plasma length when the driving few-cycle pulse energy of 400 μJ is used. The inversion of THz emission appears at the position of near 4 mm. By varying the initial CEP of the few-cycle pulses stepwise by 0.2π, the THz waveforms are recorded and plotted in Fig. 13b–d, respectively. It is noted that there are two inversions in the evolving THz waveform in Fig. 13b, c but only one in Fig. 13a, d. The positions of inversion are moving toward the upstream of air plasma as the initial CEP of the driving pulses increases. When the initial CEP shift increases to $\delta\phi_0 = 0.6\pi$, the THz waveform inversion position changes back to one as the front inversion moving out of the filament, shown in Fig. 13d. The observation is consistent with the calculation result shown in Fig. 12c, f. It is reasonable to consider that the THz modulation by initial CEP of $\phi_0 = 0.2\pi$ and 0.4π is corresponding to the conditions of the experimental observation of $\delta\phi_0 = 0.2\pi$ and 0.4π, as shown in Fig. 13b, c.

The accuracy of measuring the initial CEP is determined by how well the transition from two inversions to one is decided. Even though experimentally it is limited by the CEP stability of few-cycle laser pulses, we can calculate the accurate initial CEP value where the inversion disappears at the beginning of filament. By increasing the resolution of the initial CEP to 0.001π in calculation, we find that the first inversion of THz emission disappears at the initial CEP of $\phi_0 = 0.500\pi$–0.408π depending on the input energy (200–900 μJ), as shown in Fig. 14. It is a slope with the fitting derivative of −0.00013 (πuJ^{-1}). This variation

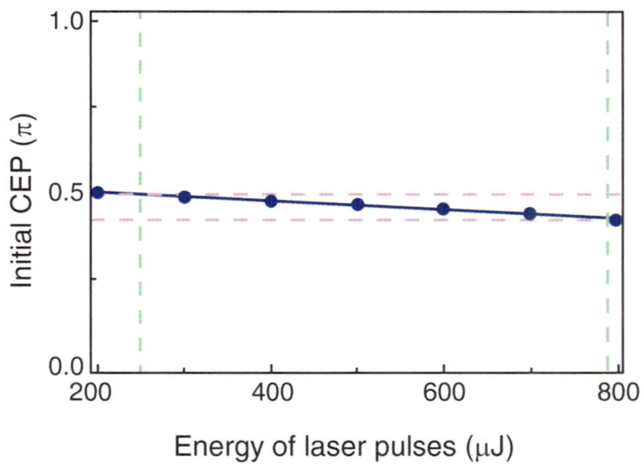

Fig. 14 The calculated initial CEP value by which the inversion of THz emission disappears at the beginning of air plasma under different input energies of 200–900 μJ

is due to the self-focusing effect and SPM that distorts the envelope of few-cycle pulses in plasma, leading to a little shift off the initial CEP of 0.5π.

There are three regions separated by green dashed lines as shown in Fig. 14, for the energy of few-cycle pluses that correspond to different THz inversion phenomena. When the energy is larger than 790 μJ (right region), three inversions of THz waveform can be seen for some initial CEP from our calculation result. In this case, the input energy is so high that the pulse envelope is split into two. It is therefore not realistic to discuss CEP related phenomena. For the input energy of 253–790 μJ (middle region), the number of inversions of THz waveforms along the filament is one or two, corresponding to the cases in Fig. 12c, f. The first inversion disappears when the initial CEP is 0.427π–0.495π. For the input energy smaller than 253 μJ (left region), the number of inversions of THz waveforms along the filament is one or none for all values of the initial CEP. In this case, the disappearance of THz waveform inversion at the upstream of air plasma indicates the initial CEP is close to 0.5π. It is a good strategy to measure the initial CEP in the experimental condition shown in the two regions of lower energy. In the middle region, it is easier to measure the THz waveform from different length of plasma since the pulse energy is higher, but the determined CEP has an error of 73 mrad. On the other hand, in left region, the accuracy of initial CEP can be less than 10 mrad if one identifies the disappearance of an inversion in the beginning of filament to be 0.5π. Therefore in practice, the measurement is only limited by the accuracy of the CEP of few-cycle laser pulses.

One may note that the accuracy of the value of 0.5π also is also determined by the completeness of the model to calculate the THz emission. In our model, even though the propagation of the few-cycle intense field is included, the other effects, such as scattering effect [9] and Coulomb repulsion effect [61], have not been taken into account. The experimental observation and theoretical refinement of these effects may demand further exploring.

6 Summary

Linearly and circularly polarized few-cycle laser pulses at the center wavelength of 1.8 μm are generated by a home-built OPA system followed by a hollow fiber compressor. The compressed pulses have a pulse duration of are CEP stabilized with ~250 mrad rms CEP fluctuations, making this source a suitable driver for the study of phenomena in intense laser fields. A scheme to generate waveform-controlled THz radiation from air plasma is demonstrated by using such CEP-stabilized few-cycle laser pulses. The generated THz waveform can be controlled by varying the length of the filament and the CEP of the driving laser pulses. The waveform evolution of THz radiation in the filament is due to the characteristic carrier phase shift and variation of laser envelope during propagation. Due to the ability to control the spatial location of femtosecond laser filamentation using prechirped laser pulses, the demonstrated scheme of waveform-controlled THz radiation generation is suitable for remote sensing exploring the phase property of THz radiation. The knowledge about the phase shift in filament is valuable for optimizing the generation of phase-matched attosecond pulses driven by intense few-cycle laser pulses. We further investigate the polarization of the generated THz emission driven by circularly polarized few-cycle pulses, and the generation and control of elliptically polarized terahertz waves from air plasma is demonstrated, without using any external electric apparatus. Experimental and simulation results reveal that electric field asymmetry in rotating directions of the few-cycle laser pulses produces the enhanced broadband transient currents, and their phase difference is partially inherited in the generation process of elliptically polarized THz emission. The ellipticity and direction of terahertz polarization is all-optically controlled. Finally the inversion of THz emission in the air plasma driven by the intense few-cycle laser fields is investigated. The variation of the number and positions of the inversions are found dependent on the initial CEP of laser pulses. The simulation indicates the first inversion vanishing point is 0.5π in the accuracy that is only limited by the CEP stability of few-cycle laser pulses. For a range of input energies of the driving pulses, the initial CEP by which the inversion disappearing at the beginning of air plasma is little energy-dependent and can be used to determine the initial CEP accurately.

Acknowledgements We acknowledge the support from National Natural Science Foundation of China (Grant Nos. 11274326, 1127901, 61221064 and 11134010), the 973 Program of China (2011CB808103), and the State Key Laboratory of High Field Laser Physics of China.

References

1. T. Kampfrath, A. Sell, G. Klatt, A. Pashkin, S. Mahrlein, T. Dekorsy, M. Wolf, M. Fiebig, A. Leitenstorfer, R. Huber, Nat. Photon. **5**(1), 31 (2011)
2. R. Ulbricht, E. Hendry, T.F.H. Jie Shan, M. Bonn, Rev. Mod. Phys. **83**(2), 543 (2011)
3. J. Liu, J. Dai, S.L. Chin, X.C. Zhang, Nat. Photon. **4**(9), 627 (2010)
4. C. D'Amico, A. Houard, M. Franco, B. Prade, A. Mysyrowicz, A. Couairon, V.T. Tikhonchuk, Phys. Rev. Lett. **98**(23), 235002 (2007)
5. T.I. Oh, Y.J. Yoo, Y.S. You, K.Y. Kim, Appl. Phys. Lett. **105**(4), 041103 (2014)
6. X. Xie, J. Dai, , X.C. Zhang, Phys. Rev. Lett. **96**(7), 075005 (2006)
7. K.Y. Kim, J.H. Glownia, A.J. Taylor, G. Rodriguez, Opt. Express **15**(8), 4577 (2007)
8. K.Y. Kim, A.J. Taylor, J.H. Glownia, G. Rodriguez, Nat. Photon. **2**(10), 605 (2008)
9. H.G. Roskos, M.D. Thomson, M. Kreß, T.L. Löffler, Laser Photon. Rev. **1**(4), 349 (2007)
10. F. Théberge, M. Châteauneuf, G. Roy, P. Mathieu, J. Dubois, Phys. Rev. A **81**(3), 033821 (2010)
11. H. Wen, A.M. Lindenberg, Phys. Rev. Lett. **103**(2), 023902 (2009)
12. J. Dai, N. Karpowicz, X.C. Zhang, Phys. Rev. Lett. **103**(2), 023001 (2009)
13. M. Kreß, T. Löffler, M.D. Thomson, R. Dörner, H. Gimpel, K. Zrost, T. Ergler, R. Moshammer, U. Morgner, J. Ullrich, H.G. Roskos, Nat. Phys. **2**(5), 327 (2006)
14. A. Baltuska, Th. Udem, M. Uiberacker, M. Hentschel, E. Goulielmakis, Ch. Gohle, R. Holzwarth, V.S. Yakovlev, A. Scrinzi, T.W. Hansch, F. Krausz, Nature **421**(6923), 611 (2003)
15. A. Apolonski, P. Dombi, G.G. Paulus, M. Kakehata, R. Holzwarth, Th. Udem, Ch. Lemell, K. Torizuka, J. Burgdörfer, T.W. Hänsch, F. Krausz, Phys. Rev. Lett. **92**(7), 073902 (2004)
16. T.M. Fortier, P.A. Roos, D.J. Jones, S.T. Cundiff, R.D.R. Bhat, J.E. Sipe, Phys. Rev. Lett. **92**(14), 147403 (2004)
17. T. Nakajima, S. Watanabe, Phys. Rev. Lett. **96**(21), 213001 (2006)
18. Y. Wu, X. Yang, Phys. Rev. A **76**(1), 013832 (2007)
19. M. Hentschel, R. Kienberger, Ch. Spielmann, G.A. Reider, N. Milosevic, T. Brabec, P. Corkum, U. Heinzmann, M. Drescher, F. Krausz, Nature **414**(6863), 509 (2001)
20. F. Krausz, M. Ivanov, Rev. Mod. Phys. **81**(1), 163 (2009)
21. M.F. Kling, Ch. Siedschlag, A.J. Verhoef, J.I. Khan, M. Schultze, Th. Uphues, Y. Ni, M. Uiberacker, M. Drescher, F. Krausz, M.J.J. Vrakking, Science **312**(5771), 246 (2006)
22. H.C. Wu, J.M. ter Vehn, Z.M. Sheng, New J. Phys. **10**(4), 043001 (2008)
23. A.A. Silaev, N. Vvedenskii, Phys. Rev. Lett. **102**(11), 115005 (2009)
24. I. Babushkin, W. Kuehn, C. Köhler, S. Skupin, L. Bergé, K. Reimann, M. Woerner, J. Herrmann, T. Elsaesser, Phys. Rev. Lett. **105**(05), 053903 (2010)
25. L. Bergé, S. Skupin, C. Köhler, I. Babushkin, J. Herrmann, Phys. Rev. Lett. **110**(7), 073901 (2013)
26. N.V. Vvedenskii, A.I. Korytin, V.A. Kostin, A.A. Murzanev, A.A. Silaev, A.N. Stepanov, Phys. Rev. Lett. **112**(5), 055004 (2014)
27. I. Babushkin, S. Skupin, A. Husakou, C. Köhler, E. Cabrera-Granado, L. Bergé, J. Herrmann, New J. Phys. **13**(12), 123029 (2011)
28. A. Baltuška, T. Fuji, T. Kobayashi, Phys. Rev. Lett. **88**(13), 133901 (2002)
29. C. Vozzi, C. Manzoni, F. Calegari, E. Benedetti, G. Sansone, G. Cerullo, M. Nisoli, S. De Silvestri, S. Stagira, J. Opt. Soc. Am. B **25**(7), B112 (2008)
30. X. Gu, G. Marcus, Y. Deng, T. Metzger, C. Teisset, N. Ishii, T. Fuji, A. Baltuska, R. Butkus, V. Pervak, H. Ishizuki, T. Taira, T. Kobayashi, R. Kienberger, F. Krausz, Opt. Express **17**(1), 62 (2009)
31. O.D. Mücke, S. Ališauskas, A.J. Verhoef, A. Pugžlys, A. Baltuška, V. Smilgevičius, J. Pocius, L. Giniūnas, R. Danielius, N. Forget, Opt. Lett. **34**(16), 2498 (2009)
32. B.E. Schmidt, P. Béjot, M. Giguère, A.D. Shiner, C. Trallero-Herrero, É. Bisson, J. Kasparian, J.P. Wolf, D.M. Villeneuve, J.C. Kieffer, P.B. Corkum, F. Légaré, Appl. Phys. Lett. **96**(12), 121109 (2010)

33. P. Béjot, B.E. Schmidt, J. Kasparian, J.P. Wolf, F. Legaré, Phys. Rev. A **81**(6), 063828 (2010)
34. C. Li, D. Wang, L. Song, J. Liu, P. Liu, C. Xu, Y. Leng, R. Li, Z. Xu, Opt. Express **19**(7), 6783 (2011)
35. C. Vozzi, F. Calegari, E. Benedetti, S. Gasilov, G. Sansone, G. Cerullo, M. Nisoli, S.D. Silvestri, S. Stagira, Opt. Lett. **32**(20), 2957 (2007)
36. T. Kobayashi, A. Baltuška, Meas. Sci. Technol. **13**(11), 1671 (2002)
37. B.E. Schmidt, P. Béjot, M. Giguère, A.D. Shiner, C. Trallero-Herrero, É. Bisson, J. Kasparian, J.P. Wolf, D.M. Villeneuve, J.C. Kieffer, P.B. Corkum, F. Légaré, Appl. Phys. Lett. **96**(12), 121109 (2010)
38. R. Trebino, K.W. DeLong, D.N. Fittinghoff, J.N. Sweetser, M.A. Krumbügel, B.A. Richman, D. Kane, Rev. Sci. Instrum. **68**(9), 3277 (1997)
39. A. Baltuška, M.S. Pshenichnikov, D.A. Wiersma, IEEE J. Quant. Electron. **35**(4), 459 (1999)
40. W.L. Chan, J. Deibel, D.M. Mittleman, Rep. Prog. Phys. **70**(8), 1325 (2007)
41. K. Kitano, N. Ishii, J. Itatani, Phys. Rev. A **84**(5), 053408 (2011)
42. Q. Wu, X.C. Zhang, Appl. Phys. Lett. **67**(24), 3523 (1995)
43. Y. Bai, L. Song, R. Xu, C. Li, P. Liu, Z. Zeng, Z. Zhang, H. Lu, R. Li, Z. Xu, Phys. Rev. Lett. **108**(25), 255004 (2012)
44. J. Liu, R. Li, Z. Xu, Phys. Rev. A **74**(4), 043801 (2006)
45. C. Köhler, E. Cabrera-Granado, I. Babushkin, L. Bergé, J. Herrmann, S. Skupin, Opt. Lett. **36**(16), 3166 (2011)
46. M.A. Porras, Phys. Rev. E **65**(2), 026606 (2002)
47. N.C.J. van der Valk, W.A.M. van der Marel, P.C.M. Planken, Opt. Lett. **30**(20), 2802 (2005)
48. E. Castro-Camus, J. Lloyd-Hughes, M.B. Johnston, Appl. Phys. Lett. **86**(25), 254102 (2005)
49. N. Kanda, K. Konishi, M. Kuwata-Gonokami, Opt. Express **15**(18), 11117 (2007)
50. S. Hughes, D.S. Citrin, J. Opt. Soc. Am. B **17**(1), 128 (2000)
51. J.L. McHale, *Molecular Spectroscopy* (Prentice Hall, Englewood Cliffs, 1999)
52. J. Xu, J. Galan, G. Ramian, P. Savvidis, A. Scopatz, R.R. Birge, S.J. Allen, K. Plaxco, in *Proceedings of SPIE Conference on Chemical and Biological Standoff Detection*, vol. 5268, ed. by J.O. Jensen, J.M. Thériault (SPIE, Providence, 2004), pp. 19–26
53. X. Lu, X.C. Zhang, Phys. Rev. Lett. **108**(12), 123903 (2012)
54. F. Jahangiri, M. Hashida, S. Tokita, T. Nagashima, M. Hangyo, S. Sakabe, Appl. Phys. Lett. **99**(16), 161505 (2011)
55. L. Song, Y. Bai, R. Xu, C. Li, P. Liu, R. Li, Z. Xu, Appl. Phys. Lett. **103**(26), 261102 (2013)
56. Y. Chen, C. Marceau, W. Liu, Z.D. Sun, Y. Zhang, F. Théberge, M. Châteauneuf, J. Dubois, S.L. Chin, Appl. Phys. Lett. **93**(23), 231116 (2008)
57. G.G. Paulus, F. Lindner, H. Walther, A. Baltuška, E. Goulielmakis, M. Lezius, F. Krausz, Phys. Rev. Lett. **91**(25), 253004 (2003)
58. T. Wittmann, B. Horvath, W. Helml, M.G. Schatzel, X. Gu, A.L. Cavalieri, G.G. Paulus, R. Kienberger, Nat. Phys. **5**(5), 357 (2009)
59. C.A. Haworth, L.E. Chipperfield, J.S. Robinson, P.L. Knight, J.P. Marangos, J.W.G. Tisch, Nat. Phys. **3**(1), 52 (2007)
60. R. Xu, Y. Bai, L. Song, P. Liu, R. Li, Z. Xu, Appl. Phys. Lett. **103**(6), 061111 (2013)
61. F. Lindner, G.G. Paulus, H. Walther, A. Baltuška, E. Goulielmakis, M. Lezius, F. Krausz, Phys. Rev. Lett. **92**, 113001 (2004)

Lasing Actions Inside a Femtosecond Laser Filament in Air

Tie-Jun Wang, Shuai Yuan, Jingjing Ju, Heping Zeng, Ruxin Li, Zhizhan Xu, and See Leang Chin

Abstract Lasing action inside a femtosecond laser filament in air has recently attracted much attention due to its important application on enhanced remote sensing. In this chapter, we summarize our recent experimental work together with physical understanding on various lasing actions inside femtosecond laser filaments in air, namely, amplified spontaneous emission (ASE) type of lasing actions from OH in a water vapor filament in air, of NH in a mixture of air and water vapor, of CH in a mixture of air and hydrocarbon gas. Coherent lasing action of N_2^+ in an air filament is also presented.

T.-J. Wang (✉) • J. Ju
State Key Laboratory of High Field Laser Physics, Shanghai Institute of Optics and Fine Mechanics, Chinese Academy of Sciences, Shanghai 201800, China

Centre d'optique, photonique et laser and Département de physique, de génie physique et d'optique, Université Laval, Québec, QC, G1V 0A6 Canada
e-mail: tiejunwang@siom.ac.cn

S. Yuan
Centre d'optique, photonique et laser and Département de physique, de génie physique et d'optique, Université Laval, Québec, QC, G1V 0A6 Canada

Shanghai Key Laboratory of Modern Optical System, Engineering Research Center of Optical Instrument and System, Ministry of Education, School of Optical-Electrical and Computer Engineering, University of Shanghai for Science and Technology, Shanghai 200093, China

State Key Laboratory of Precision Spectroscopy, East China Normal University, Shanghai 200062, China

H. Zeng
Shanghai Key Laboratory of Modern Optical System, Engineering Research Center of Optical Instrument and System, Ministry of Education, School of Optical-Electrical and Computer Engineering, University of Shanghai for Science and Technology, Shanghai 200093, China

State Key Laboratory of Precision Spectroscopy, East China Normal University, Shanghai 200062, China

R. Li • Z. Xu
State Key Laboratory of High Field Laser Physics, Shanghai Institute of Optics and Fine Mechanics, Chinese Academy of Sciences, Shanghai 201800, China

S.L. Chin
Centre d'optique, photonique et laser and Département de physique, de génie physique et d'optique, Université Laval, Québec, QC, G1V 0A6 Canada

A.D. Bandrauk et al. (eds.), *Laser Filamentation*, CRM Series in Mathematical Physics, DOI 10.1007/978-3-319-23084-9_5

1 Introduction

Femtosecond laser filamentation is a propagation phenomenon. When an intense ultrashort laser pulse propagates in a transparent medium, for instance, in air, an intense light induced refractive index change will lead to beam self-focusing. Once the pulse peak power is higher than a critical value, namely the critical power for self-focusing, self-focusing will overcome natural linear diffraction and collapse towards a high intensity zone resulting in tunnel/multiphoton ionization of air molecules. Therefore a plasma is generated. The plasma has a negative contribution to the refractive index change. Further evolution of the pulse is mainly governed by the competition between Kerr self-focusing and plasma defocusing together with linear diffraction, group velocity dispersion and multiphoton absorption. As a consequence, a weak plasma channel, referred to as a laser filament, is formed. So far, one monograph and several review papers have been published on this subject [1–7]. The laser intensity inside air filaments produced from femtoseocnd laser pulses at 800 nm is clamped at $\sim 5 \times 10^{13}\,\mathrm{W/cm^2}$. This femtosecond laser filament with high intensity can be propagated to a long distance in the atmosphere. Femtosecond laser filamentation has stimulated many applications such as new frequency generation ranging from UV to THz [8–10], standoff spectroscopy [11, 12], filament-based weather control [13–29], etc. Lasing action in air [30] is one of the frontiers in filamentation; recently, it attracted a lot of research interest mainly due to its promising applications on standoff spectroscopy [31–45]. All the above mentioned spectroscopic applications in an air or gas filament were based upon the possibility of avoiding optical breakdown, resulting in a "clean" fluorescence zone [1, 46].

The high clamped intensity inside a laser filament not only ionizes air molecules but also excites the molecules and ions into high lying excited states leading to characteristic fluorescence. ASE (Amplified Spontaneous Emission) type of fluorescence emission from air molecules in the backward direction of the 800 nm femtosecond laser filament was discovered more than 10 years ago in Laval University [30]. This type of radiation is randomly polarized. The backward lasing was also observed from a filament in a gas mixture of N_2 (>0.3 bar) and Ar (>3 bar) in a gas chamber excited by a mid-infrared pulse (80 fs, 8 mJ, 3900 nm) or near-infrared pulse (200 fs, 6 mJ, 1030 nm) [31]. The scheme combining a femtosecond filamenting pulse and other heater pulse is also proposed to generate atmospheric lasing [32, 33]. More recently, by using femtosecond mid-infrared (MIR) laser pulses (wavelength around 2 μm) generated from an optical parametric amplifier pumped by a Ti:sapphire laser system, coherent and linearly polarized lasing actions from MIR filament-induced ionic N_2 (N_2^+) [34, 35] and ionic CO_2 (CO_2^+) [36] were observed in the forward direction. The lasing polarization is dependent on the self-generated harmonic seed (third order or fifth order harmonic of the pump pulse). Similar lasing action in the nitrogen ions (N_2^+) was also observed in pure nitrogen gas in a chamber by using Ti:sapphire laser pulses (centered at ~800 nm) together with its second harmonic [37, 38]. The second harmonic of 800 nm served as a first-order harmonic seed for the forward lasing action. Forward neutral nitrogen emission could also be generated by using 10 J, 1053 nm laser with picoseconds pulse duration [39].

The mechanism of the population inversion in N_2 and N_2^+ has been studied in the past few years. One possible mechanism was proposed based upon the high (clamped) intensity inside the filament to explain the population inversion in the excited N_2^{+*} [40]. This high laser intensity inside the filament would induce further nonlinear (multiphoton) processes such as wave mixing soon after the molecule was ionized into the ionic ground state N_2^+. This would transfer efficiently the ground state population in N_2^+ into the excited state N_2^{+*} resulting in population inversion. The pump photons together with the self-generated harmonic photons of the pump laser would also mix and contribute to the population inversion. This type of nonlinear pumping using intense femtosecond (broad bandwidth) laser pulses resulting in population inversion seems universal. Consequently, forward lasing action is explained as the stimulated amplification of harmonic seed (third or fifth harmonic) at the UV range due to the population inversion generated between two vibrational levels of N_2^{+*} and N_2^+. Therefore, coherent lasing action polarized in the direction of seed pulse polarization occurs. In the case of ASE type of lasing of neutral nitrogen N_2, the population inversion comes from collisions in the plasma in the case of pumping with linearly polarized light [31, 39, 47]. It comes from high energy electron collisional pumping when using circularly polarized light [48].

In this chapter, we shall summarize our recent experimental work on lasing actions inside femtosecond laser filaments in air. The chapter is organized as follows. ASE type of lasing action from OH in a water vapor filament in air [41] will be discussed first. Then ASE type of lasing action of NH [42] and CH [43] from the same filament will be presented in Sects. 3 and 4, respectively. In Sect. 5, coherent lasing action from an air filament [44, 45] will be described. At the end, we shall give a brief summary of the chapter.

2 Lasing Action of OH in Water Vapor

Sensing water vapor concentration in the atmosphere is particularly important because the concentration of water vapor is related to the changes of state of water content. Water vapor concentration measurement based on laser filament induced OH fluorescence spectroscopy has been successfully demonstrated [49]. Toward remote sensing the water vapor in the atmosphere, amplified fluorescence emission would promise higher detection sensitivity. Toward this aim, the experiments were performed by using a 12 mJ/50 fs, 10 Hz Ti:sapphire laser beam. The schematic of the experimental setup is described in Fig. 1. The laser beam was focused by a lens with a focal length of 42 cm to create a laser filament. A large aperture dichroic mirror (M1, 7.5 cm in diameter) with high reflectivity at 800 nm and 50 % transmission from 300 to 315 nm was used right after the lens to reflect the beam at 45° incidence angle. Then a laser filament was created roughly 35 cm after the

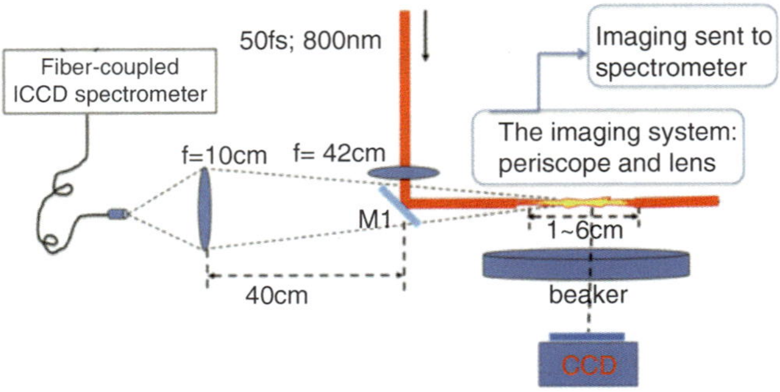

Fig. 1 Schematic of experimental setup for the gain experiment in water vapor

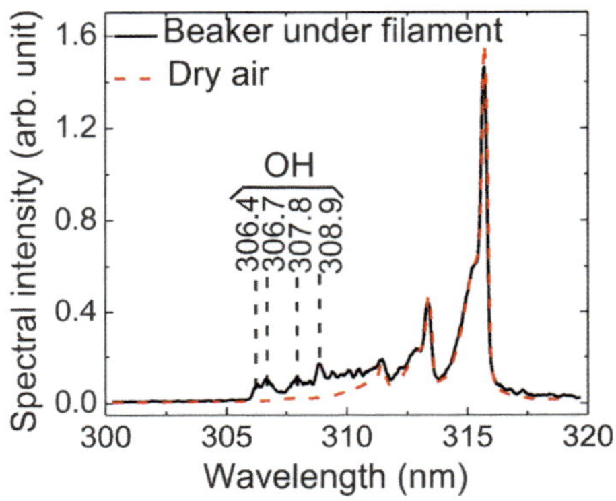

Fig. 2 Typical backward spectrum from water vapor induced by laser filament

mirror (M1) in air. A beaker with open area of 15 cm in diameter was placed 4 mm under the filament. The beaker was fully filled with distilled water, in order to provide more vapors and create an environment with a water vapor concentration of ∼2 % at room temperature (22 °C) according to our measurement in [30]. The backscattered fluorescence from the filament zone was collected by a fused-silica lens (5.08 cm in diameter, 10 cm focal length) after the dichroic mirror (M1) and sent to an ICCD-gated spectrometer (Acton Research Corporation, SpectraPro-500i) through a fiber coupler. For comparison, the fluorescence from the side was also recorded by imaging the filament into the 100 μm-wide slit of the same ICCD-gated spectrometer using two identical fused-silica lenses (ϕ5.08 cm, $f = 10$ cm) and one periscope. A grating of 1200 grooves/mm (blazed at 500 nm) was used in the experiments.

"Clean" fluorescence spectrum coming from the water vapor was found in the range of 306–309 nm as shown by the black curve in Fig. 2, at room temperature (22 °C) and with 2 mJ of the pump pulse energy. The fluorescence was identified as OH radiation from the transition of [49, 50]. The fluorescence from molecular N_2 in air was also observed around 315 nm as shown in Fig. 2. To check whether the fluorescence of N_2 has any effect on the typical OH radiation at 308.9 nm, another trial experiment was carried out. The beaker filled with water was replaced by a gas nozzle, which continuously provided a dry air blow to reduce the amount of water vapor in air. The spectrum of the back-scattered fluorescence from the filament was detected then. The result is shown as the red dashed curve in Fig. 2. In this case, the signal in the range of 306–309 nm which corresponds to the OH fluorescence decreases to the noise level. Therefore we can reasonably conclude that both N_2 and O_2 molecules in air have negligible contribution to the spectral range of the OH fluorescence. In the following measurements and discussions, the spectral peak intensities of 308.9 nm from OH were chosen as the OH fluorescence signal.

First of all, the pump energy dependence on the OH fluorescence in the backward and side directions was investigated. A CCD camera was employed to image the filament from the side (see Fig. 1) in order to characterize the length of the filament. The results of the OH fluorescence signal as a function of the filament length are shown in Fig. 3a (detected from the backward direction) and Fig. 3b (detected from the side). In Fig. 3a, the backscattered OH fluorescence increases exponentially as the filament length increases, while in Fig. 3b, the OH fluorescence from the side has a linear dependence with the filament length. If the OH fluorescence signal is incoherent, OH fluorescence emission detected in the backward and side directions should give similar results. However, in our experiment, the on-axis backward OH

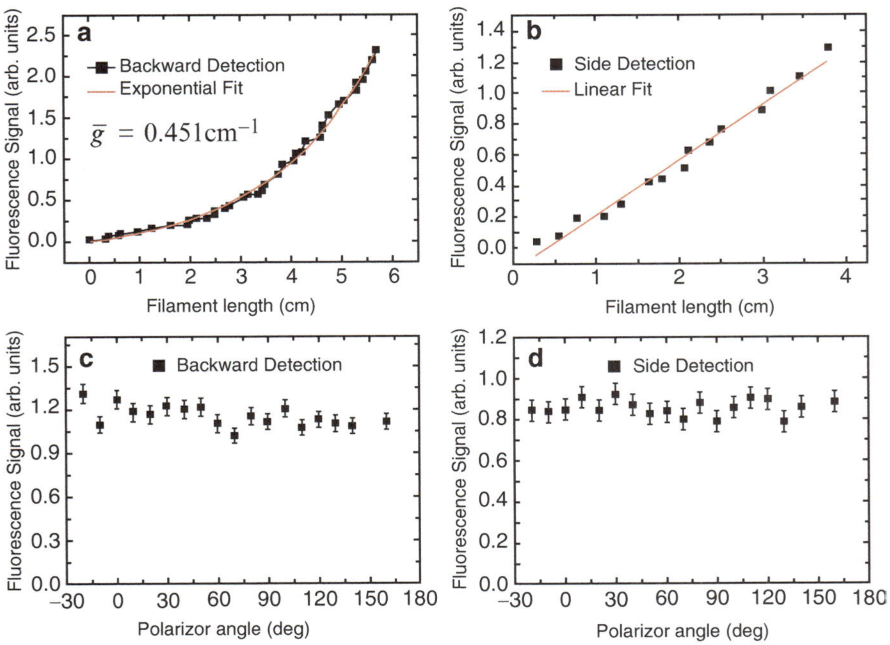

Fig. 3 The measured OH fluorescence intensity at 308.9 nm as a function of the filament length recorded in backward (**a**) and side (**b**) direction. The *red solid line* in (**a**) is the fitted gain curve calculated with (1a). The polarization property of the fluorescence in backward and side direction is shown in (**c**) and (**d**), respectively

fluorescence signal increases much faster (increase exponentially) with the filament length than the one detected from the side (increase linearly). This indicates the occurrence of ASE along the backward direction of laser filament. ASE can be simulated as:

$$I \propto P = \int_0^L P_s e^{g(l)l}\, dl = \begin{cases} \frac{P_s}{\bar{g}}(e^{\bar{g}L} - 1) & \text{(with amplification)} \quad (1a) \\ \frac{P_s}{\bar{g}} \times \bar{g}L = P_s L & \text{(when } \bar{g}L \ll 1) \quad (1b) \end{cases}$$

where I is the intensity of the spontaneous emission, P the spontaneous emission power, P_s the spontaneous emission power per unit length, L the effective length of the gain medium, which roughly equals to the length of the filament. $g(l)$ is defined as the optical gain coefficient, which is a function of the relative position l inside the filament. $\bar{g}$ is the effective gain coefficient over the filament length. In our experiment, the maximum length of the filament is 6 cm. In this sense, due to intensity clamping, we assume the intensity inside the filament is uniform and consider the filament zone approximately as a cylinder, meaning that the optical gain $g(l)$ has little deviation from the effective gain coefficient $\bar{g}$ over the filament length. In (1a), when the spontaneous fluorescence is amplified along the gain medium, the integrated fluorescence shows an exponential dependence with the length L of the gain medium (filament), which corresponds to the result in Fig. 3a. By applying (1a), the experimental data in Fig. 3a can be fitted with an effective gain coefficient of 0.451 cm^{-1}. In Fig. 3b, the OH fluorescence is proportional to the filament length, since the fluorescence detected from the side only sees the gain along the very small filament cross section (roughly 100 μm), which is a short distance. The value of the gain along the cross section should have the same magnitude as the effective gain coefficient detected from the backward direction (0.451 cm^{-1}). Therefore, in this case the product of the gain (assuming $\bar{g} = 0.451\ cm^{-1}$) and the length of the gain medium ($L = 100\ \mu m$) is negligibly small ($\bar{g}L \ll 1$), which corresponds to the case of (1b). As a result, the integrated OH fluorescence from the side is just the sum of the emission from all the small unit lengths along the filament, in accordance with our results in Fig. 3b. The polarization of the fluorescence in the backward/side direction was measured by inserting a polarizer just before the fiber head (backward direction) or the slit of the spectrometer (side). Both of them are randomly polarized as shown in Fig. 3c, d. The isotropic polarization and the difference between the gain in backward and side directions provide strong evidences for ASE lasing.

In order to understand the physics of the lasing action and the dissociation process, an energy diagram was plotted and shown in Fig. 4. The ionization potential (IP) of H_2O molecule is 12.6 eV. The OH* fluorescence at 306–309 nm in this experiment was identified as the transition from $A^2\Sigma^+ \rightarrow X^2\Pi$ [50]. From the energy diagram, there are several distinctive dissociation channels for H_2O molecule below and above the ionization potential related to OH*(A) generation: below the IP level, there is one channel dissociating into H(^2S) + OH($A^2\Sigma$),

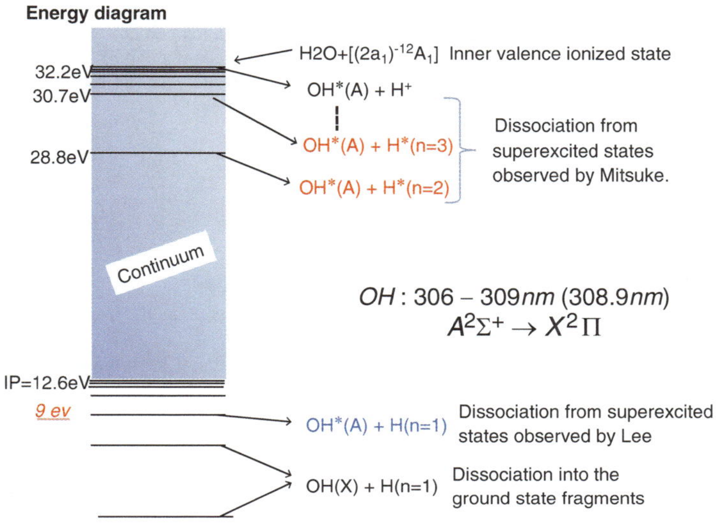

Fig. 4 The schematic of energy diagram for molecular H_2O

resulting in H in the ground state and OH in the excited state. The threshold energy of 9.136 eV for exciting this channel comes from [51–53] where single VUV photons were used in the dissociation of H_2O. If the dissociation channels above the ionization potential (12.6 eV) happened, the excited state of H, namely H*, would be generated together with OH*(A) (see Fig. 4). As a consequence, there should be emission from H* (Balmer series) together with OH emission. However, during the experiment, we did not observe any spectral line from the Balmer series of H*. Thus, it can be assumed that inside the filament the water vapor molecules can be dissociated through the channel.

During laser filamentation, the dissociation of H_2O molecule through H(^2S) + OH($A^2\Sigma^+$) by infrared photon at 800 nm is a multiphoton or highly nonlinear field interaction process (9 eV, roughly 6 photon energy). In order to prove our assumption, one supplementary experiment was carried out by measuring the intensity inside the filament and OH fluorescence emission at different input pump pulse energy. Note that although the intensity is supposed to be uniformed along the filament, the intensity inside the filament would change by varying external focusing condition [54]. In this experiment, the lens $f = 42$ cm in Fig. 1 was replaced by another four lenses with focal length of $f = 18$, 30, 50, and 80 cm. At different pump pulse energy, the effective filament length was imaged into the ICCD-gated spectrometer from the side. The intensity at every small unit lengths (0.15 mm by calibration) along the effective filament length was calculated according to [55, (13)], by measuring the ratio of the spectral intensity of two nitrogen fluorescence lines, 391 and 337 nm. Then the peak intensity was selected and the intensity of OH fluorescence at this unit length was recorded. We plotted the OH fluorescence versus the peak laser intensity over this unit length. The result is shown in a log – log plot in

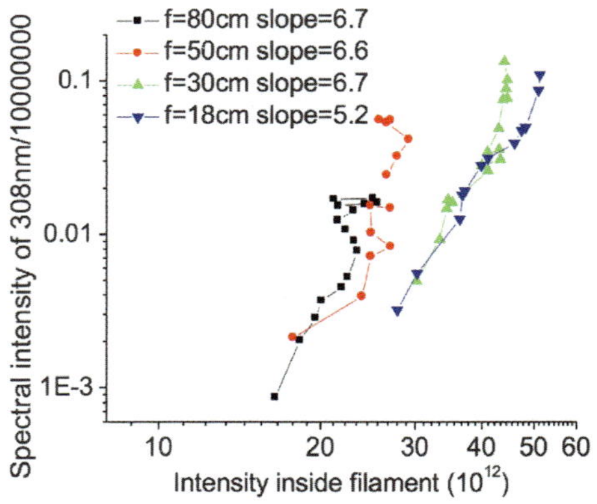

Fig. 5 Spectral intensity of OH fluorescence as a function of laser peak intensity inside filament under different external focusing conditions (with focal length of 18, 30, 50 and 80 cm)

Fig. 5, with an averaged slope of 6.3, which agrees very well with our assumption. The slope does not correspond to an exact number of photons. According to the literature [52] around 9 eV (roughly 6 photon energy) a high density of states exists. These states could be coupled by the strong field and dissociate into the channel $H(^2S)+OH(A^2\Sigma^+)$. The fluorescence from $OH(A^2\Sigma^+)$ is attributed to the pumping to these excited states. The excitation probability into these coupled excited states may have the slope of 6 or 7 (or some other non-integer values). In this sense, the slope 6.3 which is observed in the experiment is the mixture of the contribution from various excited states.

Thus, it is reasonable to conclude that the gain in the fluorescence of OH comes from the dissociation channel $H_2O + nh\nu \rightarrow H(^2S) + OH(A^2\Sigma^+)$ (see Fig. 4). Here, n indicates the mixture of the contribution from various excited states and not an exact number of photons. However, if the other lower lying channel $H_2O + n'h\nu \rightarrow H(^2S) + OH(X^2\Pi)$ which would give rise to OH (X) fragments in the ground state were also possible, it would reduce the population inversion in the ensemble of OH population. We would argue that such latter channel would be less probable. Under strong broadband femtosecond laser field excitation, many molecular states might have been coupled by the field. The higher the density of states is, the more probable the coupling could be. In the experiment, the strong field was set by intensity clamping in the air filament where the intensity was of the order of 5×10^{13} W/cm^2. According to [52], the region of higher density of states is around 9 eV which would decay into the channel $H_2O + nh\nu \rightarrow H(^2S) + OH(A^2\Sigma^+)$. Hence, the excitation of this channel would be more probable. Since OH* is created directly, its population is naturally inverted, and thus ASE lasing occurs [41].

3 Lasing Action of NH in a Mixture of Air and Water Vapor

By using a similar experimental setup as shown in Fig. 1, except that the laser filament was created by using 1.8 mJ/400 nm femtosecond laser pulse, another interesting fluorescence signal at 336 nm in a mixture of air and water vapor was

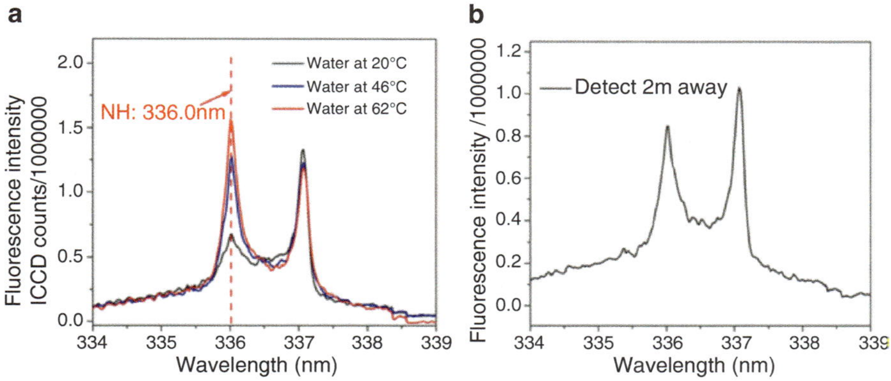

Fig. 6 Typical spectra in the range of 334–339 nm for filament-induced fluorescence of NH in air, with fluorescence collected at a distance of 50 cm (**a**) and 2 m (**b**) from the side of the filament

also observed [42]. In this experiment, the beaker was heated by a heating plate underneath to provide more water vapor.

Figure 6a depicts the typical spectra in the range of 334–339 nm for filament-induced fluorescence of NH in air. At 337 nm there was a very strong fluorescence emission which is the known emission from molecular N_2 in air. Besides this emission, there was another emission at 336 nm. The 336 nm signal became stronger when the water was heated from 20 to 62 °C. This fluorescence was identified as NH* radiation through the transition of$A^3\Pi \rightarrow X^3\Sigma^-$ [50, 55]. Here, the NH radicals might be ascribed to the complicated recombination process of the dissociated N_2 and water vapor molecules. When the water temperature increased, the water vapor concentration became higher so that the NH fluorescence got stronger. Especially at 62 °C, the intensity of the NH fluorescence at 336 nm was even stronger than that of N_2 fluorescence at 337 nm. "Clean" spectrum of NH was even clearly detected when the fluorescence was collected at a distance of 2 m away from the side of the filament, with the water temperature at 43 °C (see Fig. 6b), which shows the possibility for long distance application.

Since air is mainly composed of nitrogen and oxygen molecules, to verify our assumption that NH radicals originate from the recombination of the dissociated N_2 and water vapor molecules, another trial experiment was carried out. The beaker of water was placed inside a plastic box with input and exit windows open. A rectangular opening was cut from the side for fluorescence detection. Inside the box, a mounted gas nozzle continuously provided a flow of nitrogen or oxygen into the filament zone in order to increase the number of N_2/O_2 molecules at a constant pressure. The fluorescence was detected from the side opening at room temperature (20 °C). The spectral lines at 336 nm for NH and at 337 nm for N_2 were observed as continuous nitrogen, oxygen or no external gas blow was applied to the filament. The experimental results are shown in Fig. 7. The fluorescence intensities of these two lines are strongest when nitrogen gas was blown into the filament zone.

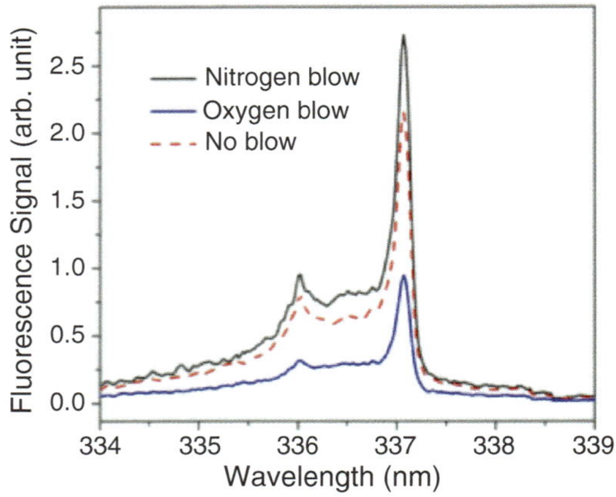

Fig. 7 The spectra in the ranges of 334–349 nm obtained with the water temperature of 20 °C, when nitrogen, oxygen, or no external gas blow was applied to the filament region

This is in good agreement with our previous assumption that NH was a consequence of the dissociation of nitrogen and water molecules followed by recombination. Since N_2 is one of the reactants, the increase of the number of N_2 molecules will definitely increase the production of NH radical. In the case of blowing oxygen into the filament zone, both the spectra decreased. Molecular O_2 have no contribution to NH generation; it just displaces N_2 and water molecules in air at a constant pressure resulting in a weak signal. The above observation indicates that the generated NH radical did originate from the nitrogen; i.e., it should have come from the dissociated product, namely, nitrogen atom. We could also say that the product quantity of the product NH is proportional to the quantities of N_2 and water molecules (NH $\propto$ N_2 and NH $\propto$ H_2O). The possibility of making use of NH fluorescence to characterize water vapor concentration has been successfully demonstrated [42].

In order to check the gain effect and have a better physical picture of the filament-induced chemical reaction, the NH fluorescence in the backward and side directions were recorded under different input pump pulse energy. The water temperature was fixed at 43 °C and the distance between the filament and the water surface was at 6 mm. The backward NH fluorescence as a function of laser filament length is shown in Fig. 8a. A clear exponential increase was observed as the filament length increases. However, as shown in Fig. 8b, the NH fluorescence from the side has a linear dependence on the filament length. These results confirm that the gain effect of NH fluorescence in the backward direction does exist. By assuming that the intensity inside the filament is uniform because of intensity clamping [1–7], the gain coefficient (g) over the filament is obtained by fitting the experimental data in Fig. 8a using (1a) to get $g = 0.556\,\text{cm}^{-1}$. We note that the value of gain from our measurement ($0.556\,\text{cm}^{-1}$ from NH*) and that from OH ($0.451\,\text{cm}^{-1}$ from OH* for fluorescence measured at 308.9 nm) are not comparable. The difference may come from the different complicated dissociation and recombination processes during laser filamentation. The polarizations of the fluorescence from the backward and side direction were measured by rotating a polarizer placed just before the fiber head which couples to the ICCD. Both of them are randomly polarized as

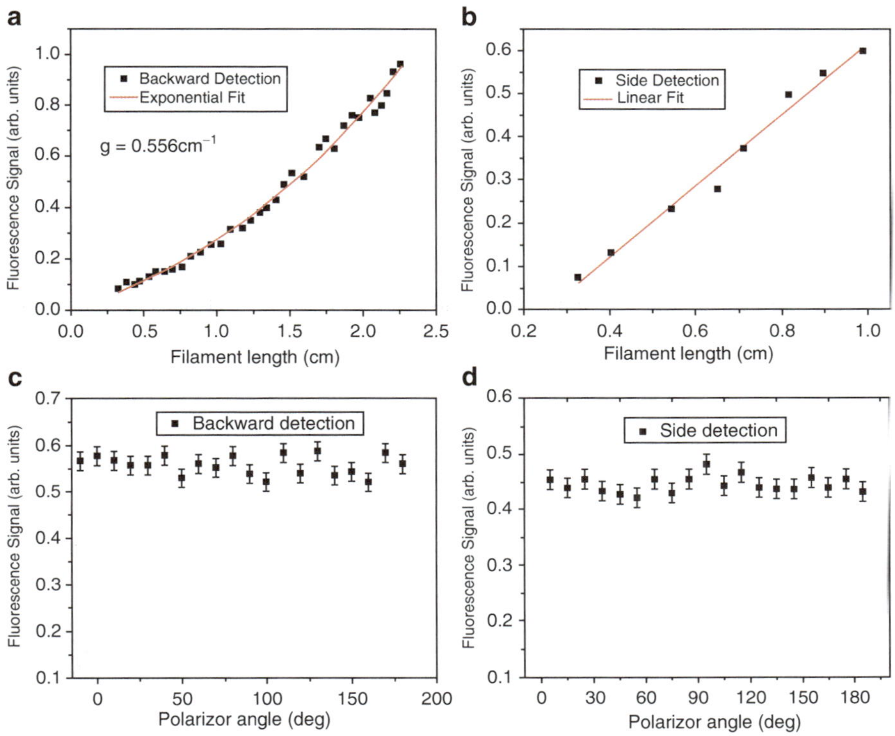

Fig. 8 The measured NH fluorescence intensity at 336 nm as a function of the filament length recorded in backward (**a**) and side (**b**) direction. The *red solid line* in (**a**) is the fitted gain curve calculated with (1a). The polarization property of the fluorescence in backward and side direction is shown in (**c**) and (**d**), respectively

shown in Fig. 8c, d. The isotropic polarization and the difference between the gain in backward and side directions confirm the occurrence of ASE lasing [42].

The ASE lasing confirms that the NH free radical, the product of filament-induced chemical reaction, is in an excited state and the population is inverted. Therefore, the translation of NH $A^3\Pi \rightarrow X^3\Sigma^-$ gives strong radiation at 336 nm, which can be used for the humidity measurement based on filament-induced spectroscopy since the fluorescence intensity from NH is dependent on the relative humidity [42]. To calibrate the relative humidity, in Fig. 9a the fluorescence intensities of NH at 336 nm and N_2 at 337 nm were measured by varying the water temperature from 20 to 66 °C. The distance between the filament and the water surface was 6 mm. By increasing the water temperature, the NH fluorescence increases linearly, while the N_2 fluorescence keeps roughly constant. As a reference for humidity calibration, a hydrometer was fixed 6 mm above the water surface. The relative humidity was recorded by changing the water temperature and linearly fitted in Fig. 9b. By combining the experimental curve in Fig. 9a and the reference in

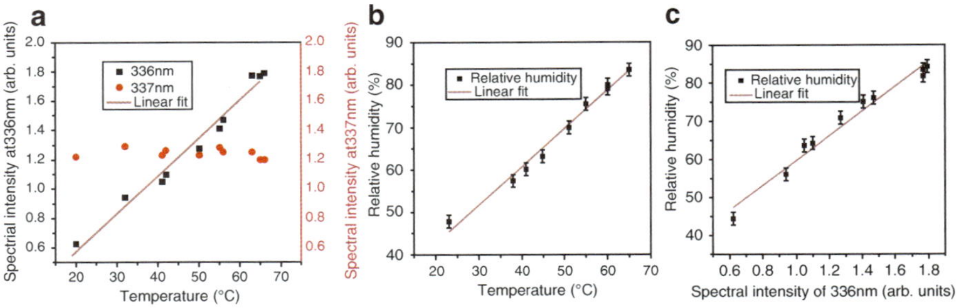

Fig. 9 (**a**) Fluorescence intensities of NH at 336 nm and N_2 at 337 nm versus the water temperature, with filament-to-water surface at 6 mm. (**b**) The relative humidity at 6 mm above the water surface versus the water temperature. The relative humidity is obtained from hygrometer. (**c**) The retrieved relative humidity by combining (**a**) and (**b**) as a function of the NH fluorescence signal at 336 nm

Fig. 9b, we obtain the relative humidity as a function of NH fluorescence (Fig. 9c). The relative humidity is linearly proportional to the fluorescence signal of NH free radical. This result provides a significant way to measure relative humidity by monitoring NH fluorescence from filament-induced chemical reaction.

Although the detailed formation mechanism of NH is not quite clear yet, it is reasonable to conclude that the high intensity laser filament induced the dissociation of the nitrogen and water molecules. After that the collision of the highly excited fragments generates the excited NH free radical with population inversion. The strong emission at 336 nm occurs which is sensitive to the number of H_2O molecules in air. Besides, lasing action in NH radical indicates that once we increase the length of the gain medium by the increasing input pulse energy, it is expected to achieve stronger emission at 336 nm in the backward direction. The strong fluorescence emission at 336 nm works as an effective approach for calibrating the humidity, even remotely [42].

4 Lasing Action of CH in a Mixture of Air and Hydrocarbon Gas [43]

Filament induced fluorescence spectroscopy has shown the capability to remotely sense the pollutants including hydrocarbon gases. If there is a gain effect of the fluorescence from the pollutants in the filament, the sensitivity of the detection of the pollutants in air will be strongly improved. Experiments were performed to look into the gain effect from hydrocarbon gases. 800 nm laser filament was formed inside a gas tube filled with a gas mixture, air with 2.0 % hydrocarbon molecular species. The total pressure was set at 1 atm. Three species, CH_4, C_2H_2, and C_2H_4 have been tested, respectively. The fluorescence signals both from the backward direction and the side have been collected using the similar detection system as

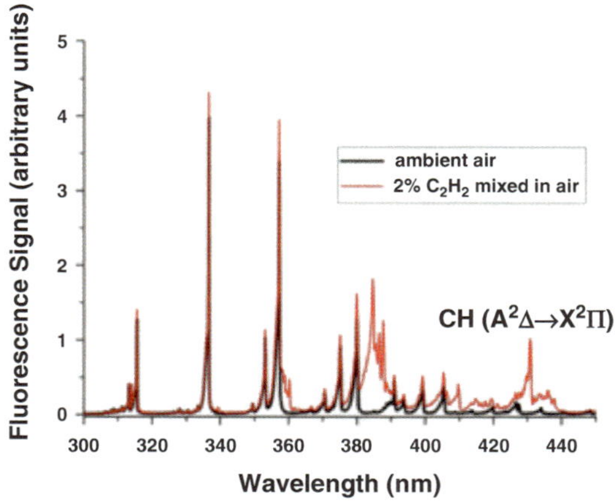

Fig. 10 The typical fluorescence emission spectrum recorded from the side in air (*black*) and in a mixture of air-acetylene (2 %) (*red*) induced by 800 nm laser filament

described in Sect. 2. The fluorescence emission spectrum recorded from the side in air (black) and in a mixture of air-acetylene (2 %) (red) is shown in Fig. 10. The fluorescence signal around 431 nm corresponds to the transition of $A^2\Delta - X^2\Pi(0,0)$ from CH fragments [50]. An interference filter that transmits 431 nm light with a spectral bandwidth of 2.5 nm (FWHM) is placed in front of the fiber head to avoid introducing the strong white light into the fiber. The fluorescence mechanism in hydrocarbon was described as a neutral dissociation through superexcited states during filamentation [56]. Highly lying neutral excited states above the ionization potential, namely superexcited states, were created through multiphoton excitation. The superexcited states could decay into different channels. Among them, there is one channel through neutral dissociation. As a consequence the CH($A^2\Delta$) fragment created in the strong field of filamenting pulse decays to the ground state of CH($X^2\Pi$) resulting in the fluorescence signal at around 431 nm.

In the spectral measurement, the white light continuum was found around 431 nm in the backward fluorescence while it was not found from the side. In order to remove the contribution of the white light continuum around 431 nm, the emission spectrum from the sample gas (air with 2 % hydrocarbon gases) was recorded first (the black squares in Fig. 11). Then the emission spectrum from air with the same partial pressure (0.98 atm) was measured (the red stars in Fig. 11). By subtracting the emission spectrum of air from the gas mixture, the white light continuum contribution was removed. The subtracted fluorescence signal (the blue triangles in Fig. 11) integrated over the spectral range from 429 to 432 nm was defined as the final CH fluorescence signal for the following analysis and discussion.

As shown in Fig. 12, the CH fluorescence signals from backward and side emission as a function of filament length are plotted for a gas mixture of air with 2 % CH_4, C_2H_2, and C_2H_4, respectively. It is clear to see the backward signal increases nonlinearly as the filament length increases. But the fluorescence signal from the side direction is linearly proportional to the filament length. By using the same numerical method to fit the experimental results [43], the gain factors of 3.19,

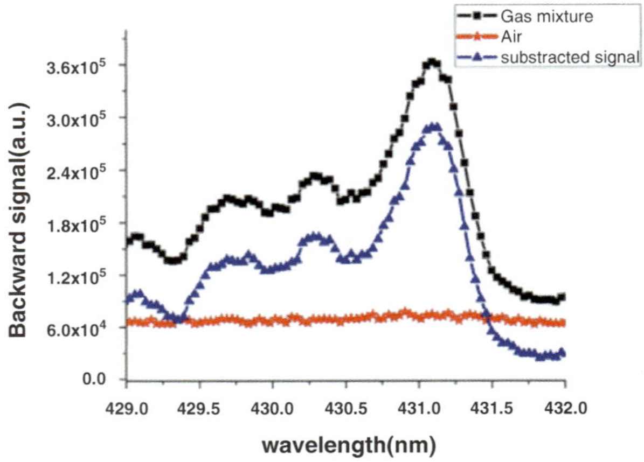

Fig. 11 The spectrum of the fluorescence emission from laser filament in 2 % CH_4 mixed with air (*black squares*), the spectrum of the background white light continuum in air (*red stars*), and the difference spectrum (*blue triangles*) obtained by subtracting the white light continuum from the emission spectrum

3.41, and 11.48 cm^{-1} were obtained for the gas mixture of 2 % CH_4, C_2H_2, and C_2H_4, respectively. Note that the gain factors are larger than the factors obtained in Sects. 2 and 3. It is because of the integration of 3 nm spectral bandwidth in the experimental data analysis. While in the cases of Sects. 2 and 3, the peak signals were used for the analysis. The important difference of the fluorescence signal's behaviors from the side and the backward direction proves the existence of ASE type of gain from the backward through filamentation. Although the details of the formation mechanism of the superexcited states and their decay processes are not well known yet, it could be concluded that only 2 % hydrocarbon gases in air could behave as a gain medium in the CH fragments when a laser filament is formed inside it [43]. The gain effect promises high sensitivity for remote sensing of pollutants in the atmosphere.

5 Lasing Actions of N_2^+ in Air

Besides the ASE type of lasing action inside the laser filament, another lasing action, namely stimulated amplification of harmonic seed inside the laser filament attracts a lot of research interest recently because of its much stronger radiation and high coherence. The stimulated emission has been observed from the transition between two vibrational levels from two different electronic states of N_2 [31], N_2^+ [34] and CO_2^+ [36]. Figure 13 depicts the energy diagram of N_2, N_2^+ together with the transitions of possible lasing actions. Most lasing actions were observed in pure N_2 and CO_2 gases inside a gas cell. Towards remote spectroscopy applications,

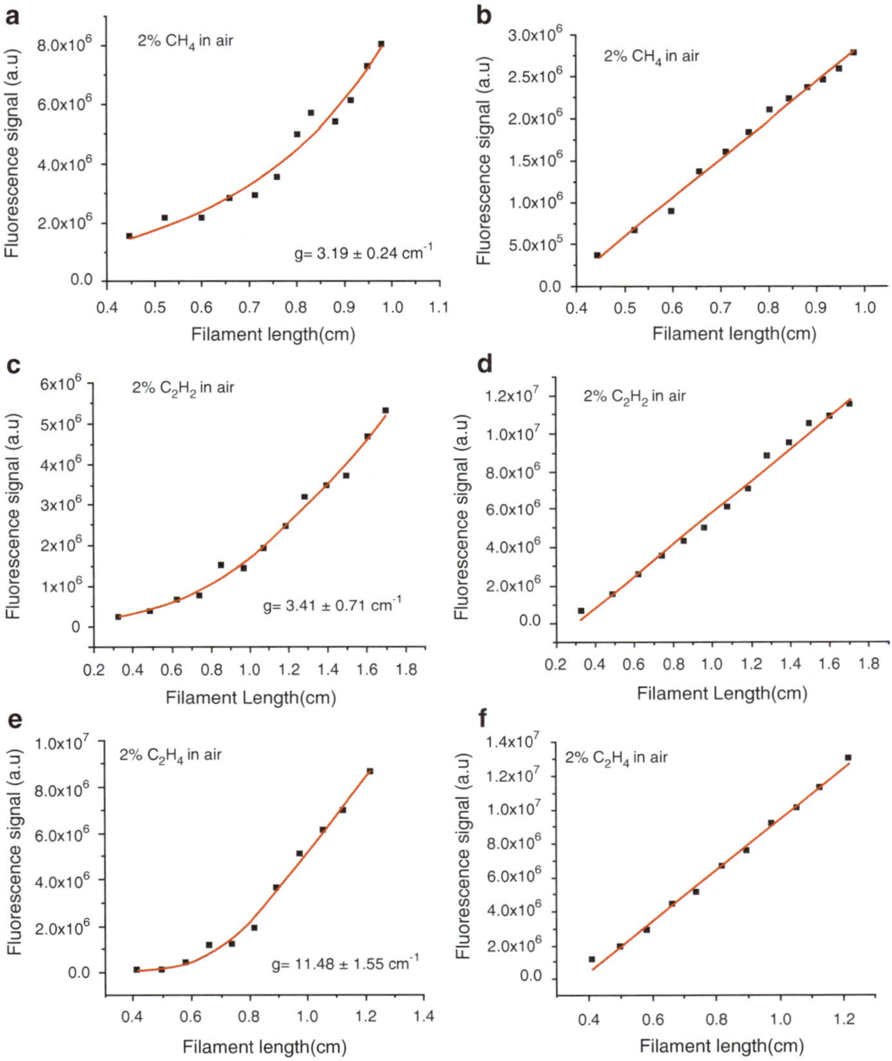

Fig. 12 The CH fluorescence signals from backward (**a**, **c**, **e**) and side emission (**b**, **d**, **f**) as a function of filament length for a gas mixture of air with 2 % CH_4 (**a**, **b**), C_2H_2 (**c**, **d**), and C_2H_4 (**e**, **f**), respectively

the lasing action occurred inside atmospheric air is much more interesting and important. In this section, we summarize our recent activities toward this latter aim.

Once the intense filament creates the population inversion between two vibrational levels from two different electronic states of N_2^+, the seed could be either self-generated from the laser pulse itself or from another external source. The only requirement of the seed is to cover the transition wavelength of two vibrational

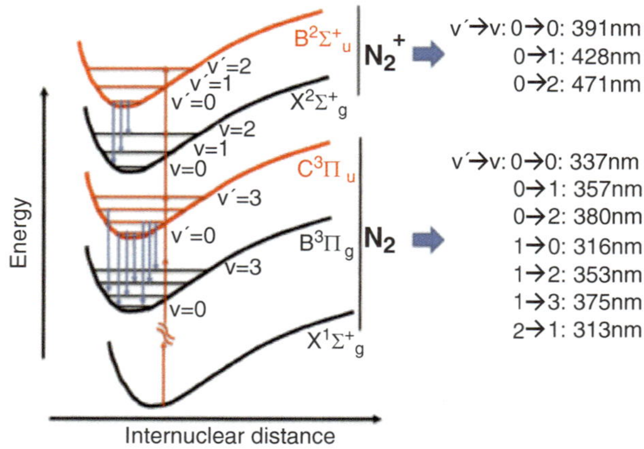

Fig. 13 Energy diagram of N_2 and N_2^+. The corresponding transitions are shown on the *right-hand side*

levels of ionic nitrogen in air. With the spectral broadening effect occurring with the filament, namely white light, can one use it as the seed to create the forward lasing, in particular to achieve simultaneous multiple wavelength lasing? In order to answer those questions, we performed systematical investigation on the coherent forward lasing action from a femtosecond laser filament in air. The experimental results confirm that this lasing action can be achieved from a single laser filament either from 800 nm [44, 45] or 400 nm femtosecond laser pulses [45]. Besides that, multiple-wavelength lasing action from 800 nm laser filament in air could also be operated, by using laser induced white light as an external seed [45].

The experiments were performed by using a 10 Hz/45 fs/800 nm laser pulse from a commercial Ti:sapphire laser system. Laser pulse energy can be tuned up to 15 mJ through the combination control of a thin film polarizer and a half-wave plate inserted before a two-pass amplifier. The beam diameter was ~1 cm. Femtosecond blue pulse (400 nm) of a few mJ was generated by frequency-doubling the fundamental Ti:sapphire laser pulse using a 100 μm thick BBO crystal. In the experiment using the self-generated white light seed (Fig. 14a, b), a femtosecond laser filament was created in air after the laser pulse passed through a focusing lens (60 cm in Fig. 14a and 22 cm in Fig. 14b). All the transparent optics used in the experiments was UV-grade and had a good transmission in the UV. In the case of "red" (800 nm) filament (Fig. 14a), a 45° 800 nm mirror was used to filter out the 800 nm filamenting pulses and let the UV emission pass through. This UV emission was collected by a 10 cm-focal length lens and sent to the integrating sphere and fiber coupled ICCD spectrometer. In the blue (400 nm) filament case (Fig. 14b), the forward spectrum was measured at 60 cm after the filament by directly sending the pulse into an integrating sphere that was connected to an ICCD spectrometer through a fiber. The laser filament induced fluorescence in both cases (Fig. 14a, b) was also collected, perpendicular to the laser beam propagation axis,

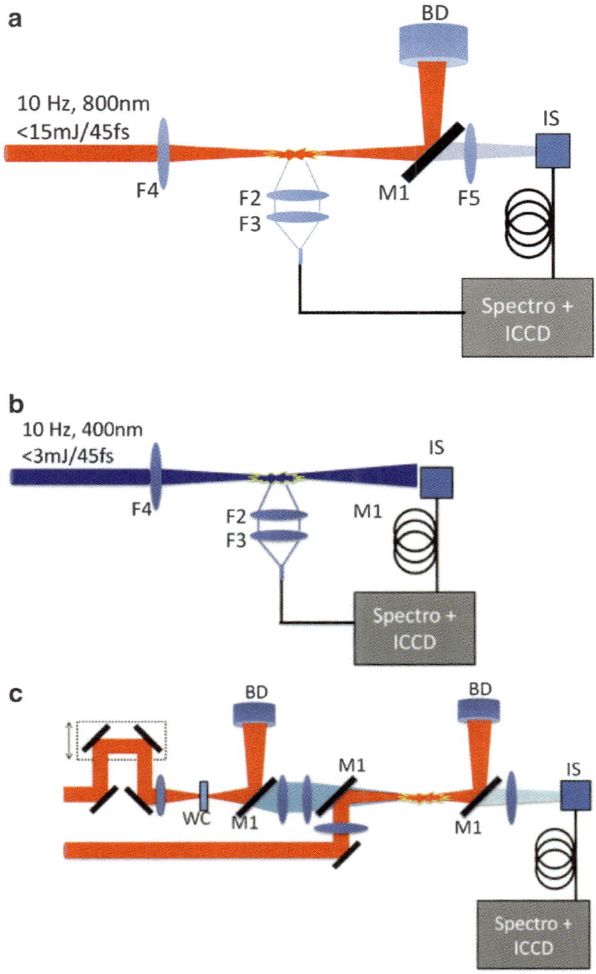

Fig. 14 Schematic illustrations of experimental setups on forward lasing action from white light seeded laser filament in air. (**a**) and (**b**) lasing action from self-seeded *red* (800 nm) and *blue* (400 nm) filament, respectively. (**c**) for externally generated white light seeding (see text for details)

by a pair of lenses and focused to a fiber link connected to the spectrometer. In the experiment of Fig. 14c, an external white light seed was generated by focusing part of the 800 nm laser pulses through a 1 cm-long water cell. The water cell was followed by an 800 nm 45° mirror (shown in Fig. 14c) to filter out the strong 800 nm fundamental pulse and let the white light from IR to near UV pass through. Another 800 nm 45° mirror was used to couple the filament and white light for collinear propagation. Spatial overlap between laser filament and focused white light was controlled by a pair of lenses. A translation stage was built in the white light path to control the temporal overlap. The UV emission was collected by using the same method that has been used in the experiment of Fig. 14b. A cubic polarizer was inserted in front of the integrating sphere/fiber head for the emission polarization measurement.

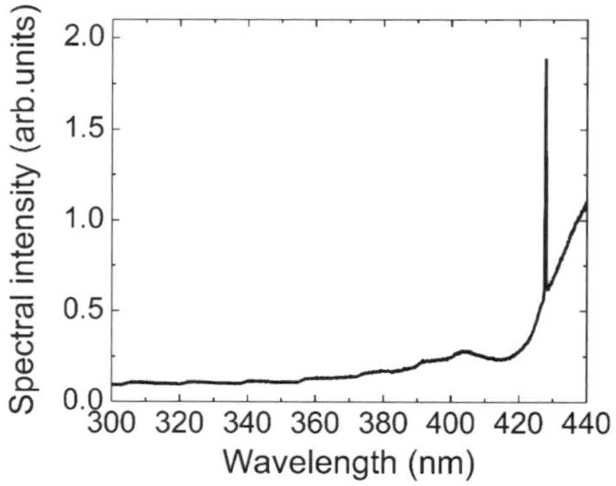

Fig. 15 Typical spectrum from forward detection in the case of only one 800 nm laser filament in air. The pump energy was ~8 mJ

5.1 Lasing Action from an 800 nm *Filament in Air by a Self-seeded White Light* [44]

Since most lasing actions observed from air molecules are from ionic nitrogen (N_2^+) in the wavelength range of 300–500 nm as shown in Fig. 13, looking for a suitable seed in this spectral range is very crucial. As it is well known, the white light spectrum generated from 800 nm air filaments can also cover the UV range of the lasing actions. The significance of using only 800 nm air filaments to achieve the lasing action is very important because one can simplify the generation method by using powerful Ti:sapphire laser system which will provide much stronger laser pulses leading to more population inversions of N_2^+.

The forward spectrum in the blue-UV part after the laser filament is shown in Fig. 15. The pulse energy was fixed at 8 mJ in the measurements. There is only one very strong ~428 nm (427.8 nm) spike sitting on the plateau of the filament-induced white light. This 428 nm emission is from the transition of $B^2\Sigma_u^+(0)$ to $X^2\Sigma_g^+(1)$ of N_2^+ [50]. We do not see the typical fluorescence spectrum covering the signals from the first negative band system of N_2^+ ($B^2\Sigma_u^+ - X^2\Sigma_g^+$ transition) with the principal wavelengths at 391 and 423 nm and the second positive band system of N_2 ($C^3\Pi_u - B^3\Pi_g$ transition) with the principal wavelengths at 316, 337, 357, and 380 nm. This is because those fluorescence signals are too weak to detect in the forward direction as compared with the generated strong white light and lasing emission accompanying the filament using the integrating sphere. The white light signal at many of the shorter wavelengths must be too weak or non-existent to be amplified. The behavior of the forward emission in the lasing window of 415–435 nm is shown in Fig. 16 obtained using two different pump energies of 2 and 8 mJ.

In the case of using the 2 mJ pulses, one can only see the white light background without any lasing signal. When the pump energy was increased to 8 mJ, not only the white light signal was enhanced, but also lasing action was induced. The peak signal around 428 nm was considered the lasing signal in the following analysis. The intensity of the white light at 428 nm was also estimated by using a fitting line

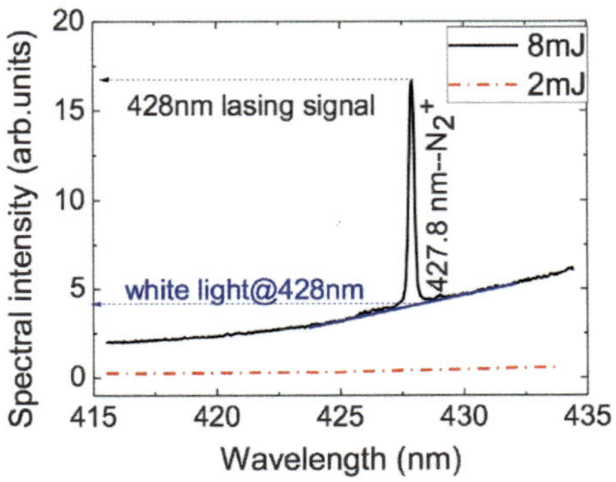

Fig. 16 Forward lasing signals in the spectral range of 415–435 nm under different pump energy (2 mJ in *dashed dotted red line* and 8 mJ in *solid black line*)

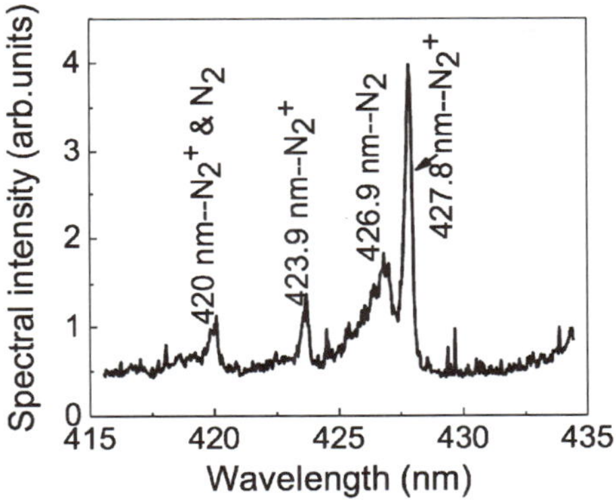

Fig. 17 Fluorescence signal detected from the side in the same spectral window in Fig. 16

as shown in Fig. 16. The spectral spike in Fig. 16 is narrow and symmetrical. There are no other lines observed. This is clearly different from the typical fluorescence signal obtained from side detected with fiber detector. The side fluorescence signal is shown in Fig. 17.

Both N_2 fluorescence and N_2^+ fluorescence are simultaneously detected over a long time (ns) as shown in Fig. 17. In the forward direction (Fig. 16), only the 428 nm transition went through stimulated amplification. The polarization property is a very useful parameter to distinguish the seeded lasing from ASE or fluorescence.

As shown in Fig. 18, the polarization of this 428 nm lasing line has the same polarization property as the pump pulse providing seeding pulse, which well agrees with previous observation using MIR filamenting wavelength [34–36]. Here the polarization of the filamenting pulse (800 nm) was characterized before the laser filaments. This again is an evidence that 800 nm filament induced white light seeding. The bandwidth of this lasing pulse is ~0.26 nm at full-width of half maximum (FWHM), which can support the Fourier-Transform-limited pulse duration of 1.03 ps. The pump energy dependence of lasing and white light at 428 nm is depicted in Fig. 19. Under low filamenting pulse energy (<5 mJ in our case), there is no lasing action detected. The signal intensities of lasing and white light were considered at the same level. Above a certain value (~5 mJ in our case),

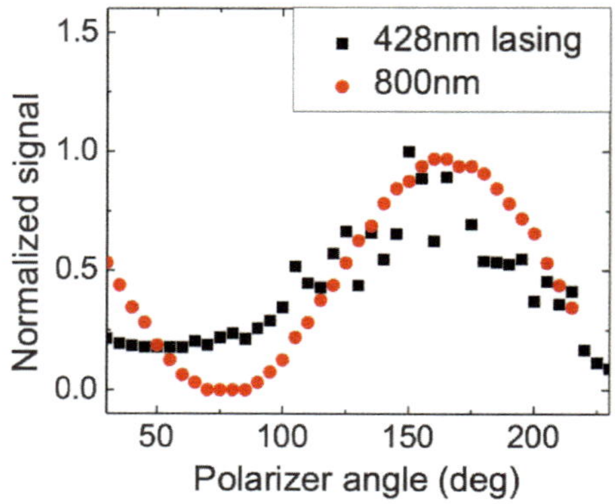

Fig. 18 Polarization property for forward 428 nm lasing (*black square*) and 800 nm filamenting pulse (*red circular*)

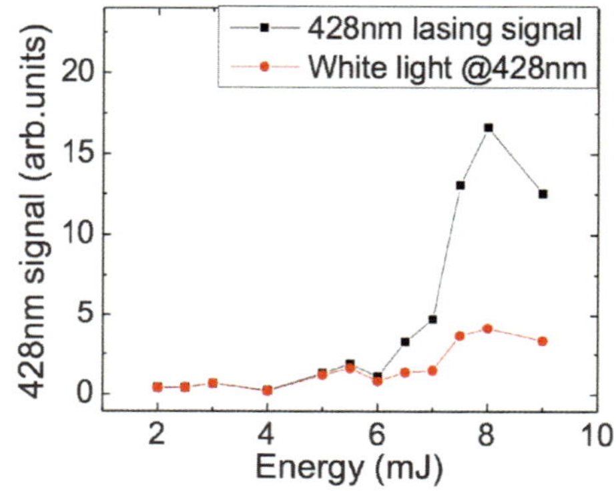

Fig. 19 Pump energy dependence of 428 nm forward lasing action and white light

the 428 nm lasing signal appeared and grew nonlinearly with increasing 800 nm pulse energy. The higher the filamenting pulse energy was, the stronger the white light signal at 428 nm was found; the 428 nm lasing line became much stronger. The 428 nm lasing pulse energy was around sub-μJ level under ~8 mJ pump through the estimation of the spectral intensity.

5.2 *Lasing Action from a White Light Self-seeded* 400 *nm Filament in Air* [45]

Except 800 nm femtosecond laser filamentation, using an intense 400 nm femtosecond laser pulse to create a filament, the generated white light spectra could also reach the UV part (<400 nm), with which self-seeded lasing action could be expected. With this idea in mind, we performed the blue filament experiment. 2 mJ blue pulses (400 nm) were used to create a filament in air. The forward spectra are shown in Fig. 20a. The spectrum measured after the focusing is much broader for the 2 mJ pump pulse, which produced a filament, as compared to the 1 μJ pump pulse. There is a clear lasing action at 391 nm sitting on the UV extension of the spectrum.

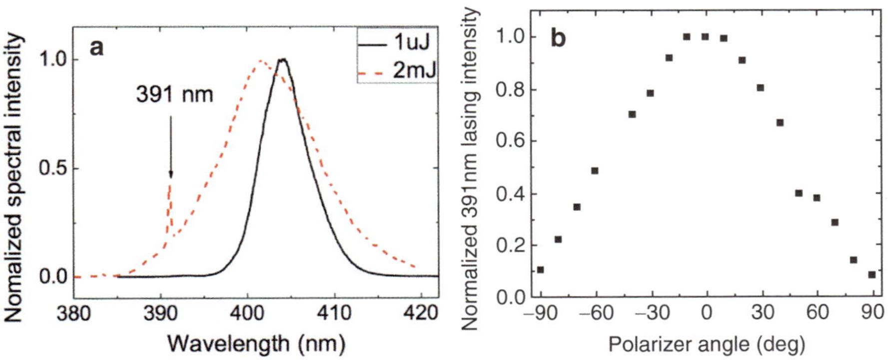

Fig. 20 (**a**) Forward spectra of blue pulses under different energy (1 μJ in *solid black line* and 2 mJ in *dashed red line*) and (**b**) polarization measurement of 391 nm lasing

Note the forward detection has to be set slightly off-axis (~1.4°) in order to avoid sending the strong central part of the beam to the integrating sphere. Polarization property is one of the most important parameters to characterize the coherence of optical emission. An emission with certain polarized direction is strong evidence on coherent lasing action because both spontaneous emission and ASE hold isotropic polarization. The polarization property of the 391 nm has been characterized in this case as shown in Fig. 20b. An emission linearly polarized in the direction parallel to that of the pump pulse was observed which is totally different from the randomly polarized fluorescence measured from the side. This also provides strong evidence on filament induced white light seed being amplified. This lasing action is from the transition of two vibrational levels of N_2^+, which are $B^2\Sigma_u^+(0)$ to $X^2\Sigma_g^+(0)$ [50]. The bandwidth of this lasing pulse is ~0.33 nm in FWHM, which can support the Fourier-Transform-limited pulse duration of 0.68 ps.

5.3 *Lasing Action from an External White Light Seeded* 800 *nm Filament in Air* [45]

Single lasing actions at 391 and 428 nm have been observed by using only 400 nm and only 800 nm air filaments, respectively. The self-generated white light seed was rather weak and also impossible to be independently controlled. The resulting lasing emission cannot be well controlled. In order to generate a strong, well-controlled white light for seeding the lasing action, we focused part of 800 nm pulse energy through a 1 cm-long water cell. Then the strong 800 nm pulse was filtered out. The rest forward white light spectrum detected is showned in Fig. 21a.

This white light seed is much stronger than the one from air filament in Fig. 16. Under the same filamenting pulse energy (~8 mJ), multiple-wavelength emission, namely, 388, 391, 428, and 471 nm, was experimentally demonstrated in this case which are depicted in Fig. 21b.

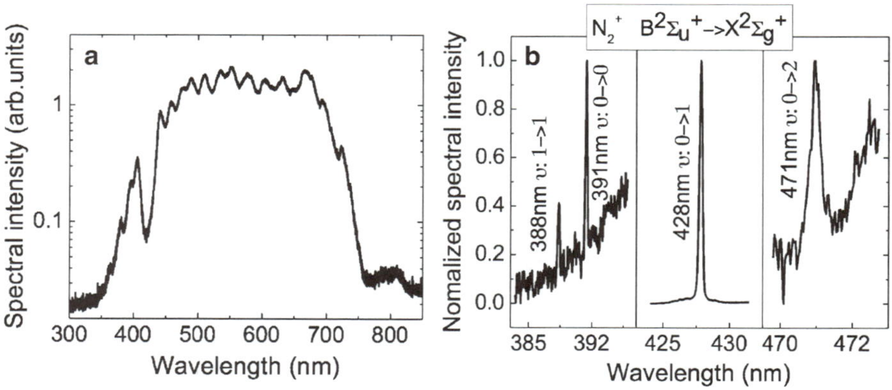

Fig. 21 (**a**) White light seed spectrum and (**b**) simultaneously generated four lasing wavelengths from an external white light seeded laser filament in air

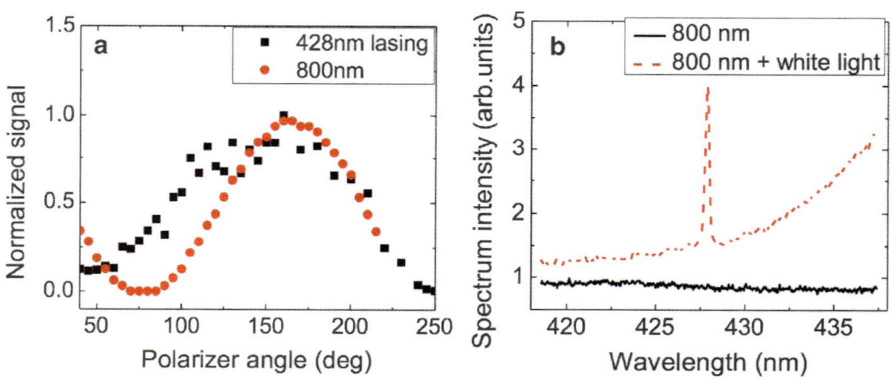

Fig. 22 (**a**) Polarization measurement of 428 nm lasing from external white light seeded 800 nm laser filament in air together with the polarizations of 800 nm pump. (**b**) Forward spectra around 428 nm lasing action from 5 mJ 800 nm laser filament with (*dashed red line*) and without (*solid black line*) external white light

The polarization of 428 nm emission shown in Fig. 22a was linear and followed the polarization direction of the 800 nm pump pulses, which once again confirms the coherent lasing actions. The 428 nm lasing is also one order of magnitude stronger than that from the self-seeded air filament. The lasing pulse energy could be μJ level under ~8 mJ pump with external white light seeding through the estimation of the spectral intensity. Among these four detected lasing wavelengths, 428 nm was the strongest, 2–3 orders of magnitude stronger than others. The reason why there was only 428 nm lasing observed in white light self-seeded 800 nm air filament is mostly due to the weak self-generated white light seed. These lasing wavelengths are from the first negative system $B^2\Sigma_u^+$ to $X^2\Sigma_g^+$ of N_2^+ [50] as shown in Fig. 13.

The bandwidths (FWHM) of those lasing pulses are 0.26 nm (388 nm), 0.24 nm (391 nm), 0.28 nm (428 nm), and 0.26 nm (471 nm), which correspond to the

Fourier-Transform-limited pulse durations of 0.85 ps, 0.93 ps, 0.96 ps, and 1.25 ps, respectively. In order to further confirm the external seeding effect, the forward spectra around 428 nm lasing action with and without external white light have been recorded as shown in Fig. 22b. With external white light seeded into 5 mJ 800 nm air filament, a clear lasing action has been observed. While without the external white light, the lasing action was absent. This is a strong evidence on white light seeded stimulated emission. This demonstration of using externally controllable white light seed in filament based laser actions provides more feasibilities to investigate this new frontier.

5.4 Physical Discussion on the Lasing Actions of N_2^+ in Air

The forward lasing actions of 388, 391, 428, and 471 nm observed above in three different experiments are identified as the transitions between two vibrational levels of N_2^+. White light as a seed is confirmed, which could be generated by laser filament itself or externally. This white light seeded forward lasing action can be ascribed to stimulated amplification process.

In the case of self-generated white light seed, the laser filament (at 800 and 400 nm) plays two roles. On the one hand, population inversion between two vibrational levels of ionic nitrogen ($B^2\Sigma_u^+ - X^2\Sigma_g^+$) is created inside the femtosecond laser filament due to the high clamping intensity [1]. On the other hand, filament-induced white light at 391 or 428 nm seeds into these transitions, which stimulated the laser emission at 391 or 428 nm in the forward direction. The white light feature with its spectral broadening to the blue and UV part has been clearly confirmed in Figs. 16 and 20a. Because of seeding effect, the polarization property of the forward lasing should be the same as that of the white light seed from 800 nm (400 nm) pump pulses. This is also confirmed by the polarization measurement (Figs. 18 and 20b). Since this lasing effect is very sensitive to the intensity of self-generated white light seeds, which is pump energy dependent, the forward lasing should depend on the pump energy also. At low pump energies, filament-induced white light is rather weak so that this white light cannot cover the seeding wavelength at 300–500 nm. At the same time the population inversion effect is also limited even though population inversion is created once a filament is formed. This forward lasing would be hard to detect. As the pump energy is increased, one can get a longer filament (and also multiple filaments) which will result in the increase of the number of population inverted ionic molecules. That will lead to the forward lasing occurring much more easily. This forward lasing action is also a highly nonlinear process. Once the filamenting energy is high enough, the lasing signal will be much stronger. This also agrees with our observation (Fig. 19). Once a strong white light covering many transitions of N_2^+ vibrational levels was seeded into air filament, multiple-wavelength lasing action should be expected and also successfully demonstrated experimentally (Fig. 21b). The white light seeding technique can be extended to operate the filament based lasing action in other gas species.

6 Summary

Although the first observation on lasing phenomenon inside a femtosecond laser filament was observed in Laval University more than 10 years ago, the topic gains a lot of research interest only in the past few years and becomes a new frontier in the community of filamentation. The laser filament based lasing action can be ASE type or seeded type. The former type of gain was observed in the fragments of OH, NH, and CH from a mixture of air with water vapor, or hydrocarbon gas. It looks like a universal phenomenon. The latter type of gain is much higher as compared with the ASE type of gain. This lasing action was observed in air by using powerful Ti:sapphire laser system, which can find more opportunity toward application.

From the theoretical point of views the challenge ahead will be a clear understanding of the physical mechanism. Although there are many successful experiments showing the lasing behaviors [48, 57, 58], it is still absent in demonstrating the desired spectroscopy application on remote sensing.

Acknowledgements This work was supported in part by National Natural Science Foundation of China (Grant Nos. 61221064, 11127901), National 973 Project (Grant No. 2011CB808103), Chinese Academy of Sciences and the State Key Laboratory of High Field Laser Physics, 100 Talents Program of Chinese Academy of Sciences, Shanghai Pujiang Program, NSERC, Canada Research Chair, the Canada Foundation for Innovation, the Canadian Institute for Photonics Innovation, and le Fonds Québécois pour la Recherche sur la Nature et les Technologies.

References

1. S.L. Chin, *Femtosecond Laser Filamentation* (Springer, New York, 2010)
2. S.L. Chin, S.A. Hosseini, W. Liu, Q. Luo, F. Théberge, N. Aközbek, A. Becker, V.P. Kandidov, O.G. Kosareva, H. Schroeder, Can. J. Phys. **83**(9), 863 (2005)
3. A. Couairon, A. Mysyrowicz, Phys. Rep. **441**(2–4), 47 (2007)
4. L. Bergé, S. Skupin, R. Nuter, J. Kasparian, J.P. Wolf, Rep. Prog. Phys. **70**(10), 1633 (2007)
5. J. Kasparian, J.P. Wolf, Opt. Express **16**(1), 466 (2008)
6. V.P. Kandidov, S.A. Shlenov, O.G. Kosareva, Quantum Electron. **39**(3), 205 (2009)
7. S.L. Chin, T.J. Wang, C. Marceau, J. Wu, J.S. Liu, O. Kosareva, N. Panov, Y.P. Chen, J.F. Daigle, S. Yuan, A. Azarm, W.W. Liu, T. Saideman, H.P. Zeng, M. Richardson, R. Li, Z.Z. Xu, Laser Phys. **22**(1), 1 (2012)
8. F. Théberge, M. Châteauneuf, V. Ross, P. Mathieu, J. Dubois, Opt. Lett. **33**(21), 2515 (2008)
9. T.J. Wang, S. Yuan, Y. Chen, J.F. Daigle, C. Marceau, F. Théberge, M. Châteauneuf, J. Dubois, S.L. Chin, Appl. Phys. Lett. **97**(11), 111108 (2010)
10. T.J. Wang, S. Yuan, Y. Chen, S.L. Chin, Chin. Opt. Lett. **11**(1), 011401 (2013)
11. S.L. Chin, H.L. Xu, Q. Luo, F. Théberge, W. Liu, J.F. Daigle, Y. Kamali, P.T. Simard, J. Bernhardt, S.A. Hosseini, M. Sharifi, G. Méjean, A. Azarm, C. Marceau, O. Kosareva, V.P. Kandidov, N. Aközbek, A. Becker, G. Roy, P. Mathieu, J.R. Simard, M. Châteauneuf, J. Dubois, Appl. Phys. B **95**(1), 1 (2009)
12. H.L. Xu, S.L. Chin, Sensors **11**(1), 32 (2011)
13. K. Yoshihara, Chem. Lett. **34**(10), 1370 (2005)
14. K. Yoshihara, Y. Takatori, K. Miyazaki, Y. Kajii, Proc. Jpn. Acad. Ser. B **83**(9–10), 320 (2007)
15. K. Yoshihara, Y. Takatori, Y. Kajii, Bull. Chem. Soc. Jpn. **85**(10), 1155 (2012)

16. K. Yoshihara, Y. Sakamoto, M. Kawasaki, Y. Takatori, S. Kato, Y. Kajii, Bull. Chem. Soc. Jpn. **87**(5), 593 (2014)
17. P. Rohwetter, J. Kasparian, K. Stelmaszczyk, Z. Hao, S. Henin, N. Lascoux, W.M. Nakaema, Y. Petit, M. Queisser, R. Salamé, E. Salmon, L. Wöste, J.P. Wolf, Nat. Photon. **4**, 451 (2010)
18. Y. Petit, S. Henin, J. Kasparian, J.P. Wolf, Appl. Phys. Lett. **97**(2), 021108 (2010)
19. Y. Petit, S. Henin, J. Kasparian, J.P. Wolf, P. Rohwetter, K. Stelmaszczyk, Z.Q. Hao, W.M. Nakaema, L. Woste, A. Vogel, T. Pohl, K. Weber, Appl. Phys. Lett. **98**(4), 041105 (2011)
20. M. Petrarca, S. Henin, K. Stelmaszczyk, S. Bock, S. Kraft, U. Schramm, C. Vaneph, A. Vogel, J. Kasparian, R. Sauerbrey, K. Weber, L. Woste, J.P. Wolf, Appl. Phys. Lett. **99**(14), 141103 (2011)
21. S. Henin, Y. Petit, P. Rohwetter, K. Stelmaszczyk, Z.Q. Hao, W.M. Nakaema, A. Vogel, T. Pohl, F. Schneider, J. Kasparian, K. Weber, L. Woste, J.P. Wolf, Nat. Commun. **2**, 456 (2011)
22. H. Saathoff, S. Henin, K. Stelmaszczyk, M. Petrarca, R. Delagrange, Z. Hao, J. Lüder, O. Möhler, Y. Petit, P. Rohwetter, M. Schnaiter, J. Kasparian, T. Leisner, J.P. Wolf, L. Wöste, Atmos. Chem. Phys. **13**, 4593 (2013)
23. T. Leisner, D. Duft, O. Möhler, H. Saathoff, M. Schnaiter, S. Henin, K. Stelmaszczyk, M. Petrarca, R. Delagrange, Z. Hao, J. Lüder, Y. Petit, P. Rohwetter, J. Kasparian, J.P. Wolf, L. Wöste, Proc. Natl. Acad. Sci. U. S. A. **110**(25), 10106 (2013)
24. J. Ju, J. Liu, C. Wang, H. Sun, W. Wang, X. Ge, C. Li, S.L. Chin, R. Li, Z. Xu, Opt. Lett. **37**(7), 1214 (2012)
25. J. Ju, J. Liu, C. Wang, H. Sun, W. Wang, X. Ge, C. Li, S.L. Chin, R. Li, Z. Xu, Appl. Phys. B **110**(3), 375 (2013)
26. H. Sun, J. Liu, C. Wang, J. Ju, Z. Wang, W. Wang, X. Ge, C. Li, S.L. Chin, R. Li, Z. Xu, Opt. Express **21**(8), 9255 (2013)
27. J. Ju, H. Sun, A. Sridharan, T.J. Wang, C. Wang, J. Liu, R. Li, Z. Xu, S.L. Chin, Phys. Rev. E **88**(6), 062803 (2013)
28. A. Sridharan, T.J. Wang, A. Gregoire, J. Ju, J. Liu, R. Li, Z. Xu, D. Boudreau, S.L. Chin, Laser Phys. Lett. **10**(12), 125301 (2013)
29. J. Ju, T. Leisner, H. Sun, A. Sridharan, T.J. Wang, J. Wang, C. Wang, J. Liu, R. Li, Z. Xu, S.L. Chin, Appl. Phys. B **117**(4), 1001 (2014)
30. Q. Luo, W. Liu, S.L. Chin, Appl. Phys. B **76**(3), 337 (2003)
31. D. Kartashov, S. Alisauskas, G. Andriukaitis, A. Pugzlys, M. Shneider, A. Zheltikov, S.L. Chin, A. Baltuska, Phys. Rev. A **86**(3), 033831 (2012)
32. P.R. Hemmer, R.B. Miles, P. Polynkin, T. Siebert, A.V. Sokolov, P. Sprangle, M.O. Scully, Proc. Natl. Acad. Sci. U. S. A. **108**(8), 3130 (2011)
33. P. Sprangle, J. Penano, B. Hafizi, D. Gordon, M. Scully, Appl. Phys. Lett. **98**(21), 211102 (2011)
34. J.P. Yao, B. Zeng, H.L. Xu, G.H. Li, W. Chu, J.L. Ni, H.S. Zhang, S.L. Chin, Y. Cheng, Z.Z. Xu, Phys. Rev. A **84**(5), 051802 (2011)
35. J.L. Ni, W. Chu, H.S. Zhang, C.R. Jing, J.P. Yao, H.L. Xu, B. Zeng, G.H. Li, C.J. Zhang, S.L. Chin, Y. Cheng, Z.Z. Xu, Opt. Express **20**(19), 20970 (2012)
36. W. Chu, B. Zeng, J.P. Yao, H.L. Xu, J.L. Ni, G.H. Li, H.S. Zhang, F. He, C.R. Jing, Y. Cheng, Z.Z. Xu, Europhys. Lett. **97**(6), 64004 (2012)
37. J.P. Yao, G.H. Li, C.R. Jing, B. Zeng, W. Chu, J.L. Ni, H.S. Zhang, H.Q. Xie, C.J. Zhang, H.L. Li, H.L. Xu, S.L. Chin, Y. Cheng, Z.Z. Xu, New J. Phys. **15**(2), 023046 (2013)
38. J.L. Ni, W. Chu, C.R. Jing, H.S. Zhang, B. Zeng, J.P. Yao, G.H. Li, H.Q. Xie, C.J. Zhang, H.L. Xu, S.L. Chin, Y. Cheng, Z.Z. Xu, Opt. Express **21**(7), 8746 (2013)
39. D. Kartashov, S. Alisauskas, A. Baltuska, A. Schmitt-Sody, W. Roach, P. Polynkin, Phys. Rev. A **88**(4), 041805 (2013)
40. S.L. Chin, H.L. Xu, Y. Cheng, Z.Z. Xu, K. Yamanouchi, Chin. Opt. Lett. **11**(1), 013201 (2013)
41. S. Yuan, T.J. Wang, Y. Teranishi, A. Sridharan, S.H. Lin, H. Zeng, S.L. Chin, Appl. Phys. Lett. **102**(22), 224102 (2013)
42. S. Yuan, T.J. Wang, P. Lu, S.L. Chin, H. Zeng, Appl. Phys. Lett. **104**(9), 091113 (2014)
43. S. Hosseini, A. Azarm, J.F. Daigle, Y. Kamali, S.L. Chin, Opt. Commun. **316**, 61 (2014)

44. T.J. Wang, J. Ju, J.F. Daigle, S. Yuan, R. Li, S.L. Chin, Laser Phys. Lett. **10**(12), 125401 (2013)
45. T.J. Wang, J.F. Daigle, J. Ju, S. Yuan, R. Li, S.L. Chin, Phys. Rev. A **88**(5), 053429 (2013)
46. A. Talebpour, M. Abdel-Fattah, S.L. Chin, Opt. Commun. **183**(5–6), 479 (2000)
47. H.L. Xu, A. Azarm, J. Bernhardt, Y. Kamali, S.L. Chin, Chem. Phys. **360**(1–3), 171 (2009)
48. S. Mitryukovskiy, Y. Liu, P. Ding, A. Houard, A. Mysyrowicz, Opt. Express **22**(11), 12750 (2014)
49. T.J. Wang, H. Xu, J.F. Daigle, A. Sridharan, S. Yuan, S.L. Chin, Opt. Lett. **37**(10), 1706 (2012)
50. R.W.B. Pearse, A.G. Gaydon, *The Identification of Molecular Spectra*, 4th edn. (Chapman and Hall, New York, 1976)
51. L.C. Lee, L. Oren, E. Phillips, D.L. Judge, J. Phys. B **11**(1), 47 (1978)
52. L.C. Lee, J. Chem. Phys. **72**(8), 4334 (1980)
53. O. Dutuit, A. Tabche-Fouhaile, I. Nenner, H. Frohlich, P.M. Guyon, J. Chem. Phys. **83**(2), 584 (1985)
54. F. Théberge, W. Liu, P.T. Simard, A. Becker, S.L. Chin, Phys. Rev. E **74**(3), 036406 (2006)
55. S.Q. Xu, X. Sun, B. Zeng, W. Chu, J. Zhao, W. Liu, Y. Cheng, Z.Z. Xu, S.L. Chin, Opt. Express **20**(1), 299 (2012)
56. F. Kong, Q. Luo, H. Xu, M. Sharifi, D. Song, S.L. Chin, J. Chem. Phys. **125**(13), 133320 (2006)
57. W. Chu, G. Li, H. Xie, J. Ni, J. Yao, B. Zeng, H. Zhang, C. Jing, H. Xu, Y. Cheng, Z. Xu, Laser Phys. Lett. **11**(1), 015301 (2014)
58. Y. Liu, Y. Brelet, G. Point, A. Houard, A. Mysyrowicz, Opt. Express **21**(19), 22792 (2013)

Filamentation and Pulse Self-compression in the Anomalous Dispersion Region of Glasses

A. Couairon, V. Jukna, J. Darginavičius, D. Majus, N. Garejev, I. Gražulevičiūtė, G. Valiulis, G. Tamošauskas, A. Dubietis, F. Silva, D.R. Austin, M. Hemmer, M. Baudisch, A. Thai, J. Biegert, D. Faccio, A. Jarnac, A. Houard, Y. Liu, A. Mysyrowicz, S. Grabielle, N. Forget, A. Durécu, M. Durand, K. Lim, E. McKee, M. Baudelet, and M. Richardson

A. Couairon (✉) • V. Jukna
Centre de Physique Théorique, CNRS, Ecole Polytechnique, 91128 Palaiseau, France
e-mail: couairon@cpht.polytechnique.fr

J. Darginavičius • D. Majus • N. Garejev • I. Gražulevičiūtė • G. Valiulis • G. Tamošauskas
A. Dubietis
Department of Quantum Electronics, Vilnius University, Saulėtekio Avenue 9, Building 3, 10222 Vilnius, Lithuania

F. Silva • D.R. Austin • M. Hemmer • M. Baudisch • A. Thai
ICFO—Institut de Ciències Fotòniques, Mediterranean Technology Park, 08860 Castelldefels, Barcelona, Spain

J. Biegert
ICFO—Institut de Ciències Fotòniques, Mediterranean Technology Park, 08860 Castelldefels, Barcelona, Spain

ICREA—Institució Catalana de Recerca i Estudis Avançats, 08010 Barcelona, Spain

D. Faccio
School of Engineering and Physical Sciences, Heriot-Watt University, Edinburgh EH14-4AS, UK

A. Jarnac • A. Houard • Y. Liu • A. Mysyrowicz
Laboratoire d'Optique Appliquée, ENSTA ParisTech, Ecole Polytechnique, CNRS, 91762 Palaiseau, France

S. Grabielle • N. Forget
FASTLITE, Centre Scientifique d'Orsay, Plateau du Moulon, 91401 Orsay, France

A. Durécu,
Onera-The French Aerospace Lab, BP 80100, 91123 Palaiseau Cedex, France

M. Durand
Laboratoire d'Optique Appliquée, ENSTA ParisTech, Ecole Polytechnique, CNRS, 91762 Palaiseau, France

Townes Laser Institute, CREOL-The College of Optics and Photonics, University of Central Florida, Orlando, FL 32816, USA

K. Lim • E. McKee • M. Baudelet • M. Richardson
Townes Laser Institute, CREOL-The College of Optics and Photonics, University of Central Florida, Orlando, FL 32816, USA

A.D. Bandrauk et al. (eds.), *Laser Filamentation*, CRM Series in Mathematical Physics, DOI 10.1007/978-3-319-23084-9_6

Abstract The propagation of near-infrared ultra-short laser pulses in the regime of anomalous dispersion of transparent solids is associated with a host of self-induced effects including a significant spectral broadening extending from the ultraviolet into the infrared region, pulse self-compression down to few-cycle pulse durations, free and driven third harmonic generation, conical emission and the formation of stable filaments over several cm showing the emergence of conical light bullets. We review measurements performed in different experimental conditions and results of numerical simulations of unidirectional propagation models showing that the interpretation of all these phenomena proceeds from the formation of non-spreading conical light bullets during filamentation.

1 Introduction

Laser pulses with central wavelengths in the near and mid-infrared region are available in many laboratories. All sources are exclusively based on optical parametric amplifiers since there are no femtosecond lasers operating in this range. As shown in Fig. 1, which illustrates the variation of the second order dispersion coefficient k_0'' as a function of wavelength for different materials, near- and mid-infrared laser sources operating at 2 or 3 μm lie within the region of anomalous dispersion for fused silica, calcium fluoride, and YAG.

This opens up new possibilities for filament based applications [1]: for instance, nonlinear pulse propagation in this regime leads to pulse compression rather than pulse splitting and the possibility for ultrabroadband generation from supercontinuum generation. The generation of odd harmonics in the lowest orders falls into the visible region providing a direct probing of higher order nonlinearities. The development of near and mid-infrared laser sources also led to new investigations on the fundamentals of filamentation in transparent solids. Several publications in this area during the last 2 years [2–9] uncovered filamentation physics specific to the anomalous dispersion regime and clarified the links between pioneering results [10–12].

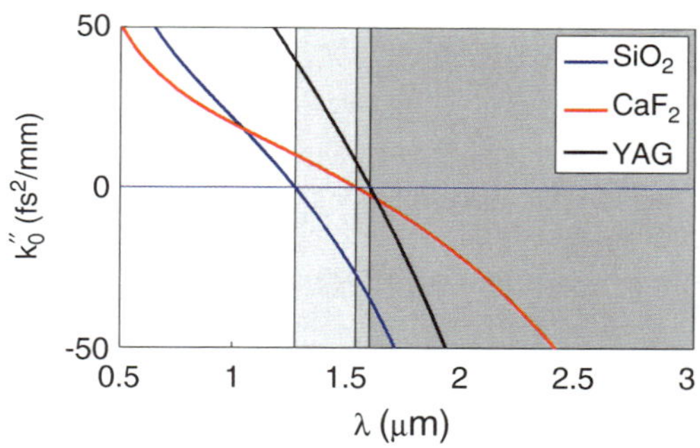

Fig. 1 Group velocity dispersion of fused silica, CaF_2 [13] and YAG [14]. For each material, the *gray area* indicates the anomalous dispersion region

In 2003, Moll and Gaeta investigated the role of dispersion in *multiple collapse dynamics* [10]. On the basis of experiments and numerical simulations, they have shown that the filamentation length is extended in the anomalous dispersion regime: collapse events can occur at locations in the medium many diffraction lengths beyond the initial collapse point, in contrast with the normal dispersion regime in which multiple collapse occurs within a diffraction length.

In 2005, Kolesik, Wright, and Moloney proposed an interpretation of conical emission of femtosecond laser pulses propagating in bulk media [11], valid in the normal as well as in the anomalous dispersion regime. They have shown that the input pulse with initially localized spectrally resolved far-field is scattered off a *material wave*, which is nothing but the self-generated nonlinear polarization wave associated with nonlinear propagation of the pulse. This results in preferential locations for populating the far-field, which can be predicted by an effective three wave mixing model and phase matching arguments.

The same year, Porras et al. have discovered a new type of spatiotemporal light bullet, which is nonsolitary and exhibits a conical nature [12]. Concretely, they have found a stationary solution to an extended nonlinear Schrödinger equation describing propagation in a Kerr medium, in the regime of anomalous dispersion, and in the presence of nonlinear losses. This solution represents a nonlinear and stable statiotemporal wavepacket that does not spread in space or time and withstands nonlinear losses. For this reason, it differs from a soliton or a light bullet as defined by Silberberg [15]. Figure 2 shows a typical profile for this solution, which takes the form of a nonlinear Bessel function along a reduced coordinate $R \equiv \sqrt{r^2 + t^2/k_0 k_0''}$ combining the transverse spacial coordinate r and the pulse local time t, where k_0 and k_0'' denote the wavenumber and the second order dispersion coefficient in the medium at the central wavelength of the pulse. These solutions have a ring shaped spectrum indicating that wavenumbers are distributed along the

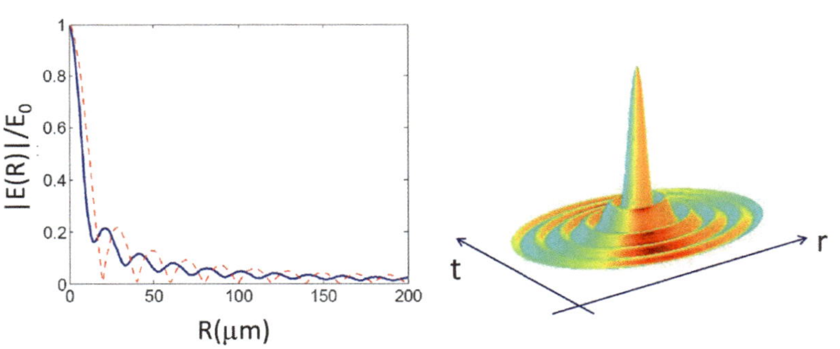

Fig. 2 (**a**) Amplitude profile for conical light bullets (nonlinear O-waves from the theory in[12]) supported by anomalous dispersion with normalized intensity. Solutions exist as a continuous family parametrized by the maximum amplitude. In this example, the *dashed red curve* corresponds to a linear solution. The continuous *blue curve* is typical for nonlinear solutions and corresponds to the a peak intensity of 9×10^{12} W/cm^2. (**b**) Space-time representation of the amplitude of conical light bullets

Fig. 3 Side view photography of the blue fluorescence of fused silica induced by a near infrared laser pulse at 1.9 µm with 8 µJ energy. The laser pulse is focused on a 70 µm spot (FWHM) at the entrance surface of the 3 cm thick sample. *Vertical dash-dotted lines* indicate the positions of the entrance and exit surfaces of the sample. Adapted from [5]

surface of a cone. This is typical for Bessel-like profiles. This solution can therefore be viewed as a nonlinear polychromatic Bessel beam, i.e., a light bullet with a conical nature.

The goal of this chapter is to show the links between these three results in light of recent experimental results which led us to discover a new filamentation regime, featured by pulse self-compression, stationarity, and the formation of similar light bullets as those predicted by Porras et al. [12]. Figure 3 illustrates this new filamentation regime, obtained by recording with a CCD camera the side image of the blue luminescence after focusing the beam of a 40 fs, 8 µJ laser pulse at wavelength of 1.9 µm on the entrance surface of a 3 cm-thick block of fused silica. The measurement reveals a thin and regular filamentary track extending over a large part of the sample in the anomalous dispersion regime, whereas a significantly shorter filament with multiple refocusing cycles is obtained in the regime of normal dispersion [5].

The outline of this chapter is the following. Section 2 briefly presents the model used in this work for numerical simulations of nonlinear pulse propagation and filamentation. Section 3 deals with investigations on ultrabroadband supercontinuum generation. Section 4 shows that these spectra are a signature of efficient pulse self-compression. Section 5 shows that a visible spectral peak appears in the ultrabroadband spectra, corresponding to the axial part of conical emission. Section 6 deals with free and driven third harmonic generation, and finally Sect. 7 shows measurements and simulations on a new filamentation regime where dissipative spatiotemporal light bullets emerge.

2 Numerical Simulation Model

For interpreting the measurements, numerical simulations were performed on the basis of unidirectional propagation equations [16, 17]. The Maxwell Wave Equation (MWE) describes the evolution of the pulse electric field along z, the propagation axis.

$$\frac{\partial \hat{E}(z,\omega,r)}{\partial z} = i(k(\omega) - k_0'\omega)\hat{E} + \frac{i}{2k(\omega)}\nabla_\perp^2 \hat{E} + \frac{1}{2\epsilon_0 c n(\omega)}[i\omega \hat{P}(z,\omega,r) - \hat{J}(z,\omega,r)], \quad (1)$$

where $\hat{E}(z, \omega, r) = \mathcal{F}[E(z, t, r)]$ denotes the frequency components (Fourier transform $t \to \omega$) of the laser electric field $E(z, t, r)$ and the transverse coordinate r indicates that revolution symmetry is assumed. The effects accounted for include material dispersion via a Sellmeier dispersion relation for the medium $k(\omega)$, diffraction, and the nonlinear response of the medium described by the nonlinear polarization

$$P(z, t, r) \equiv \epsilon_0 \chi^{(3)} E^3(z, t, r) \tag{2}$$

and current $J(z, t, r)$ source terms. These terms in turn represent the main effects playing a role in filamentation: the optical Kerr effect with nonlinear index coefficient n_2, ($\chi^{(3)} \equiv (4/3)\epsilon_0 c n_0^2 n_2$), plasma effects including ionization, plasma defocusing and absorption and nonlinear absorption governed by the system of Eqs. (3)–(5)

$$\frac{\partial J_e}{\partial t} + \frac{J_e}{\tau_c} = \frac{q_e^2}{m_e} \rho E(z, t, r) \tag{3}$$

$$\frac{\partial \rho}{\partial t} = W(E)(1 - \rho/\rho_b) + \frac{\sigma}{U_i} \rho E^2 \tag{4}$$

$$J(z, t, r) = J_e + \epsilon_0 c n_0 \frac{W(E)}{E^2} U_i (1 - \rho/\rho_b) E, \tag{5}$$

where q_e and m_e denote the electron charge and mass, τ_c the effective collision time, σ the cross section for inverse Bremsstrahlung, U_i the bandgap of the material, and ρ_b the background neutral atom density. Optical field ionization rates $W(E)$ are described by the Keldysh formulation. The model is valid for describing the propagation of few-cycle pulses with broad spectra. A formally analogous formulation based on an envelope propagation equation will be also used when explicitly specified. In the latter case, ionization rates and nonlinearity depend on intensity rather than on the field squared.

3 Supercontinuum Generation

3.1 *Supercontinuum Generation in Glasses at* 2 μm

Figure 4 shows a series of measurements for the supercontinuum spectra obtained by propagation of 15 fs (2.3 optical-cycle), CEP-stable pulses with central wavelength of 2 μm in different solid media, fused silica, CaF_2 and YAG. The corresponding conical emission is shown as a cross section of the emerging beam taken at 15 cm distance from the output face of the nonlinear medium.

Different features are observed: (1) generation of supercontinuum spectra; (2) presence of a spectral peak in the visible region; (3) emergence of a third harmonic

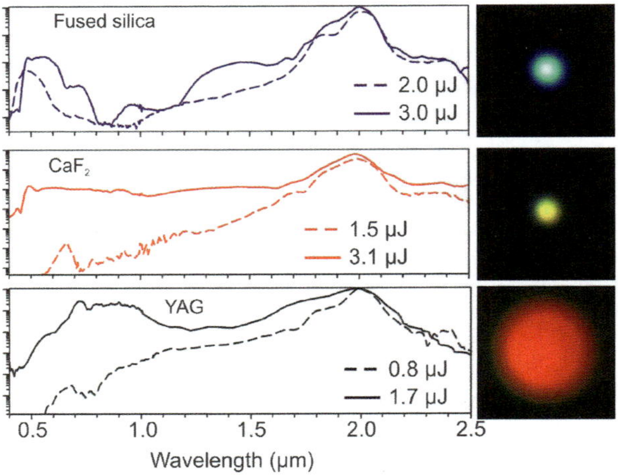

Fig. 4 Supercontinuum spectra generated in (**a**) fused silica, (**b**) CaF_2, (**c**) YAG. The *dashed* and *solid curves* represent supercontinuum spectra in the transient and saturation regimes of the spectral broadening, respectively. Curve labels stand for the input-pulse energy. Images on the *right side* show the corresponding far-field patterns of the conical emission in the visible range, recorded in the saturated regime of the spectral broadening. Adapted from [4]

peak for moderate pulse energy, and (4) conical emission in the form of angular spreading of visible light. These features are quite generic and reproducible in different media, at other pulse wavelengths and durations.

In fused silica, the supercontinuum spectrum has a deep extended minimum around 1 μm and a distinct intense peak in the visible (blue peak), which shows red-shifted broadening with increase of the input-pulse energy. CaF_2 exhibits lower dispersion than fused silica: a broad and flat supercontinuum spectrum from 450 nm to more than 2.5 μm was generated. In YAG which has higher nonlinearity, a smooth supercontinuum spectrum with elevated spectral intensity in the visible near-infrared 600–1000 nm range was generated using low input-pulse energy. The differences in conical emission bear upon the characteristic coloring and angular divergence of the supercontinuum radiation in the saturated regime of the spectral broadening, corresponding to higher values of the input-pulse energy. Among the tested materials YAG has the highest nonlinear index of refraction, leading to a strong laser–matter interaction, and the largest dispersion, therefore visible extension of supercontinuum in YAG has the largest angular spread. This feature will be further commented in Sect. 5.

In all cases, prior to supercontinuum generation, a characteristic third harmonic peak centered at 660 nm was observed. Its nature is discussed in Sect. 6. It is best visible in CaF_2 and YAG. In fused silica, it becomes rapidly masked by the strong and broad visible peak when energy is increased, which is distinct from the third harmonic as will be thoroughly explained in Sect. 5.

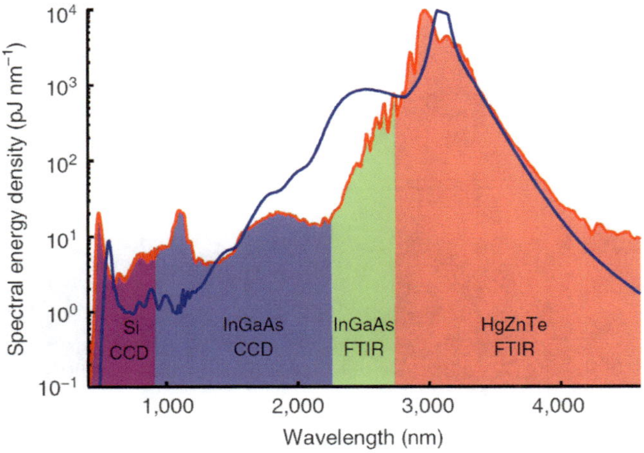

Fig. 5 Supercontinuum generated by 3100 nm, 2.6 μJ pulses in YAG (*red*). The ranges of each spectrometer/detector are indicated and superimposed is the angle-integrated spectrum from our simulation (*blue*). Two different detectors (InGaAs and HgCdZnTe) were used to cover the long wavelength region. FTIR, Fourier Transform Infrared Spectrometer. Adapted from [2]

3.2 Supercontinuum Generation in YAG at 3 μm

Recent experimental results have shown ultrabroadband supercontinuum generation in a YAG crystal from mid-IR laser pulses with central wavelength of 3 μm [2]. Pulses of 7 μJ energy were focused in a 2 mm-thick YAG plate placed in the focal region. It produced a clearly visible filament. After the exit plane of the YAG plate, we recorded spectra by imaging the supercontinuum into four spectrometer fibers. Each fiber delivered light to four different spectrometers working in different spectral regions, as indicated by the colored regions in Fig. 5. Intensity calibrations were applied for ensuring consistency in their regions of overlap (925, 2300 and 2700 nm) and with the measured pulse energy.

The result is shown in Fig. 5, along with a corresponding numerical simulation. The spectrum shows that the long wavelength edge decays smoothly, reaching the noise level at 4.5 μm. The short wavelengths extend down to 450 nm. The highest spectral energy density is found around the mid-infrared pump wavelength. The smallest energy content is observed in the 750–1000 nm range at a few picojoules per nanometer.

We observed two spectral peaks at 600 and 1100 nm seemingly close to the third and fifth harmonics. The blue curve corresponds to the result of a numerical simulation. It does reproduce well the peak in the visible region but not the third harmonic which merges in the supercontinuum. Another simulation with an envelope formulation, which does not account for harmonic generation, produced the same features in the supercontinuum spectrum, in particular a prominent visible peak, suggesting that this peak at 600 nm is not related to fifth harmonic generation.

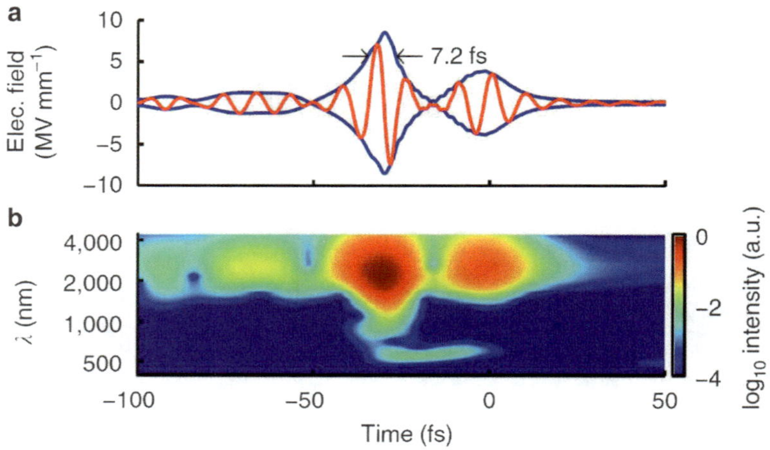

Fig. 6 Temporal profile and spectrogram of the output pulse after filamentation in YAG at 3 μm. (**a**) Temporal envelope (*blue*) and field oscillations (*red*) and (**b**) spectrogram of the on-axis simulated pulse emerging from the YAG plate; note the logarithmic intensity colour scale in the spectrogram. Adapted from [2]

Simulations with the MWE (1) and (3)–(5) detailed how the spectrally resolved far-field is populated along propagation [2]. The visible peak appeared in the last 400 μm of the YAG plate and is not affected by harmonic generation. Its generation is due to a scattering process explained in Sect. 5. It coincides with the loci of phase-matched angle and wavelength lying in the visible region of the spectrum, i.e., this peak is nothing but the axial part of conical emission. The presence of this spectral peak is also a strong signature of the partial reshaping of the pulse into O- or fishlike waves in media with anomalous dispersion [18–20].

We used the numerics to examine the potential of our setup for pulse compression. Figure 6a shows a typical output temporal on-axis profile obtained by numerical simulation. We found that the pulse duration has been reduced by a factor of 10 (the satellite peaks are 50 % lower in peak intensity), and the main peak has near-single-cycle duration. A more complete picture of the output pulse is given by the spectrogram shown in Fig. 6b. The short wavelength components, below 1500 nm, emerge from the output face nearly simultaneously with the main peak. The overall time-frequency distribution is simple, containing only a few distinct peaks. Therefore, the numerics indicate the possibility of achieving in an extremely simple setting, near-single-cycle and carrier-envelope phase stable pulses using mid-infrared pump lasers.

4 Pulse Self-compression in the Regime of Anomalous Dispersion in Glasses

We used the simple experimental setup sketched in Fig. 7a to investigate direct pulse self-compression. The parameters of this setup consist of a 10 cm focal length CaF_2 lens and a 3-mm thick, Yttrium Aluminum garnet (YAG) plate which was moved 6 mm across the focal plane in steps of 0.5 mm to initiate nonlinear propagation inside the plate at different peak intensities. The emerging beam was collimated with a silver-coated mirror and the pulse was characterized with frequency-resolved optical gating (FROG). The FROG device was specifically designed to handle ultrabroad bandwidths in the mid-IR region [21]. Mid-IR spectra were also measured using a Fourier transform infrared spectrometer.

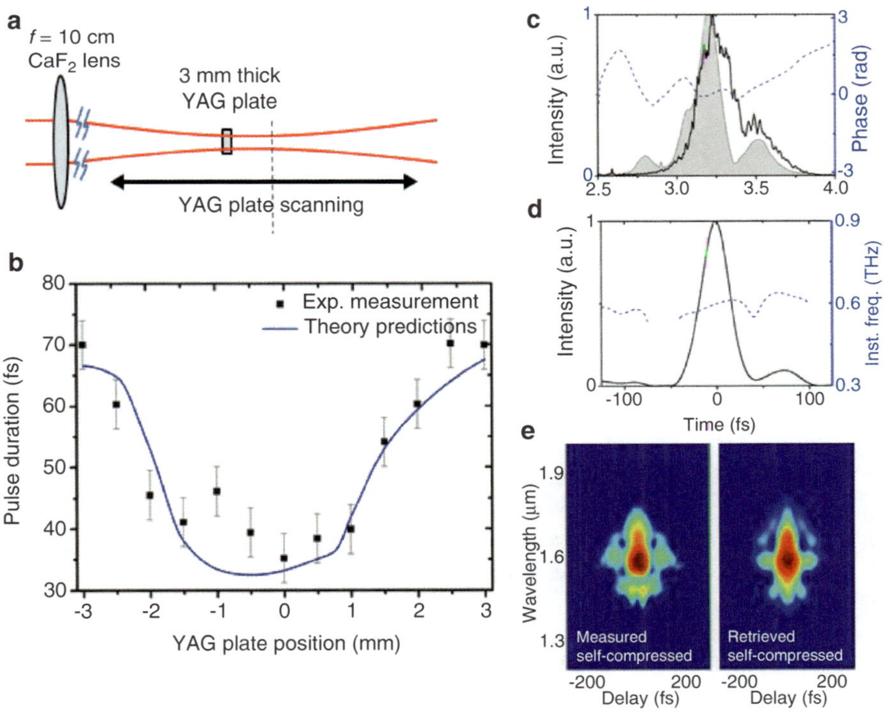

Fig. 7 (**a**) Layout of the self-compression setup. The input beam is focused by a 10 cm focal length CaF_2 lens into a 3 mm-thick YAG plate. The YAG plate is moved along the beam propagation axis, through focus. (**b**) Measured (*black dots*) and simulated (*blue line*) evolution of the pulse duration as a function of YAG plate position through focus. (**c**) Measured (*black line*) and retrieved (*shadow*) spectra of the self-compressed pulses. (**d**) Retrieved normalized 32 fs (2.9-cycle) duration intensity profile of the self-compressed pulses. (**e**) Measured and retrieved SHG-FROG traces of the self-compressed pulses on logarithmic color scale. Adapted from [3]

The results in Fig. 7c show measured and simulated durations of the emerging pulse as a function of the YAG plate position, representing the distance between the front of the YAG plate and the linear focus. Simulations and experiments are in good agreement. They indicate that self-compression to 32 fs (2.9 cycle) occurs and that no pulse splitting is observed. Optimum self-compression conditions for 3 μJ seed pulses were identified and the optimally self-compressed pulses were carefully characterized by sampling the inner part of the beam. Figure 7c–e show measured and retrieved FROG traces with a maximum 0.5 % error. Figure 7d shows the optimally self-compressed pulse. The measured spectrum in Fig. 7c for the optimally self-compressed pulse shows an extent of over 1000 nm and the retrieved spectrum from the FROG measurement is in good agreement with the measured spectrum confirming the consistency of the measurement. We retrieve a FWHM pulse duration of 32 fs which corresponds to 2.9-cycles of electric field at the slightly red-shifted output of 3200 nm. The spatial profile of the beam after self-compression was measured using a knife-edge technique and showed that the process left the seed spatial profile undisturbed.

5 Conical Emission

Nonlinear propagation of an intense laser pulse generates new frequencies leading to the supercontinuum. The angular spread in the visible wavelength range is regarded as conical emission, whose angles are set by dispersion-related phase matching conditions. In other words, for a given material, the phase matching condition entirely determines the loci of the far-field that can be populated in priority, i.e., the dependence of angles as a function of wavelength in the conical emission pattern. The effective scattering of a specific color at a given angle also depends on the nonlinear pulse–matter interaction ensuring that the phase matching condition is fulfilled over a certain propagation distance.

As shown in Figs. 4 and 5, the third harmonic does not always emerge from the supercontinuum and cannot always be distinguished from the spectral peak in the visible region. For instance, near-infrared pump pulses with central wavelength close to 2 μm usually undergo spectral broadening and frequency shift during nonlinear propagation leading to a loosely defined third harmonic wavelength in the visible range.

Figure 8 (right column) indicates the physical origin of the visible spectral peak. We compared the spectrally resolved far-fields obtained by numerical simulation of filamentation in fused silica for different pulse wavelengths ($\lambda_0 = 800, 1200, 1900$ nm, see Fig. 8). We found that the generation of the visible peak is consistent with the effective three wave mixing model proposed by Kolesik et al. [11]. Due to the reshaping of the pulse during nonlinear propagation, the peak field and therefore the nonlinear polarization wave induced by the pulse (2)–(5) travels at velocity v_p, usually slightly lower than the pulse group velocity in the conditions of anomalous dispersion [6, 11]. This constitutes a material wave, the velocity of

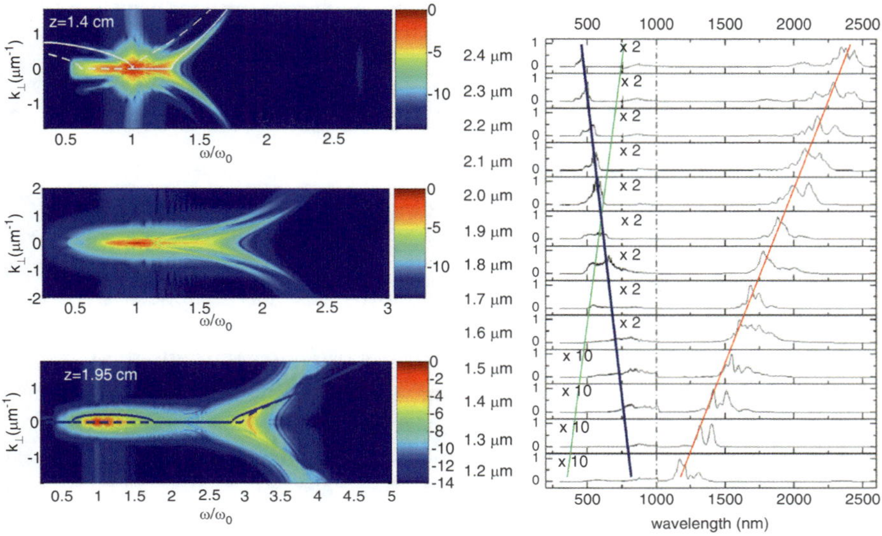

Fig. 8 *Left column*: Spectrally resolved far-fields $k_\perp - \omega$ obtained by numerical simulation of filamentation of 50 fs laser pulses in fused silica, for different pump wavelengths: *First row*: $\lambda_0 =$ 800 nm (normal dispersion). *Second row*: $\lambda_0 =$ 1200 nm (close to zero dispersion). *Third row*: $\lambda_0 =$ 1900 nm (anomalous dispersion). Units on the logarithmic colorbars correspond to decades in intensity. *Right column*: Supercontinuum spectra measured after filamentation in a 4 cm thick fused silica sample, with variable pump central wavelengths from 1.2 to 2.4 μm. For legibility, magnification factors were applied in the region of the spectra up to 1 μm (10× for $1.2 < \lambda_0 <$ 1.5 μm) and 2× for $1.6 < \lambda_0 < 2.4$ μm). Adapted from [6]

which can be quantified from the results of numerical simulations by monitoring the temporal profile of the pulse along propagation (see example below in Fig. 12). Conical emission is essentially interpreted as a scattering process: the pump pulse is scattered off the material wave. Newly generated frequencies will populate the spectrum in regions where the longitudinal phase matching condition (6) between the pump pulse (axial wavenumber k_0), the material wave (axial wavenumber $(\omega - \omega_0)/v_p$), and the scattered wave (k_z) is fulfilled.

$$\Delta k \equiv k_0 + \frac{\omega - \omega_0}{v_p} - k_z(\omega, k_\perp) \simeq 0, \tag{6}$$

$$k_\perp^2 + k_z^2 = k^2(\omega). \tag{7}$$

Equation (7) links transverse and longitudinal components for dispersive waves in the medium, and $k(\omega)$ denotes a Sellmeier like dispersion relation characterizing the medium. The angular distribution of frequencies (or wavelengths) where the far-field is preferentially populated along propagation is obtained by the combination of (6) and (7):

$$k_{\perp} = \pm\sqrt{k^2(\omega) - [k_0 - \Delta k + \alpha(\omega - \omega_0)]^2} \tag{8}$$

where $\alpha \equiv v_p^{-1}$ and Δk is a free parameter (zero for perfect phase matching).

In Fig. 8, we plotted the phase matching condition (8) for both cases of normal and anomalous dispersion. For $\lambda_0 = 800$ nm, we used two different values of the α parameter, corresponding to the two velocities of the split pulses arising in filaments with normal dispersion [22, 23]. This leads to two pairs of hyperbolic branches (one pair for each velocity highlighted by the white continuous and dashed curves). For $\lambda_0 = 1900$ nm, the same procedure leads to two branches: one is elliptic around the pump, a second one is hyperbolic and marks the conical emission in the visible range. Note that measurements of the spectrally resolved far-field are very well fitted by these branches both in the normal and the anomalous dispersion regimes [11, 23–26], which confirms the physical origin of conical emission and highlights the nature of the spectral peak in visible region of the supercontinuum.

To further confirm that spectral peaks in the visible region originate from dispersive waves populating the supercontinuum and distinguish them from third harmonic generation, another set of experiments was performed on filamentation in fused silica, where the pulse wavelength was varied from 1.2 to 2.4 μm [6]. The experiments clearly confirmed that the visible spectral peak is not related to the third harmonic but corresponds to the axial part of conical emission. The laser source was an OPA pumped by a titanium:sapphire chirped-pulse amplified laser system. The pulses at the output of the OPA had energies up to 100 μJ, at 100 fs. The laser beam was focused with a 50 cm focal length lens inside a 3 cm-long fused silica hexagonal rod. The measurements taken by three spectrometers were combined to cover the spectral range from 300 to 2500 nm. Results are shown in the right column of Fig. 8.

The visible spectral peak, highlighted by a blue line, is clearly shifted to shorter wavelengths when the initial pulse wavelength (red line) is varied to longer wavelengths, ruling out third-harmonic generation as the mechanism at the origin of the visible emission. The green line marks the position of the third harmonic, showing clearly that it is well separated from the visible peak except for a short range of pump wavelengths around 2 μm.

We have finally applied this model to determine the position of the visible peak as a function of the input pulse wavelengths. The results plotted in Fig. 9. The comparison of the model prediction, in solid black curve, with the measurements in red dots, shows very good agreement in the range of pump wavelengths 1500–2000 nm and a slight redshift with respect to the prediction in the range 2000–2400 nm. The figure also shows the extent of the visible spectral peak.

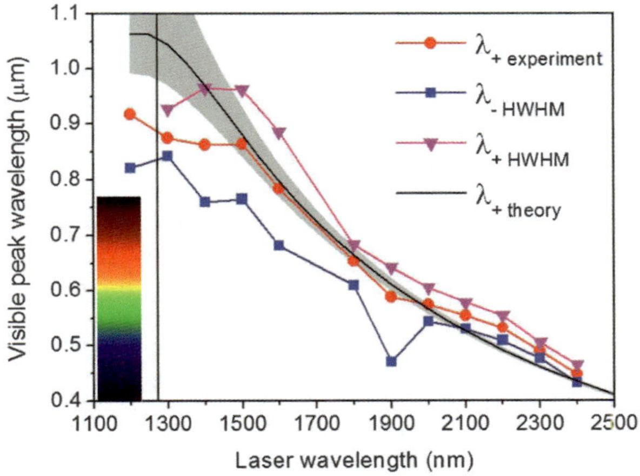

Fig. 9 Comparison of the wavelength of the spectral peak in the visible region obtained from the theory ($\lambda_{+,\text{theory}}$, *black curve*) with the measurements ($\lambda_{+,\text{experiment}}$, *red dotted curve*). The visible peak is generated during filamentation in a 3 cm-thick fused silica sample, as a function of the input pulse central wavelength. The curves $\lambda_{-,\text{HWHM}}$ and $\lambda_{+,\text{HWHM}}$ mark the extent of the visible peak. Adapted from [6]

6 Third Harmonic Generation

As shown in Fig. 4, third harmonic generation is clearly observed before the onset of spectral broadening and supercontinuum generation, in the transient regime of supercontinuum generation [4, 7]. Typically, the third harmonic radiation was detected with half the lowest pulse energies indicated in Fig. 4 in fused silica and in YAG, and the measured third harmonic efficiency varied from 10^{-6} to 10^{-4}, depending on the input-pulse energy.

A closer inspection of third harmonic spectra revealed a fast periodic modulation, whose frequency changed with the nonlinear material and its length. An example of third harmonic spectrum generated with a 15 fs input-pulse with 0.70 μJ energy, before the onset of supercontinuum generation, is shown in Fig. 10a, b for YAG and fused silica, respectively. This spectral modulation is a signature of a double-peaked third harmonic pulse, which occurs naturally, without the splitting of the input pulse. We interpret our results in the framework of large phase and group-velocity mismatched third harmonic generation [27]. The operating conditions impose that the third harmonic radiation consists of two pulses, the free and driven waves, which are solutions to the homogeneous and the inhomogeneous wave equations, respectively. The free third harmonic wave propagates at the velocity u_f set by the material dispersion, and walks-off from the pump pulse, i.e., the input-pulse at the fundamental frequency. The driven third harmonic wave propagates at the same velocity u_d as the nonlinear polarization and appears as locked under the pump, i.e., travels at the pump pulse group velocity (see Fig. 10a, b).

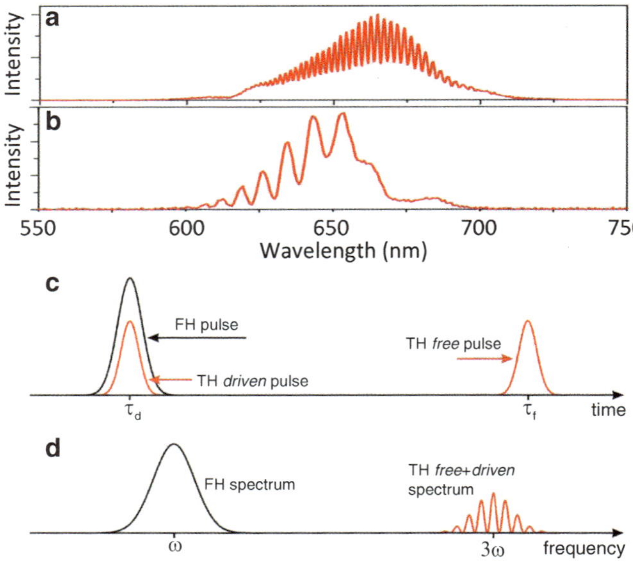

Fig. 10 (**a**), (**b**) Fringe pattern in the third harmonic spectra after propagation of a 15 fs pulse in (**a**) 6 mm of YAG, suggesting two pulses separated by 670 fs and (**b**) 5 mm of fused silica. (**c**), (**d**) Sketch for the generation of third harmonic radiation in the conditions of large phase and group velocity mismatch. (**c**) Time domain and (**d**) frequency domain

The emerging third harmonic radiation consists of two pulses separated in time by a duration $(u_f^{-1} - u_d^{-1})L$, proportional to the group velocity mismatch, where L denotes the medium length. This produces beating in the third harmonic spectrum with frequency inversely proportional to the temporal separation between the free and driven components. By inserting the relevant values for YAG: $L = 6$ mm and group velocity mismatch $u = 115$ fs/mm, the estimated temporal separation between the free and driven third harmonic pulses is 690 fs, which is very close to that of 670 fs, as retrieved from the fringe pattern shown in the spectrum.

In order to verify our interpretation, we performed numerical simulations with the model described in Sect. 2. Figure 11a shows the result of numerical simulation for the spectrally resolved far-field in the transient regime of spectral broadening in YAG. The inset shows the magnified portion of the third harmonic spectrum in the 550–750 nm range, highlighting the spectral beatings. The profiles in Fig. 11b show the respective temporal profiles of the main (pump) pulse (black curves) and pulses at the third harmonic wavelength (red curves), retrieved by using a super-Gaussian filter with full width at half maximum of 150 nm and centered at 660 nm. The retrieved third harmonic temporal profile consists of two distinct pulses temporally separated by 700 fs. They correspond to the *free* and *driven* third harmonic pulses, in fair agreement with our analytical interpretation and experimental results. The driven third harmonic pulse remains short as it is locked under the envelope of the fundamental (pump) pulse, which does not broaden much due to the interplay

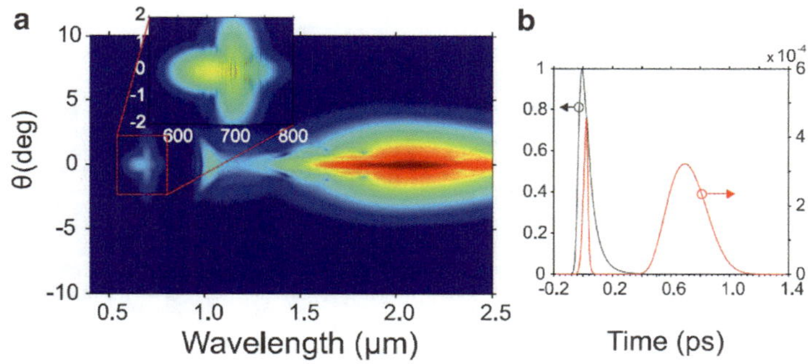

Fig. 11 (**a**) Spectrally resolved far-fields in YAG in the transient regime (**a**) of spectral broadening. The *inset* highlights the modulation around the third harmonic spectral range. (**b**) shows the normalized temporal profiles of the pump (*black curve*) and third harmonic (*red curves*) pulses after spectral filtering. Additional *arrows* in (**b**) indicate driven (*on the left*) and free (*on the right*) third harmonic pulses. Adapted from [4]

of self-phase modulation and anomalous dispersion, while the free third harmonic pulse moves away and experiences considerable temporal broadening due to normal dispersion.

7 Conical Light Bullets Generated by Filamentation in the Anomalous Dispersion

Figure 3 presented in introduction constitutes a typical result for filamentation in the anomalous dispersion regime of transparent solids. In an energy range above the threshold, the filament is very regular, featured by an absence of focusing-defocusing cycle sustaining filaments in the normal dispersion regime, at shorter wavelengths. These cycles, however, appear in the anomalous dispersion regime when the input pulse energy is increased and for slightly different focusing conditions [28] but after a few cycles, the filament regularizes and becomes regular as in Fig. 3 [5].

Numerical simulation results with the parameters of the experiments of Fig. 3 are shown in Fig. 12. The beam shrinks and forms the filament with a fairly constant diameter over more than 2 cm (Fig. 12a). Figure 12b shows the evolution of the axial temporal profile of the pulse, in the local frame, along the propagation distance. Both results show the emergence of a stable non-spreading structure in space of time at the origin of the filamentary track extending over a large part of the sample. This structure possesses some attributes of light-bullets, as introduced by Silberberg, referring to *a pulse that propagates without changing its temporal or spatial shape, held together by the nonlinear forces* [15]. Our measurements and simulations show quasi-stationary pulses, which do not change temporal or spatial shape for several diffraction lengths.

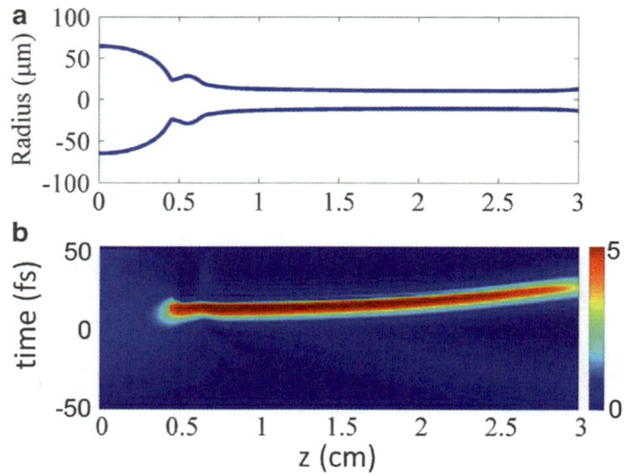

Fig. 12 Results of numerical simulation for the propagation through fused silica of an optical pulse at 1.9 μm of 110 μm diameter and 40 fs duration (FWHM) and 5×10^{11} W/cm^2 intensity collimated on the front sample surface. (**a**) Beam radius as a function of distance z. (**b**) Axial temporal profile of the laser pulse. Numbers on the colorbar denote TW / cm^2. Adapted from [5]

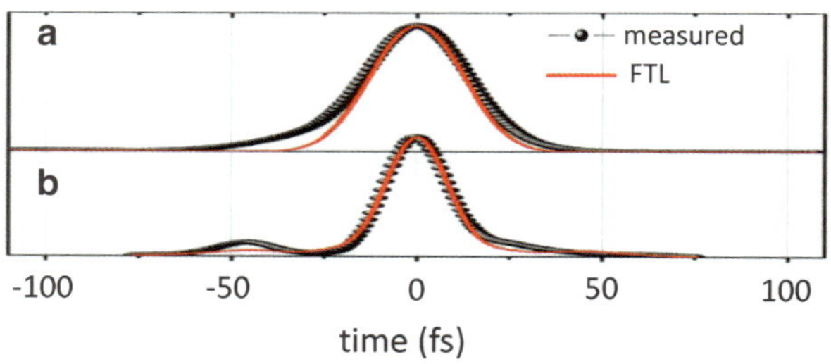

Fig. 13 Temporal pulse shape measured with the Wizzler for the initial pulse (40 fs FWHM) (**a**) and for filament lengths of 0.5 cm (20 fs FWHM) (**b**). Also shown in *red* is the Fourier Transform Limited (FTL) pulse. Adapted from [5]

For instance, the duration of the pulse emerging from the sample was characterized by a single-shot, self-referenced spectral interferometry technique called Wizzler operating in the near infrared region [29]. The measurement shown in Fig. 13 corresponds to a fused silica thickness of 5 mm and an integrated profile over 20 shots. A pulse duration of 20 fs was obtained, with a fair reproducibility. Propagation of the infrared pulse has led to self-compression by a factor of two. However, the self-compressed structure does not qualify as light-bullets in Silberberg's sense for two reasons: (1) they continuously lose energy due to multiphoton absorption and (2) they are stable whereas Silberberg's light bullets are unstable in more than one dimension.

In contrast, the pulse undergoing filamentation in the anomalous dispersion regime was shown in [8] to possess self-reconstruction properties and all the attributes of conical light bullets as proposed by Porras et al. [12]. The most important property of these structures is their conical nature allowing them to remain propagation invariant while undergoing multiphoton absorption. This is possible because the high intensity core where energy is lost is continually refueled by an energy flux from the low intensity energy reservoir of the wave. Invariant pulses in this form are found as mathematical solutions to a nonlinear Schrödinger equation accounting for the effects of diffraction, Kerr self-focusing, anomalous dispersion, and multiphoton absorption for any peak intensity up to a certain limit [12]. They are stable provided multiphoton absorption is sufficiently strong, and can therefore be observed in real experiments.

A confirmation of the conical nature of the filament was obtained from numerical simulation results, showing the evolution of the energy density flux along the propagation in the fused silica sample. The complex field (intensity and phase) allows us to reconstruct the energy density flux in the full spatial and temporal coordinates [30]:

$$\mathbf{F} = k_0^{-1} I[(\partial_r \phi)\mathbf{u}_r + \beta(\partial_\zeta \phi)\mathbf{u}_z], \tag{9}$$

where $I(r, \zeta)$ and $\phi(r, \zeta)$ denote the intensity and phase distributions along transverse r and longitudinal $\zeta \equiv k_0'^{-1} t - z$ axes, $\mathbf{u}_r$ and $\mathbf{u}_\eta$ are the unit vectors in the corresponding directions,

Figure 14 shows the energy flux components as arrows, along with an overlaid color plot of the intensity profiles at six distances along the propagation axis. Figure 14a–c evidence the pulse reshaping (compression in space and time) before the propagation invariant stage in Fig. 14d–f, sustained by a density flux essentially directed towards the leading and central part of the pulse from its periphery. A slow shift to the right corresponding to the subluminal motion of the pulse is the only visible change. The spatiotemporal distribution of the currents is almost identical to that of a conical light bullet [30]. We underline that Silberberg's bullets

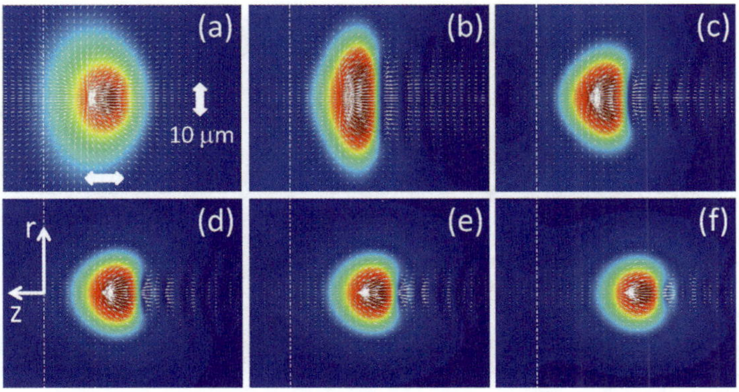

Fig. 14 Energy flux (*white arrows*) superimposed to intensity distribution (in *color*) obtained by numerical simulation at six propagation distances within a fused silica sample. (**a**) $z = 4.5$ mm. (**b**) $z = 6$ mm. (**c**) $z = 7.5$ mm. (**d**) $z = 12$ mm. (**e**) $z = 15$ mm. (**f**) $z = 19.5$ mm. The *dash-dotted lines* indicate the center of the input pulse (local frame center). The *white arrows* in (**a**) indicate the scale of 10 μm valid for all figures. The axis in (**f**) indicate the transverse and the longitudinal coordinates r, z, with the longitudinal direction corresponding to the propagation direction

and solitons exhibit a flat phase and hence no transverse energy density flow in the spatiotemporal domain. Conversely, the energy density flux unveiled by the simulations is inherent to conical light bullets. Together with detailed measurements further highlighting the reconstruction properties of these bullets [8, 9], these results unambiguously show the conical nature of the light bullets generated by self-focusing in a bulk dielectric medium with anomalous dispersion.

8 Conclusion

Laser pulse filamentation in transparent solids for near- and mid-infrared wavelengths leads to a significant spectral broadening and supercontinuum generation spanning more than three octaves. The pulse is efficiently self-compressed down to few-cycle pulse durations. Signatures in the form of free and driven third harmonic generation, conical emission and typical spectral peaks in the visible region, energy flux pointing toward the high intensity peak, propagation invariant O-shaped spatiotemporal intensity profile and characteristic angularly resolved spectrum all point out established properties of conical light bullets.

Acknowledgements We acknowledge support from the French National Agency for Research, the French DGA, the ONERA, the European Social Fund under the Global Grant measure (Grant No. VP1-3.1-ŠMM-07-K-03-001), the Lithuanian Science Council, the Spanish Ministerio De Economia Y Competitividad (MINECO) through "Plan Nacional" (FIS2011-30465-C02-01) and the Catalan Agencia de Gestió d'Ajuts Universitaris i de Recerca (AGAUR) with SGR 2014-2016. This research has been supported by Fundació Cellex Barcelona, LASERLAB-EUROPE grant agreement 284464 and COST Action MP1203. F.S. was partially supported by FCT-Fundaçãso para a Ciência e a Tecnologia grant SFRH/BD/69913/2010 and D.R.A. from the Marie Curie Intra-European Fellowship program, the European Research Council under the European Union's Seventh Framework Programme (FP/2007-2013)/ERC GA 306559. We acknowledge support from the US Army Research Office MURI Program and the State of Florida.

References

1. A. Couairon, A. Mysyrowicz, Phys. Rep. **441**(2–4), 47 (2007)
2. F. Silva, D.R. Austin, A. Thai, M. Baudisch, M. Hemmer, D. Faccio, A. Couairon, J. Biegert, Nat. Commun. **3**, 807 (2012)
3. M. Hemmer, M. Baudisch, A. Thai, A. Couairon, J. Biegert, Opt. Express **21**(23), 28095 (2013)
4. J. Darginavičius, D. Majus, V. Jukna, N. Garejev, G. Valiulis, A. Couairon, A. Dubietis, Opt. Express **21**(21), 25210 (2013)
5. M. Durand, A. Jarnac, A. Houard, Y. Liu, S. Grabielle, N. Forget, A. Durécu, A. Couairon, A. Mysyrowicz, Phys. Rev. Lett. **110**(11), 115003 (2013)
6. M. Durand, K. Lim, V. Jukna, E. McKee, M. Baudelet, A. Houard, M. Richardson, A. Mysyrowicz, A. Couairon, Phys. Rev. A **87**(4), 043820 (2013)
7. N. Garejev, I. Gražulevičiūtė, D. Majus, G. Tamošauskas, V. Jukna, A. Couairon, A. Dubietis, Phys. Rev. A **89**(3), 033856 (2014)

8. D. Majus, G. Tamošauskas, I. Gražulevičiūtė, N. Garejev, A. Lotti, A. Couairon, D. Faccio, A. Dubietis, Phys. Rev. Lett **112**(19), 193901 (2014)
9. I. Gražulevičiūtė, G. Tamošauskas, V. Jukna, A. Couairon, D. Faccio, A. Dubietis, Opt. Express **22**(25), 30613 (2014)
10. K.D. Moll, A.L. Gaeta, Opt. Lett. **29**(9), 995 (2004)
11. M. Kolesik, E.M. Wright, J.V. Moloney, Opt. Express **13**(26), 10729 (2005)
12. M.A. Porras, A. Parola, P. Di Trapani, J. Opt. Soc. Am. B **22**(7), 1406 (2005)
13. M.J. Weber, *Handbook of Optical Materials* (CRC Press, New York, 2003)
14. M. Bass, C. DeCusatis, J. Enoch, V. Lakshminarayanan, G. Li, C. MacDonald, V. Mahajan, E.V. Stryland, *Handbook of Optics*, vol. 4, 3rd edn. (McGraw Hill, New York, 2009)
15. Y. Silberberg, Opt. Lett. **15**(22), 1282 (1990)
16. M. Kolesik, J.V. Moloney, Phys. Rev. E **70**(3), 036604 (2004)
17. A. Couairon, E. Brambilla, T. Corti, D. Majus, O. de J. Ramírez-Góngora, M. Kolesik, Eur. Phys. J. Spec. Top. **199**(1), 5 (2011)
18. M.A. Porras, A. Dubietisand, E. Kučinskas, F. Bragheri, V. Degiorgio, A. Couairon, D. Faccio, P. Di Trapani, Opt. Lett. **30**(24), 3398 (2005)
19. D. Faccio, A. Averchi, A. Couairon, A. Dubietis, R. Piskarskas, A. Matijosius, F. Bragheri, M.A. Porras, A. Piskarskas, P. Di Trapani, Phys. Rev. E **74**(4), 047603 (2006)
20. M.A. Porras, A. Dubietis, A. Matijošius, R. Piskarskas, F. Bragheri, A. Averchi, P. Di Trapani, J. Opt. Soc. Am. B **24**(3), 581 (2007)
21. P.K. Bates, O. Chalus, J. Bieger, Opt. Lett. **35**(9), 1377 (2010)
22. M. Kolesik, E.M. Wright, J.V. Moloney, Phys. Rev. Lett. **92**(25), 253901 (2004)
23. D. Faccio, M. Porras, A. Dubietis, F. Bragheri, A. Couairon, P. Di Trapani, Phys. Rev. Lett. **96**(19), 193901 (2006)
24. D. Faccio, A. Matijosius, A. Dubietis, R. Piskarskas, A. Varanavičius, E. Gaižauskas, A. Piskarskas, A. Couairon, P. Di Trapani, Phys. Rev. E **72**(3), 037601 (2005)
25. D. Faccio, A. Averchi, A. Lotti, P. Di Trapani, A. Couairon, D. Papazoglou, S. Tzortzakis, Opt. Express **16**(3), 1565 (2008)
26. D. Faccio, M.A. Porras, A. Dubietis, G. Tamošauskas, E. Kučinskas, A. Couairon, P. Di Trapani, Opt. Commun. **265**(2), 672 (2006)
27. G. Valiulis, V. Jukna, O. Jedrkiewicz, M. Clerici, E. Rubino, P. Di Trapani, Phys. Rev. A **83**(4), 043834 (2011)
28. E.O. Smetanina, V.O. Kompanets, S.V. Chekalin, A.E. Dormidonov, V.P. Kandidov, Opt. Lett. **38**(1), 16 (2013)
29. T. Oksenhendler, S. Coudreau, N. Forget, V. Crozatier, S. Grabielle, R. Herzog, O. Gobert, D. Kaplan, Appl. Phys. B **99**(1–2), 7 (2010)
30. A. Lotti, A. Couairon, D. Faccio, P. Di Trapani, Phys. Rev. A **81**(2), 023810 (2010)

Nonperturbative Nonlinear Maxwell–Schrödinger Models for Intense Laser Pulse Propagation

E. Lorin, M. Lytova, and A.D. Bandrauk

Abstract This paper is devoted to the derivation of nonperturbative nonlinear models for intense and short laser pulses propagating in a gas. Starting from a computationally complex micro-macro Maxwell–Schrödinger–Plasma model (Lorin et al., Comput Phys Commun 177(12):908, 2007; Lorin et al., Commun Comput Phys 9(2):406, 2011) we derive an optics model, where the nonlinear response is nonperturbatively computed using a nonlinear evolution evolution equation of polarization. Only initial data of the polarization equation are computed at the microscopic level via TDSEs, thus providing for an efficient algorithm in Maxwell–Schrödinger's problems.

1 Introduction

1.1 Introductory Remarks

This paper is devoted to the derivation of non-standard nonperturbative nonlinear optics models for intense electromagnetic fields propagating in a gas. This work is motivated by the recurrent debate in the nonlinear optics community, regarding appropriate choice of models for simulating laser filaments, in particular appropriate modeling of high order nonlinearities and plasma of free electrons. In the present

E. Lorin (✉)
School of Mathematics and Statistics, Carleton University, Ottawa, ON, Canada K1S 5B6

Centre de recherches mathématiques, Université de Montréal, Montréal, QC, Canada H3T 1J4
e-mail: elorin@math.carleton.ca

M. Lytova
School of Mathematics and Statistics, Carleton University, Ottawa, ON, Canada K1S 5B6
e-mail: MariannaLytova@cmail.carleton.ca

A.D. Bandrauk
Faculté des Sciences, Laboratoire de chimie théorique, Université de Sherbrooke, Sherbrooke, QC, Canada J1K 2R1

Centre de recherches mathématiques, Université de Montréal, Montréal, QC, Canada H3T 1J4
e-mail: andre.bandrauk@usherbrooke.ca

A.D. Bandrauk et al. (eds.), *Laser Filamentation*, CRM Series in Mathematical Physics, DOI 10.1007/978-3-319-23084-9_7

paper we summarize and extend some of the ideas presented in [1], where different models and methods were derived for nonperturbative optics. These issues were already discussed in the celebrated papers [2, 3]. Over the years, several models were proposed and the interested reader can refer to complete review papers [4–6]. Although, some elaborate models (HOKE, UPPE [7, 8]) allow for accurate simulations and analysis of laser filamentation in several physical frameworks [7, 9–13] at very high intensity the accurate modeling of plasma of free electrons and nonperturbative nonlinearities, is still an open problem, see [14–20].

Our starting point is a micro-macro model constituted of Maxwell's equations, MEs, coupled with time-dependent Schrödinger equations, TDSEs, modeling the nonlinear response of a gas to an electromagnetic field, see [21–24]. From a practical point of view, this model confronts a major issue, which is the huge computational cost for computing the nonlinear polarization using TDSEs. Realistic simulations are then only possible on very short propagation distances, which makes this ab-initio model impractical for filamentation. In order to reduce the overall computational complexity, a model is proposed which is based on a second order equation, modeling the time evolution of the polarization. This additional equation allows to reduce the number of TDSEs to be solved, and then reduces the overall computational complexity of the numerical model. Although the model may not be efficient enough for large scale simulation of filamentation, it is a good candidate for a fundamental understanding of this phenomenon [4, 5]. It is shown that the general derived model reduces to a simple transport equation in its most basic version. We finally propose an slowly varying envelope approximation (SVEA) with paraxial version of the model, which is more suitable for filamentation simulation. Although it still requires the computation of the TDSE, this last model allows, in principle, for much longer laser pulse propagation.

1.2 Maxwell–Schrödinger Model

We recall that the Maxwell–Schrödinger model developed in [21, 22, 24] is based on the coupling of the 3D macroscopic MEs (Maxwell's Equations) with many TDSEs under the dipole approximation (the electric field is assumed to be constant in space at the molecular scale). This approximation is typically valid when the smallest internal wavelengths $\lambda_{\min}$ of the electromagnetic field are much larger than the molecule size ℓ, that is $\ell \ll \lambda_{\min}$ (typically $\lambda_{\min} \approx 800\,\text{nm}$ where $\ell \approx 0.1\,\text{nm}$). We then define the MEs on a bounded space domain with a boundary Γ, and $\mathbf{x}' = (x', y', z')^T$ denotes the electromagnetic field space variable. At the molecular scale, and working under the Born–Oppenheimer approximation, that is static nuclei, we denote by $\mathbf{x} = (x, y, z)^T$ the TDSE space variable (for electrons). The molecular density is assumed to be constant in time, continuous in space, and is denoted by $\mathcal{N}(\mathbf{x}')$. The complete field-molecule equations we consider are the following ones:

$$\begin{cases} \partial_t \mathbf{B}(\mathbf{x}',t) = -c\nabla \times \mathbf{E}(\mathbf{x}',t) \ , \\ \partial_t \mathbf{E}(\mathbf{x}',t) = c\nabla \times \mathbf{B}(\mathbf{x}',t) - 4\pi\big(\partial_t \mathbf{P}(\mathbf{x}',t)\big) \ , \\ \nabla \cdot \mathbf{B}(\mathbf{x}',t) = 0 \ , \\ \nabla \cdot \big(\mathbf{E}(\mathbf{x}',t) + 4\pi \mathbf{P}(\mathbf{x}',t)\big) = e(\mathcal{N}_I - \mathcal{N}_e) \ . \end{cases} \tag{1}$$

The polarization-TDSE is written as:

$$\begin{cases} \mathbf{P}(\mathbf{x}',t) = \mathcal{N}(\mathbf{x}') \sum_{i=1}^{m} \mathbf{P}_i(\mathbf{x}',t) \\ \qquad\qquad = \mathcal{N}(\mathbf{x}') \sum_{i=1}^{m} \chi_{\Omega_i}(\mathbf{x}') \int_{\mathbb{R}^3} \psi_i(\mathbf{x},t)\mathbf{x}\psi_i^*(\mathbf{x},t)\,\mathrm{d}\mathbf{x} \ , \\ \mathrm{i}\partial_t \psi_i(\mathbf{x},t) = -(\nabla_{\mathbf{x}}^2/2)\psi_i(\mathbf{x},t) + \mathbf{x}\cdot\mathbf{E}_{\mathbf{x}_i'}(t)\psi_i(\mathbf{x},t) + V_{\mathrm{C}}(\mathbf{x})\psi_i(\mathbf{x},t) \\ \qquad\qquad\qquad\qquad\qquad\qquad\qquad\qquad \forall i \in \{1,\ldots,m\} \end{cases} \tag{2}$$

where V_{C} denotes the Coulomb interaction potential. Computation of the TDSE provides complete wavefunctions, ψ, including ionization, i.e. a continuum spectrum of free electrons propagating in a laser pulse. Such electrons can recombine with the parent ion with maximum energy $I_p + 3.17U_p$, where I_p is the ionization potential and $U_p = E^2/4m\omega^2$, the ponderomotive energy acquired by a particle of mass m in a field E and frequency ω or with neighbors with energies exceeding $3U_p$, [25, 26]. In (1), Ω_i denotes the macroscopic spatial domain containing a reference molecule associated with a wavefunction ψ_i, and $\mathbf{P}_i$ denotes the macroscopic polarization in this domain. In other words, Domain Ω_i contains $\mathcal{N}(\mathbf{x}')\mathrm{vol}(\Omega_i)$ molecules represented by one single eigenfunction ψ_i (under the assumption of a unique pure state), and we have $\cup_{i=1}^{m}\Omega_i = \Omega$. We now assume that the spatial support of ψ_i is included in a domain $D_i \subset \mathbb{R}^3$, which is supposed to be *sufficiently large*. We allow free electrons to reach the boundary D_i and we impose absorbing boundary conditions on ∂D_i. In the proposed model, the part of the wavefunction absorbed at the boundary generates a plasma of free electrons. We refer to [21, 25] for a complete description of the geometry of this model. Functions χ_{Ω_i} are defined by $\chi \otimes \mathbf{1}_{\Omega_i}$ where χ is a plateau function and $\mathbf{1}_{\Omega_i}$ is the characteristic function of Ω_i. Finally $\mathbf{E}_{\mathbf{x}_i'}$ denotes the electric field (supposed constant in space) in Ω_i. The current density evolution can be integrated in the model via the classical Drude-model

$$\partial_t \mathbf{J} + \nu_e \mathbf{J} = \rho \mathbf{E} \tag{3}$$

where ν_e denotes the electron collision frequency and ρ the free electron density. In practice the time dependence of ρ is computed from laser-molecule ionization. More specifically

$$\rho(\mathbf{x}',t) = \sum_{i=1}^{m} \chi_{\Omega_i}(\mathbf{x}')\mathcal{I}_i(t) \tag{4}$$

where

$$\mathcal{I}_i(t) = 1 - \int_{\Omega_i} |\psi_i(\mathbf{x}, t)|^2 \, \mathrm{d}^3\mathbf{x} \, . \tag{5}$$

In order to describe accurately, the medium response to a laser pulse, a very large number of TDSEs is required, making huge the overall numerical complexity of the model. Details can again be found in [25, 26]. This paper's objective is to reduce considerably the complexity of the ME-TDSE model. A macroscopic wave equation is derived modeling the evolution of the polarization vector **P** (2), and is coupled to MEs. Although the model still requires TDSE computations for constructing the initial data of the polarization equation, this approach allows to reduce drastically the number of TDSEs involved in the model. The most simple polarization evolution equation is a homogeneous transport equation. More accurate models are then proposed, including, in particular, the electric field variations.

When the polarization is computed in the hypothesis of a unique pure state [27], we get at $(\mathbf{x}', t)$

$$\mathbf{P}(\mathbf{x}', t) = \mathcal{N}_0(\mathbf{x}')\mathbf{d}(\mathbf{x}', t) \, , \tag{6}$$

where $\mathbf{d}(\mathbf{x}', t)$ is the dipole moment of a molecule "located" at $\mathbf{x}'$. Assuming p pure states, we get

$$\mathbf{P}(\mathbf{x}', t) = \mathcal{N}_0(\mathbf{x}') \sum_{l=1}^{p} \mathbf{d}^{(l)}(\mathbf{x}', t) \tag{7}$$

where $\mathbf{d}^{(l)}(\mathbf{x}', t) = \langle \psi^{(l)} | \mathbf{x} | \psi^{(l)} \rangle$ and $\psi^{(l)}$ is a solution to

$$i\partial_t \psi(\mathbf{x}, t) = \big(H_0 + \mathbf{x} \cdot \mathbf{E}_\alpha(t)\big)\psi(\mathbf{x}, t) \, , \quad \psi(\mathbf{x}, 0) = \phi_l(\mathbf{x}) \, , \tag{8}$$

with $H_0\phi_l = \epsilon_l \phi_l$.

2 Polarization Evolution Equation for Ultrashort Pulses

2.1 Simple Evolution Equation for the Dipole Moment

Within the ME-TDSE model, interaction with a molecule requires the solution to the following TDSE:

$$\mathrm{i}\partial_t \psi = -\tfrac{1}{2}\Delta \psi + V_c(\mathbf{x})\psi + \mathbf{x} \cdot \mathbf{E}_{\mathbf{x}'}(t)\psi \, . \tag{9}$$

with $\mathbf{x} = (x, y)$, $\mathbf{E}_{\mathbf{x}'} = (E_{x'}, E_{y'})$. In the velocity gauge [28], this is written equivalently as:

$$i\partial_t\psi = -\frac{1}{2}\Delta\psi + V_c(\mathbf{x})\psi + i\mathbf{A}_{\mathbf{x}'}(t)\cdot\nabla\psi + \frac{\|\mathbf{A}_{\mathbf{x}'}(t)\|^2}{2}\psi\,. \tag{10}$$

with the vector potential $\mathbf{A}_{\mathbf{x}'} = (A_{x'}, A_{y'})$, $\mathbf{E}_{\mathbf{x}'} = -\partial_t\mathbf{A}_{\mathbf{x}'}$. The computation of the wavefunction ψ of a molecule "located in $\mathbf{x}'$" allows to deduce the dipole moment $\mathbf{d}$ as follows:

$$\mathbf{d}(\mathbf{x}', t) = \int_{\mathbb{R}^2}|\psi(\mathbf{x}, t)|^2\mathbf{x}\,\mathrm{d}x\,\mathrm{d}y \tag{11}$$

Following [29] and [30], we easily deduce an evolution equation for $\mathbf{d}$. Assuming that the wavefunction, ψ_1, of molecule m_1 "located" in $\mathbf{x}'_1 = (x'_1, y'_1, z'_1)$ is known, a sequence of evolution equations to estimate the dipole moment of molecule m_2 "located" in $\mathbf{x}'_2 = (x'_1, y'_1, z'_2)$ with $z'_2 > z'_1$ can be derived as follows.

If $|\Delta z'| := |z'_2 - z'_1|$ is small enough to neglect the medium effects on $\mathbf{E}$, we may assume that a circularly polarized electromagnetic field interacting with molecule m_2 is almost identical (up to a time delay) to the one which molecule m_1 is subject to. Note that overall, we do not assume that the electromagnetic field propagates as in a linear medium (or vacuum) in Maxwell's equations. In that case, let us define ψ_i the wavefunction of molecule m_i

$$i\partial_t\psi_i = -\tfrac{1}{2}\Delta\psi_i + V_c(\mathbf{x})\psi_i + \mathbf{x}\cdot\mathbf{E}_i(t)\psi_i\,. \tag{12}$$

and $\mathbf{d}(\mathbf{x}'_i, t)$ the corresponding dipole moment, with $i = 1, 2$. In addition $\mathbf{E}_i(t) = \mathbf{E}(\mathbf{x}'_i, t)$ denotes the electric field that molecule m_i is subject to. The above assumption mathematically implies: $\mathbf{E}_2(t) = \mathbf{E}_1(t - \Delta z'/v_g)$ and as a consequence

$$\mathbf{d}(\mathbf{x}'_2, t) = \mathbf{d}\left(\mathbf{x}'_1, t - \frac{\Delta z'}{v_g}\right) \tag{13}$$

where v_g is the envelope velocity (c in vacuum). Then the polarization $\mathbf{P}$ satisfies for $z' \in [z'_1, z'_2]$ the following transport equation

$$\partial_t\mathbf{P}(\mathbf{x}', t) + v_g\partial_{z'}\mathbf{P}(\mathbf{x}', t) = \mathbf{0} \tag{14}$$

with initial data $\mathbf{P}(\mathbf{x}'_1,) = \mathcal{N}_0\mathbf{d}(\mathbf{x}'_1, \cdot)$. The model is then purely macroscopic (except for the computation of the initial data for $\mathbf{P}$), see Fig. 1. *This model is applicable as long as $|\Delta z'|$ is small enough, or/and if the molecule density is small enough, that is as long as the effect of the medium on* $\mathbf{E}$ *during the pulse propagation from* $\mathbf{x}'_1$ *to* $\mathbf{x}'_2$ *is sufficiently negligible to not be included in the dipole moment calculation of molecule* m_2. In order to include the medium effects on $\mathbf{E}$ in the propagation from $\mathbf{x}'_1$ to $\mathbf{x}'_2$ an improvement of the model is necessary. In particular, this approach will allow to consider larger propagation lengths and times.

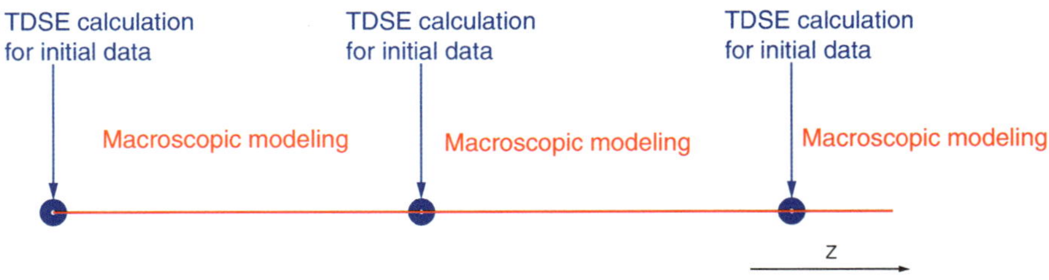

Fig. 1 Macroscopic model with initial data from TDSE

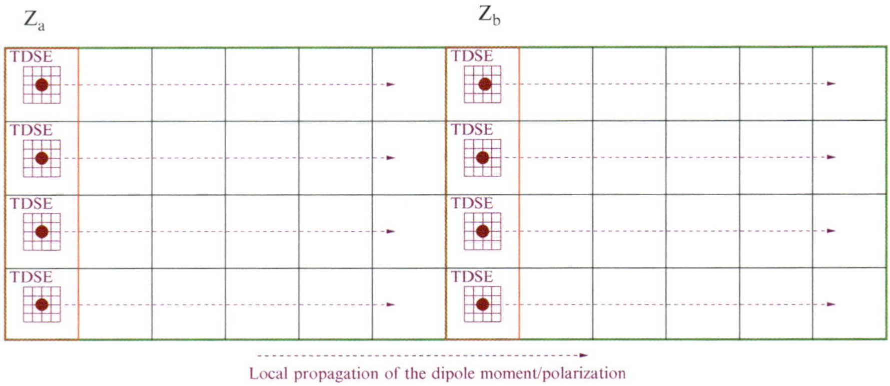

Fig. 2 Spatial evolution of polarization

The evolution equation for **P** is then coupled to MEs (1) for modeling the propagation of the laser pulse over (z'_1, z'_2). Polarization at (x', y', z') for $z' > z'_2$, is then computed again using the TDSE from (2). More specifically, the domain is decomposed in subdomains in the z' direction, and only at certain locations of each sudomain, TDSEs are computed to evaluate the dipole moment. At the other points, the macroscopic evolution equations on **P** are used. This methodology is summarized in Fig. 2, where the dipole moment/polarization is computed from TDSEs only for certain z' (Z_a, Z_b on the figure). Elsewhere, the evolution equation for **P** allows a "inexpensive" (as fully macroscopic) computational evaluation. Computational details can be found in [29], where this strategy is presented in 1-d dimension, and where it is shown that a reduction of almost *two* orders of magnitude in the computational complexity can be reached. From a practical point of view, the range of application of the presented approach is then limited only pulses on femtosecond time scale.

2.2 General Evolution Equation for the Dipole Moment

As we consider multidimensional electromagnetic field propagation, including forward and backward propagation effects, the derived polarization equation is expected to more generally be of the form:

$$\partial_{tt}\mathbf{P} - v_g^2\Delta\mathbf{P} = \mathbf{S}(\mathbf{E}, \mathbf{P}) \tag{15}$$

with initial data to be determined from quantum TDSEs. This is possible extending the arguments developed above. The main idea consists of including as much information as possible in the polarization evolution equation (14). In this goal, we assume first that $\mathbf{P} \approx \chi^{(1)}\mathbf{E} + \chi^{(3)}|\mathbf{E}|^2\mathbf{E}$, where $\chi^{(1)}$ and $\chi^{(3)}$ are estimates of the first and third instantaneous susceptibility tensors, is a good approximation of $\mathbf{P}$ over a certain propagation length. This leads to

$$\mathbf{E} \approx \frac{\mathbf{P}}{\chi^{(1)}}\frac{1}{1 + \big(\chi^{(3)}/\chi^{(1)}\big)|\mathbf{E}|^2} \approx \frac{\mathbf{P}}{\chi^{(1)}}\Big(1 - \frac{\chi^{(3)}}{\chi^{(1)}}|\mathbf{E}|^2\Big). \tag{16}$$

Then from the first and second MEs (1) we get the equation for $\mathbf{E}$:

$$\partial_t^2\mathbf{E} - c^2\Delta\mathbf{E} - c^2\nabla(\nabla\cdot\mathbf{E}) = -4\pi(\partial_t^2\mathbf{P} + \partial_t\mathbf{J}) \tag{17}$$

Furthermore for SVEA-scheme, we omit the term $\nabla(\nabla\cdot\mathbf{E})$.

Note that $\mathbf{J}$ satisfies the following evolution equation [5, 22], where ν_e is the collision frequency:

$$\partial_t\mathbf{J} + \nu_e\mathbf{J} = \rho\mathbf{E} \tag{18}$$

Then substituting $\mathbf{E}$ from (16), in the above equation for $\mathbf{E}$ we get a general equation for $\mathbf{P}$:

$$\begin{aligned}\partial_t^2\mathbf{P} - v_g^2\Delta\mathbf{P} = &-\frac{c^2\chi^{(3)}}{\chi^{(1)}(1 + 4\pi\chi^{(1)})}\Big[\Delta(\mathbf{P}|\mathbf{E}|^2) - \frac{1}{c^2}\partial_t^2(\mathbf{P}|\mathbf{E}|^2)\Big]\\ &-\frac{4\pi\chi^{(1)}}{1 + 4\pi\chi^{(1)}}\partial_t\mathbf{J}\end{aligned} \tag{19}$$

At specific locations, say $\mathbf{x}'_\alpha$, the initial data $\mathbf{P}(\mathbf{x}'_\alpha, 0)$, $\partial_t\mathbf{P}(\mathbf{x}'_\alpha, 0)$ are computed from TDSEs

$$\begin{cases} \mathbf{P}(\mathbf{x}'_\alpha, 0) = \mathcal{N}(\mathbf{x}'_\alpha) \int_{\mathbf{R}^3} |\psi(\mathbf{x}, t)|^2 \mathbf{x} \, \mathrm{d}^3\mathbf{x} \\ \partial_t \mathbf{P}(\mathbf{x}'_\alpha, 0) = \mathcal{N}(\mathbf{x}'_\alpha) \int_{\mathbf{R}^3} \partial_t |\psi(\mathbf{x}, t)|^2 \mathbf{x} \, \mathrm{d}^3\mathbf{x} \end{cases} \tag{20}$$

This equation is then coupled to the usual Maxwell equations. Note that (19) is a general wave equation for **P** from which we can derive (14), neglecting the backward propagation effect and the right-hand side in (19). The general model now reads

$$\begin{aligned} &\partial_t \mathbf{P} + v_g \partial_{z'} \mathbf{P} + c\sqrt{\frac{\chi^{(3)}}{\chi^{(1)}(1 + 4\pi\chi^{(1)})}} \partial_{z'} \left(\mathbf{P}|\mathbf{E}|^2\right) - v_g^2 \triangle_\perp \mathbf{P} \\ &= -\frac{c^2 \chi^{(3)}}{\chi^{(1)}(1 + 4\pi\chi^{(1)})} \left[\triangle_\perp (\mathbf{P}|\mathbf{E}|^2) - \frac{1}{c} \partial_t (\mathbf{P}|\mathbf{E}|^2)\right] - \frac{4\pi\chi^{(1)}}{2\left(1 + 4\pi\chi^{(1)}\right)} \mathbf{J} \end{aligned} \tag{21}$$

Except for the computation of the initial polarization, **P**, requiring TDSE computations, this general evolution equation is fully macroscopic.

In order to estimate the velocity v_g in a given medium and for which n_0, n_2 are approximately known (from $\chi^{(1)}$, $\chi^{(3)}$), we use a standard approach. From $n \approx n_0 + n_2|\mathbf{E}|^2$, where n_0 (resp. n_2) is the linear (resp. second nonlinear) refractive index and n_g, the velocity reads

$$v_g = \frac{c}{n_g(\omega)} = \frac{c}{n(\omega) + \omega(\partial n(\omega)/\partial\omega)} \tag{22}$$

In a first approximation, we can take $v_g \approx c/\sqrt{1 + 4\pi\chi^{(1)}}$, where $\chi^{(1)}$ is the instantaneous linear susceptibility.

First Example. To illustrate this model, its efficiency and its limits, we propose a numerical example which consists of the comparison of high order harmonic generation (HHG) spectra obtained from an electric field solution to a nonhomogeneous *one-dimensional* MEs, coupled with Schrödinger equations. The molecules are assumed to be aligned, and we compare the total electric field frequency spectra (using FFT on **E**), when the response (polarization) of the medium is computed from, respectively, 1024, 256, 64, 16, 4, and 2 TDSEs. The physical data are as follows: $\mathcal{N}_0 = 1.2 \times 10^{-5}\,\mathrm{mol}/a_0^3$ with $a_0 \approx 0.053\,\mathrm{nm}$. The number of cycles is ≈ 7, and the incoming pulse wavelength is 800 nm. The total propagation length is $\approx 30\,\mu\mathrm{m}$, including $\approx 10\,\mu\mathrm{m}$ in the gas. As expected the HHG spectra are quite close, even when the polarization is computed from 4 TDSEs versus 1024, see Fig. 3(left). In Fig. 3(right), we represent as logscale, the L^2-norm of the error with respect to the spectrum of reference (computed from 1024 TDSEs), as well as the estimation of the CPU/storage gain. We can therefore estimate that dividing by a factor N the number of TDSEs to solve (compared to a full Maxwell–Schrödinger model) allows to reduce by a factor N the overall computational complexity and data storage.

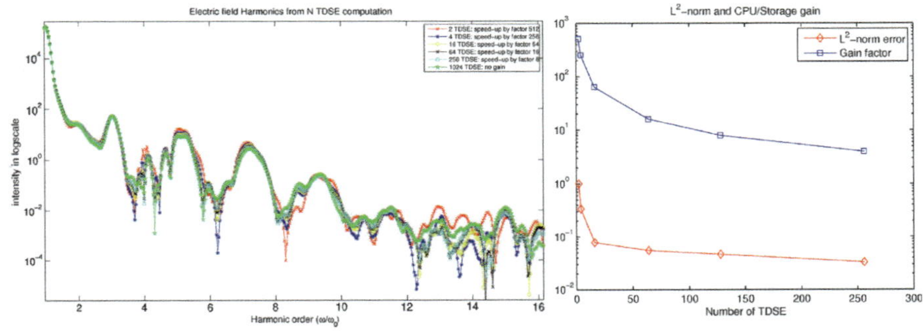

Fig. 3 (*left*) Electric field harmonic order spectrum (harmonic order N) comparison for 4, 16, 64, 256, and 1024 TDSEs; (*right*) CPU/storage gain and L^2-norm error, as function of TDSEs ($I = 10^{14}\,\mathrm{W/cm^2}$, $\lambda \approx 800\,\mathrm{nm}$, $\mathcal{N}_0 = 1.2 \times 10^{-5}\,\mathrm{mol}\,/a_0^3$)

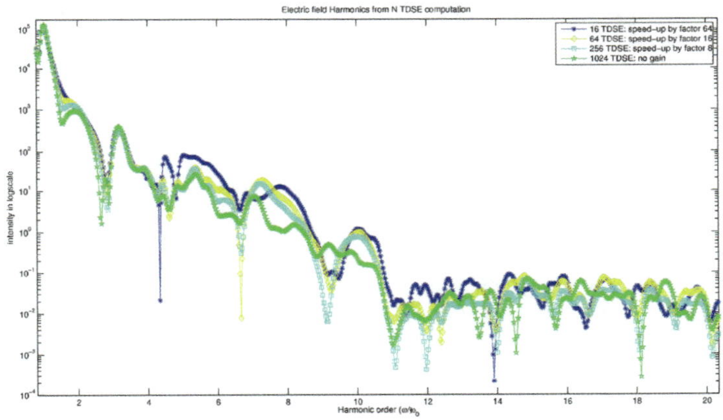

Fig. 4 Electric field harmonic spectrum order comparison, for 4, 16, 64, 256, and 1024 TDSEs ($I = 10^{14}\,\mathrm{W/cm^2}$, $\lambda \approx 800\,\mathrm{nm}$, $\mathcal{N}_0 = 5.2 \times 10^{-5}\,\mathrm{mol}\,/a_0^3$)

We now extend the same example (same data) to a medium 4.5 times denser than in the previous example. In that case, we expect that nonlinearities will deteriorate the numerical solution, see Fig. 4 where are reported harmonics up to order N with $E = N\hbar\omega_0$. A good representation for the medium requires now more TDSEs, which increases the overall complexity of the simulation. More advanced models should then derived to include nonlinearities.

Second Example. In this second example, we consider circularly polarized pulse propagation for gas models of aligned two-dimensional molecules. The incoming pulse is solution to Maxwell–Schrödinger equations,

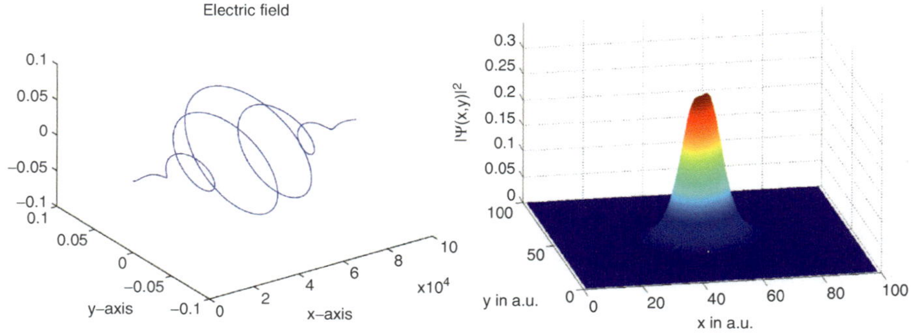

Fig. 5 (*left*) Circularly polarized laser pulse; (*right*) initial electronic wavefunction of the two-center H_2^+ molecule

$$\begin{cases} \partial_t E_{x'}(z',t) = -c\partial_{z'}B_{y'}(z',t) \\ \partial_t E_{y'}(z',t) = c\partial_{z'}B_{x'}(z',t) \\ \partial_t B_{x'}(z',t) = c\partial_{z'}E_{y'}(z',t) \\ \partial_t B_{y'}(z',t) = -c\partial_{z'}E_{x'}(z',t) \end{cases} \tag{23}$$

for which a solution to the (Maxwell) equations is

$$\begin{cases} E_{x'}(z',t) = E_0 f(kz' - \omega_0 t)\sin(kz' - \omega_0 t) \\ E_{y'}(z',t) = E_0 f(kz' - \omega_0 t)\cos(kz' - \omega_0 t) \\ B_{x'}(z',t) = -B_0 f(kz' - \omega_0 t)\cos(kz' - \omega_0 t) \\ B_{y'}(z',t) = B_0 f(kz' - \omega_0 t)\sin(kz' - \omega_0 t) \end{cases} \tag{24}$$

where f is the envelope of the initial electromagnetic field, and $k = \omega_0/c$. In Fig. 5(left), we illustrate an example of a pulse we will consider, where we use atomic units: $E_0(\text{a. u.}) = 5 \times 10^9$ V/cm corresponding to $I = cE_0^2/8\pi = 3.5 \times 10^{16}$ W/cm^2, $T(\text{a. u.}) = 24 \times 10^{-18}$ s $= 24$ asec, $a_0 = 0.0529$ nm, $e = \hbar = m_e = 1$. A gas of 2-D H_2^+ molecules is considered, with fixed internuclear equilibrium distance $R = 1.25$ a. u., Fig. 5 (right). The corresponding two-dimensional TDSEs are solved using a standard Crank-Nicolson scheme [31]. Up to 128 processors from a 4000 core computer, was used to solve the Maxwell–Schrödinger model. H_2^+ molecules are assumed to be aligned, and we compare the electric field frequency spectra, when the response (polarization) of the medium is computed from, respectively, 1024, 16 and 2 TDSEs. The other physical data are as follows: $\mathcal{N}_0 = 1.2 \times 10^{-5}$ mol/a_0^{-3} with $a_0 \approx 0.053$ nm. The number of cycles is $\approx$5, and the incoming pulse wavelength is 800 nm. The total propagation length is $\approx$30 µm, including $\approx$10 µm in the gas. One expects electronic circular orbits with radius $r_n = 2E/\omega_0^2\sqrt{1 + (n + 1/2)^2\pi^2}$ where n is an integer dependent on the time

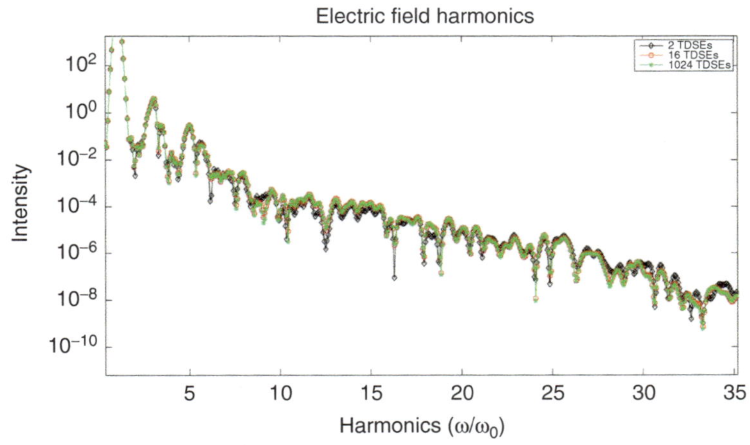

Fig. 6 CPU/storage as function of TDSEs

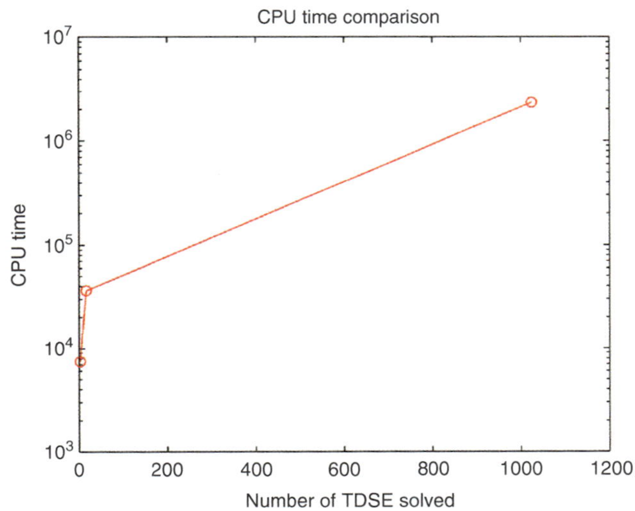

Fig. 7 Electric field harmonic order spectrum comparison for 2, 16, and 1024 TDSEs and H_2^+ density $\approx 1.2 \times 10^{-5}$ a. u.

duration and frequency of the pulse, see [32], for details. Numerical results in Fig. 6 show that at relatively low density, it is possible to reduce considerably the computational complexity of the overall model, keeping the generated electric field very accurate, see Fig. 7.

When the gas density is increased by a factor 4.5, the electric field variation must be included in the TDSE, as it is suggested in Fig. 8, where $\mathcal{N}_0 \approx 5.2 \times 10^{-5}$ a. u. It is then observed that the electric field estimation, from polarization computation with only 2 TDSEs is inaccurate.

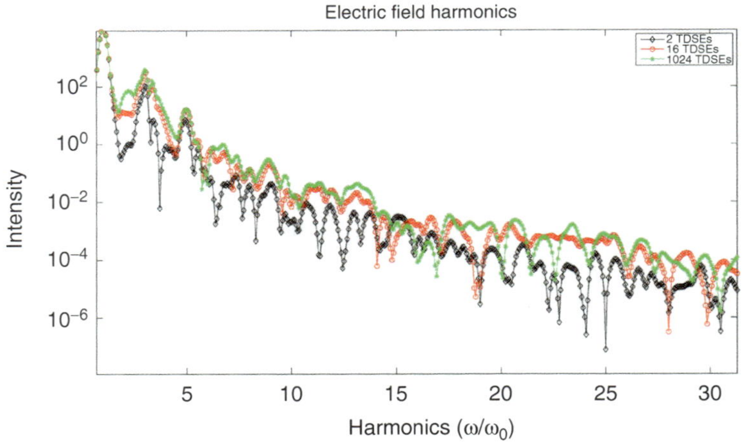

Fig. 8 Electric field harmonic order spectrum comparison for 2, 16, and 1024 TDSEs and H_2^+ density $\approx 5.2 \times 10^{-5}$ a. u.

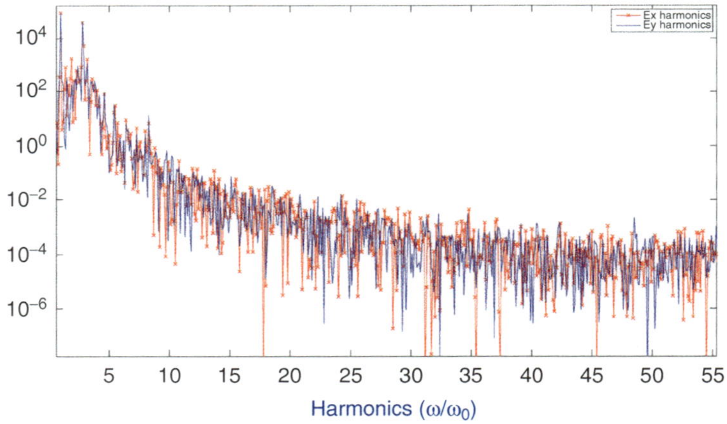

Fig. 9 Electric field harmonic order spectrum comparison a propagation length of $\approx$0.328 mm for H_2^+ molecule density $\approx 10^{-5}$ a. u. Incoming pulse intensity 5×10^{14} W/cm^2 at 800 nm

Third Example. The last example is devoted to the propagation of a circularly polarized laser pulse over a "long" propagation distance. The configuration is similar to the second example. A chain of aligned two-dimensional H_2^+ molecules is subject to a circularly polarized laser pulse of intensity 5×10^{14} W/cm^2. We solve the equivalent of 64×1024 TDSEs using the Maxwell–Schrödinger model coupled

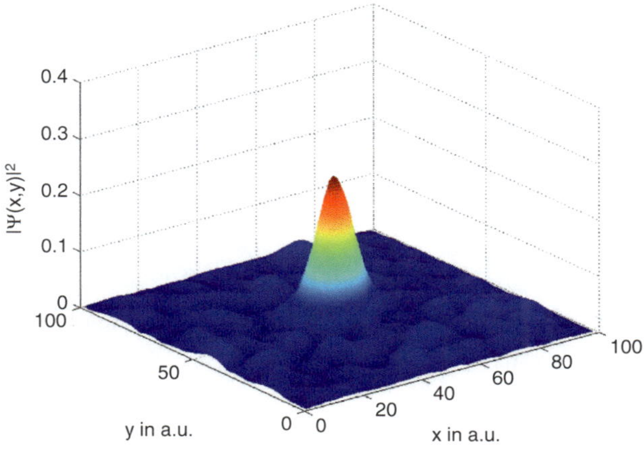

Fig. 10 Wavefunction of molecule subject to the electric field 5×10^{14} W/cm^2 and $\lambda_0 \approx 800$ nm

with the polarization equation (14). The corresponding distance of propagation is approximately $64 \times 1024 \times 100$ atomic units, that is approximately 0.328 mm. We represent the harmonic spectra of the electric field components in Fig. 9. We also represent in Fig. 10, the wavefunction $\left(x, y, |\psi(x, y, t)|^2\right)$ of a molecule which was subject to the electric field defined above.

3 Polarization Evolution Equation for Longer Pulses

The next two subsections are dedicated to the derivation of an evolution equation for the polarization under the paraxial and SVEA. This will allow for realistic macroscopic simulation of relevance to current high intensity laser experiments.

3.1 General Model Under the Paraxial Approximation

The interaction of a laser field with the medium breaks the spatio-temporal symmetry of the incoming pulse. We derive here a model in the paraxial approximation [33]. We write the electric field propagation in direction $\mathbf{e}_{z'}$ as

$$\mathbf{E}(\mathbf{x}'_{\perp}, z', t) = A(\mathbf{x}'_{\perp}, z', t)\mathrm{e}^{\mathrm{i}(k_{z'}z' - \omega t)}\mathbf{e}_{z'} \tag{25}$$

At time t_a and in $(\mathbf{x}'_{\perp,1}, z'_1)$, the initial polarization $\mathbf{P}(\mathbf{x}'_{\perp,1}, t_a)$ and its derivative $\partial_t \mathbf{P}(\mathbf{x}'_{\perp,1}, t_a)$ are computed from the TDSE

$$\mathrm{i}\partial_t \psi_1 = -\frac{1}{2}\triangle \psi_1 + V_c(\mathbf{x})\psi_1 + \mathbf{x}\cdot\mathbf{E}(t)\psi_1 \tag{26}$$

with initial data $\psi_1(t = t_a) = \phi$. Next, we search for an evolution equation for $\mathbf{P}$ starting from (19) but under the paraxial approximation. So that for $\mathbf{P}$ and $\mathbf{J}$ are in the form

$$\mathbf{P}(\mathbf{x}'_\perp, z', t) = \Pi(\mathbf{x}'_\perp, z', t)\mathrm{e}^{\mathrm{i}(k_{z'}z'-\omega t)}\mathbf{e}_{z'} \tag{27}$$

$$\mathbf{J}(\mathbf{x}'_\perp, z', t) = \Lambda(\mathbf{x}'_\perp, z', t)\mathrm{e}^{\mathrm{i}(k_{z'}z'-\omega t)}\mathbf{e}_{z'} \tag{28}$$

where A, Π, Λ are slowly varying complex amplitudes. We rewrite the ME coupled with (19) under the SVEA:

$$\begin{cases} \triangle\mathbf{E} = (\triangle_\perp A + \partial^2_{z'}A + 2\mathrm{i}k_{z'}\partial_{z'}A - k^2_{z'}A)\mathrm{e}^{\mathrm{i}(k_{z'}z'-\omega t)}\mathbf{e}_{z'}, \\ \partial^2_t\mathbf{E} = (\partial^2_t A - 2\mathrm{i}\omega\partial_t A - \omega^2 A)\mathrm{e}^{\mathrm{i}(k_{z'}z'-\omega t)}\mathbf{e}_{z'} \end{cases} \tag{29}$$

From the current $\mathbf{J}$ relation

$$\partial_t\mathbf{J} = (\partial_t\Lambda - \mathrm{i}\omega\Lambda)\mathrm{e}^{\mathrm{i}(k_{z'}z'-\omega t)}\mathbf{e}_{z'} \tag{30}$$

the continuity equation can be rewritten

$$\partial_t\Lambda = (\mathrm{i}\omega - \nu_e)\Lambda + \rho A \tag{31}$$

Generating thus a new equation

$$\partial_t\mathbf{J} = (-\nu_e\Lambda + \rho A)\mathrm{e}^{\mathrm{i}(k_{z'}z'-\omega t)}\mathbf{e}_{z'}$$

We now rewrite the nonlinearity in the right-hand side of (19), as follows:

$$\begin{aligned} (\partial^2_t - c^2\triangle)(\mathbf{P}|\mathbf{E}|^2) = \big(&\partial^2_t(\mathrm{e}^{\mathrm{i}(k_{z'}z'-\omega t)}\Pi|A|^2) - c^2\triangle_\perp(\Pi|A|^2)\mathrm{e}^{\mathrm{i}(k_{z'}z'-\omega t)} \\ &- c^2\partial^2_{z'}(\mathrm{e}^{\mathrm{i}(k_{z'}z'-\omega t)}\Pi|A|^2)\big)\mathbf{e}_{z'}, \end{aligned} \tag{32}$$

resulting

$$\begin{cases} \triangle_\perp(\Pi|A|^2) = |A|^2\triangle_\perp\Pi + \Pi\triangle_\perp|A|^2 + 2\nabla_\perp|A|^2\cdot\nabla_\perp\Pi \\ \partial^2_{z'}(\mathrm{e}^{\mathrm{i}(k_{z'}z'-\omega t)}\Pi|A|^2) \\ \quad = \mathrm{e}^{\mathrm{i}(k_{z'}z'-\omega t)}\big(-k^2\Pi|A|^2 + 2ik|A|^2\partial_{z'}\Pi + 2ik\Pi\partial_{z'}|A|^2 \\ \qquad +2(\partial_{z'}|A|^2)(\partial_{z'}\Pi) + \Pi\partial^2_{z'}|A|^2 + |A|^2\partial^2_{z'}\Pi\big) \\ \partial^2_t(\mathrm{e}^{\mathrm{i}(k_{z'}z'-\omega t)}\Pi|A|^2) \\ \quad = \mathrm{e}^{\mathrm{i}(k_{z'}z'-\omega t)}\big(-\omega^2\Pi|A|^2 - 2\mathrm{i}\omega|A|^2\partial_t\Pi - 2\mathrm{i}\omega\Pi\partial_t|A|^2 \\ \qquad +2(\partial_t|A|^2)(\partial_t\Pi) + \Pi\partial^2_t|A|^2 + |A|^2\partial^2_t\Pi\big) \end{cases} \tag{33}$$

We now apply the SVEA

$$\begin{cases} |\partial_{z'}^2 A| \ll k_{z'}|\partial_{z'} A| \ll k_{z'}^2|A| \\ |\partial_{z'}^2 \Pi| \ll k_{z'}|\partial_{z'} \Pi| \ll k_{z'}^2|\Pi| \\ |\partial_t^2 A| \ll \omega|\partial_t A| \ll \omega^2|A| \\ |\partial_t^2 \Pi| \ll \omega|\partial_t \Pi| \ll \omega^2|\Pi| \end{cases} \tag{34}$$

Then, using the dispersion relation $\omega = ck_{z'}$ we get a complete model.

$$\begin{cases} \partial_t A + c\partial_{z'} A = \dfrac{\mathrm{i}c}{2k_{z'}}\Delta_\perp A + 4\pi\Big(-\partial_t \Pi + \dfrac{\mathrm{i}\omega}{2}\Pi + \dfrac{\mathrm{i}\nu_e}{2\omega}\Lambda - \dfrac{\mathrm{i}\rho}{2\omega}A\Big) \\ \partial_t \Pi + c\partial_{z'}\Pi \\ \quad = \dfrac{\mathrm{i}c}{2k_{z'}}\Delta_\perp \Pi + 4\pi\chi^{(1)}\Big(-\partial_t \Pi + \dfrac{\mathrm{i}\omega}{2}\Pi + \dfrac{\mathrm{i}\nu_e}{2\omega}\Lambda - \dfrac{\mathrm{i}\rho}{2\omega}A\Big) \\ \quad + \dfrac{\chi^{(3)}}{\chi^{(1)}}\Big(\partial_t(\Pi|A|^2) + c\partial_{z'}(\Pi|A|^2) - \dfrac{\mathrm{i}c}{2k_{z'}}\Delta_\perp(\Pi|A|^2)\Big) \\ \partial_t \Lambda = (\mathrm{i}\omega - \nu_e)\Lambda + \rho A \end{cases} \tag{35}$$

A more simple version, where cubic effects E^3 are neglected, i.e. using the linear relation $\mathbf{P} \approx \chi^{(1)}\mathbf{E}$ reduces to

$$\begin{cases} \partial_t A + \dfrac{c}{1+4\pi\chi^{(1)}}\partial_{z'} A \\ \quad = \dfrac{1}{1+4\pi\chi^{(1)}}\Big(\dfrac{\mathrm{i}c}{2k_{z'}}\Delta_\perp A + 2\mathrm{i}\pi(\omega\chi^{(1)} - \dfrac{\rho}{\omega})A + \dfrac{2\mathrm{i}\pi\nu_e}{\omega}\Lambda\Big) \\ \partial_t \Lambda = (\mathrm{i}\omega - \nu_e)\Lambda + \rho A \end{cases} \tag{36}$$

To complete the model, we need the additional equation for ρ:

$$\partial_t \rho + v_g \partial_z \rho = 0 \tag{37}$$

The interest of this model is that it gives an accurate description of the polarization envelope (as compared to the electric field envelope), and therefore allows for long pulse propagations. Again, the evolution equation for Π is used, only in a localized spatial region, and from initial data $\mathbf{P}(\cdot, t_a)$, $\partial_t \mathbf{P}(\cdot, t_a)$, computed from TDSEs following the same technique as the one described in Sect. 2, except that we can now consider much larger propagation distances. From (35) it is possible to derive furthermore simple models neglecting certain terms on the right-hand side. This will be the subject of a forthcoming paper on the accuracy of such new models.

4 Conclusion

In this paper we have derived nonperturbative nonlinear optics models which include bound-bound, bound-continuum, and to a certain extent continuum-continuum transitions occurring in ionization. This was possible through TDSE computations. Although the present, fully nonperturbative model, still contains microscopic TDSE computations, compared to the micro–macro Maxwell–Schrödinger model, it is a great improvement in terms of computational complexity and efficiency.

In order to validate and to analyze further this model, current simulations are in progress to test it in realistic situations of intense laser–matter interactions. The inclusion of nonlinearities coming from bound states as well as continuum free electrons is now accepted to be essential in order to describe accurately filament dynamics [4, 5].

Acknowledgements The authors thank the CRM (Centre de recherches mathématiques, Montréal), the KITP (Kavli Institute of Theoretical Physics, University of California, Santa Barbara) for supporting this research via workshops and Compute Canada for access to high performance parallel computers.

References

1. E. Lorin, M. Lytova, A. Memarian, A.D. Bandrauk, J. Phys. A **48**(10), 105201 (2015)
2. T. Brabec, F. Krausz, Rev. Mod. Phys. **72**(2), 545 (2000)
3. A.V. Husakou, J. Herrmann, Phys. Rev. Lett. **87**(20), 203901 (2001)
4. A. Couairon, A. Mysyrowicz, Phys. Rep. **441**(2–4), 47 (2007)
5. L. Bergé, S. Skupin, R. Nuter, J. Kasparian, J.P. Wolf, Rep. Prog. Phys. **70**(10), 1633 (2007)
6. M. Kolesik, J.V. Moloney, Rep. Prog. Phys. **77**(1), 016401 (2014)
7. I. Babushkin, L. Bergé, J. Math. Phys. **55**(3), 032903 (2014)
8. M. Kolesik, J.V. Moloney, Phys. Rev. E **70**(3), 036604 (2004)
9. S. Skupin, L. Bergé, Physica D **220**(1), 14 (2006)
10. S. Champeaux, L. Bergé, D. Gordon, A. Ting, J. Peñano, P. Sprangle, Phys. Rev. E **77**(3), 036406 (2008)
11. A. Vinçotte, L. Bergé, Physica D **223**(2), 163 (2006)
12. M. Kolesik, E.M. Wright, J.V. Moloney, Opt. Lett. **35**(15), 2550 (2010)
13. L. Bergé, C. Gouédard, J. Schjodt-Eriksen, H. Ward, Physica D **176**(3–4), 181 (2003)
14. P. Béjot, J. Kasparian, S. Henin, V. Loriot, T. Vieillard, E. Hertz, O. Faucher, B. Lavorel, J.P. Wolf, Phys. Rev. Lett. **104**(10), 103903 (2010)
15. A. Spott, A. Jaroń-Becker, A. Becker, Phys. Rev. A **90**(1), 013426 (2014)
16. M. Richter, S. Patchkovskii, F. Morales, O. Smirnova, M. Ivanov, New J. Phys. **15**(8), 083012 (2013)
17. C. Köhler, R. Guichard, E. Lorin, S. Chelkowski, A.D. Bandrauk, L. Bergé, L., S. Skupin, Phys. Rev. A **87**(4), 043811 (2013)
18. P. Béjot, E. Cormier, E. Hertz, B. Lavorel, J. Kasparian, J.P. Wolf, O. Faucher, Phys. Rev. Lett. **110**(4), 043902 (2013)
19. P. Polynkin, M. Kolesik, E.M. Wright, J.V. Moloney, Phys. Rev. Lett. **106**(15), 153902 (2011)
20. T. Morishita, A.T. Le, Z. Chen, C.D. Lin, Phys. Rev. Lett. **100**(1), 013903 (2008)

21. E. Lorin, S. Chelkowski, A. Bandrauk, Comput. Phys. Commun. **177**(12), 908 (2007)
22. E. Lorin, S. Chelkowski, A. Bandrauk, Commun. Comput. Phys. **9**(2), 406 (2011)
23. E. Lorin, S. Chelkowski, A. Bandrauk, in *High-Dimensional Partial Differential Equations in Science and Engineering*, vol. 41, ed. by A. Bandrauk, M.C. Delfour, C. Le Bris. CRM Proc. Lecture Notes (American Mathematical Society, Providence, 2007), pp. 161–182
24. E. Lorin, S. Chelkowski, E. Zaoui, A. Bandrauk, Physica D **241**(12), 1059 (2012)
25. P.B. Corkum, Phys. Rev. Lett. **71**(13), 1994 (1993)
26. A.D. Bandrauk, S. Chelkowski, S. Goudreau, J. Mod. Opt. **52**(2–3), 411 (2005)
27. R.W. Boyd, *Nonlinear Optics*, 2nd edn. (Academic, San Diego, 2003)
28. A.D. Bandrauk, F. Fillion-Gourdeau, E. Lorin, J. Phys. B **46**(15), 153001 (2013)
29. E. Lorin, A.D. Bandrauk, J. Comput. Sci. **3**(3), 159 (2012)
30. H.S. Nguyen, A. Suda, K. Midorikawa, Phys. Rev. A **60**(3), 2587 (1999)
31. J. Crank, P. Nicolson, Adv. Comput. Math. **6**, 207 (1996)
32. K.J. Yuan, A.D. Bandrauk, J. Phys. B **45**(7), 074001 (2012)
33. B. Bidégaray-Fesquet, *Hiérarchie de modèles en optique quantique. De Maxwell-Bloch à Schrödinger non-linéaire*. Mathématiques and Applications, vol. 49 (Springer, Berlin, 2006)

Numerical Simulation of Ultra-Short Laser Pulses

Paris Panagiotopoulos, Patrick Townsend Whalen, Miroslav Kolesik, and Jerome V. Moloney

Abstract This chapter contains the numerical models used to describe ultra-short pulse propagation while taking a brief look on their derivation from the Maxwell equations. The main models described here are the unidirectional pulse propagation equation (UPPE) and the modified Kadomtsev–Petviashvili equation (MKP), while a brief look is taken at the main envelope counterpart, the nonlinear envelope equation (NEE) model. The chapter concludes with the presentation of various applications of the UPPE and the MKP in the study of carrier wave shocks in various realistically model media, and a special case of the propagation properties of the two-dimensional XY Airy beam and its radial symmetric version, the ring-Airy beam, in the linear and nonlinear regimes.

1 Numerical Models

In this section the numerical models are described.

In computational nonlinear optics, solving the full spatio-temporal Maxwell model is known to be extremely expensive in computation time and hence is impractical. Therefore, numerical studies of ultra-short pulse propagation are traditionally conducted either by using envelope type models like the nonlinear

P. Panagiotopoulos (✉) • M. Kolesik
College of Optical Sciences, University of Arizona, Tucson, AZ 85721-0094, USA

Arizona Center for Mathematical Sciences, University of Arizona, Tucson, AZ 85721-0094, USA
e-mail: paris.panagiotopoulos@gmail.com; kolesik@optics.arizona.edu

P.T. Whalen
College of Optical Sciences, University of Arizona, Tucson, AZ 85721-0094, USA

Department of Mathematics, University of Arizona, Tucson, AZ 85721-0094, USA
e-mail: pwhalen@email.arizona.edu

J.V. Moloney
College of Optical Sciences, University of Arizona, Tucson, AZ 85721-0094, USA

Arizona Center for Mathematical Sciences, University of Arizona, Tucson, AZ 85721-0094, USA

Department of Mathematics, University of Arizona, Tucson, AZ 85721-0094, USA
e-mail: jml@acms.arizona.edu

A.D. Bandrauk et al. (eds.), *Laser Filamentation*, CRM Series in Mathematical Physics, DOI 10.1007/978-3-319-23084-9_8

Schrödinger equation [1] or full electric field propagators such as the unidirectional pulse propagation equation (UPPE) [2]. In Sect. 1.1 we will discuss the most general unidirectional field propagation method, the UPPE system. In Sect. 1.2, we derive the generalized modified Kadomtsev–Petviashvili (GMKP) equation from the UPPE model, and analyze the relevant length scales of the MKP equation.

1.1 Derivation of the UPPE from the Maxwell Equations

In this section the derivation of the UPPE from Maxwell is shown.

In fully spectral solvers, the native representation of the optical field is stored in arrays representing spectral amplitudes that are functions of angular frequency ω, and transverse wavenumbers k_x, k_y. For example, $\mathbf{E}_{k_x,k_y,+}(\omega, z)$ would stand for the spectral amplitude in the plane wave expansion of the field at a given propagation distance z.

The spectral nature of the solver makes it possible to solve the linear propagation problem exactly. For a given medium permittivity $\epsilon(\omega)$, the linear propagation properties are encoded in the propagation constants of the modal fields. In the case of a homogeneous medium these are the well-known plane waves, and the propagation constant depends on both the angular frequency and on the transverse wavenumbers:

$$k_z(\omega, k_x, k_y) \equiv \sqrt{\omega^2\epsilon(\omega)/c^2 - k_x^2 - k_y^2}\ . \tag{1}$$

The permittivity is in general complex-valued, so that not only the chromatic dispersive properties, but also the absorption can be accounted for exactly. This is a clear advantage of the UPPE model to the GMKP as we will see later on.

The medium properties, other than linear ones, enter the propagation model through the constitutive relations in Maxwell equations. Thus, the polarization $\mathbf{P}(x, y, z, t)$ and the current density $\mathbf{J}(x, y, z, t)$ must be given as functionals of the electric field $\mathbf{E}(x, y, z, t)$. For the sake of brevity, we will not show the dependence of either $\mathbf{P}$ or $\mathbf{J}$ on the electric field, because from the point of view of the propagation solver, the nature of this dependence is irrelevant. Let us just note one important aspect and that is that the medium response is calculated from the history of the electric field $\mathbf{E}(x, y, z, t)$ independently at a each spatial point.

The following exact system of equations describes the evolution of modal amplitudes along the z-axis for the forward and backward propagating field. It is a coupled pair of UPPE [2]:

$$\begin{aligned}\partial_z\mathbf{E}^{\perp}_{k_x,k_y,+}(\omega, z) &= +\mathrm{i}k_z\mathbf{E}^{\perp}_{k_x,k_y,+}(\omega, z)\\ &\quad+\sum_{s=1,2}\mathbf{e}_s^{\perp}\mathbf{e}_s\cdot\left[\frac{\mathrm{i}\omega^2}{2\epsilon_0 c^2 k_z}\mathbf{P}_{k_x,k_y}(\omega, z) - \frac{\omega}{2\epsilon_0 c^2 k_z}\mathbf{J}_{k_x,k_y}(\omega, z)\right]\end{aligned} \tag{2}$$

which describes the forward propagating wave, and

$$\partial_z \mathbf{E}^{\perp}_{k_x,k_y,-}(\omega,z) = -\mathrm{i}k_z \mathbf{E}^{\perp}_{k_x,k_y,-}(\omega,z) - \sum_{s=1,2} \mathbf{e}^{\perp}_s \mathbf{e}_s \cdot \left[\frac{\mathrm{i}\omega^2}{2\epsilon_0 c^2 k_z} \mathbf{P}_{k_x,k_y}(\omega,z) - \frac{\omega}{2\epsilon_0 c^2 k_z} \mathbf{J}_{k_x,k_y}(\omega,z) \right] \tag{3}$$

which represents the backward propagating radiation.

The above UPPEs are mutually coupled through the medium response represented by the polarization and current density terms. As noted above, they require the knowledge of the full (i.e., forward *and* backward) electric field. This is where the unidirectional approximation must be adopted. It requires that the medium response can be accurately calculated from only the forward portion of the field, i.e.:

$$\mathbf{P}(\mathbf{e}_- + \mathbf{e}_+), \mathbf{J}(\mathbf{e}_- + \mathbf{e}_+) \rightarrow \mathbf{P}(\mathbf{e}_+), \mathbf{J}(\mathbf{e}_+) \tag{4}$$

If this is the case, then it is sufficient to solve only the UPPE representing the forward wave. Obviously, in practice *all* pulse propagation simulations, independently of the propagation model they utilize, require this approximation. The physical meaning of "unidirectionality" was discussed by Kinsler in [3–6] and a practical way to quantify and test its limitations is described in [7]. The canonical form of the scalar UPPE is as follows (for brevity, we consider the current term grouped with the nonlinear polarization and omit the component and direction subscripts):

$$\partial_z E_{k_x,k_y}(\omega,z) = \mathrm{i}K(k_x,k_y,\omega)E_{k_x,k_y}(\omega,z) + \mathrm{i}Q(k_x,k_y,\omega)P_{k_x,k_y}(\omega,z) \tag{5}$$

with the exact linear propagator

$$K(k_x,k_y,\omega) = \sqrt{\omega^2\epsilon(\omega)/c^2 - k_x^2 - k_y^2} \tag{6}$$

and the nonlinear response coupling

$$Q(k_x,k_y,\omega) \equiv \frac{\omega^2}{2\epsilon_0 c^2 \sqrt{\omega^2\epsilon(\omega)/c^2 - k_x^2 - k_y^2}} \,. \tag{7}$$

This is in fact the form most often utilized in practical simulations.

1.2 Derivation of the GMKP Equation from the UPPE

In this section the derivation of MKP from UPPE is shown.

While the canonical UPPE predicts carrier field shock for long-wavelength high power lasers, this effect and associated scales are hidden in the general form of the model itself. To reveal the physics in a more explicit way, we make several approximations to Eq. (5) that are valid for a large class of filamentation problems yet reduce the model to a simpler and easier to analyze form.

First, we apply the following Taylor expansions to $K(k_x, k_y, \omega)$ and $Q(k_x, k_y, \omega)$

$$k_z(\omega, k_x, k_y) = k(\omega)\left(1 - \frac{k_\perp^2}{2k^2(\omega)} - \frac{k_\perp^4}{8k^4(\omega)} - \cdots\right)$$

$$\frac{1}{k_z(\omega, k_x, k_y)} = \frac{1}{k(\omega)}\left(1 + \frac{k_\perp^2}{2k^2(\omega)} + \cdots\right),$$

with $k_\perp^2 = k_x^2 + k_y^2$, and $k(\omega) = \omega\sqrt{\epsilon(\omega)}/c$. Then our expanded UPPE is

$$\begin{aligned}\partial_z E_{k_x,k_y}(\omega, z) &= \mathrm{i}k(\omega)E_{k_x,k_y}(\omega, z) - \frac{\mathrm{i}ck_\perp^2}{2n(\omega)\omega}E_{k_x,k_y}(\omega, z)\\ &+ \frac{\mathrm{i}\omega}{2\epsilon_0 cn(\omega)}P_{k_x,k_y}(\omega, z) - \frac{\mathrm{i}ck_\perp^2}{8n(\omega)\omega}\frac{k_\perp^2}{k^2(\omega)}E_{k_x,k_y}(\omega, z)\\ &+ \frac{\mathrm{i}\omega}{4\epsilon_0 cn(\omega)}\frac{k_\perp^2}{k^2(\omega)}P_{k_x,k_y}(\omega, z) + \text{h.o.t.}\end{aligned} \quad (8)$$

where h.o.t. denotes the higher-order terms in the respective expansions of k_z and k_z^{-1}. Restricting ourselves to paraxial beam propagation implies that only the first three right-hand-side terms must be retained. Furthermore, $n^{-1}(\omega)$ in the remaining terms can be Taylor expanded around a reference frequency ω_R such that

$$\begin{aligned}\partial_z E_{k_x,k_y}(\omega, z) &= \mathrm{i}k(\omega)E_{k_x,k_y}(\omega, z)\\ &- \frac{\mathrm{i}ck_\perp^2}{2n(\omega_R)\omega}\left[1 - \frac{n'(\omega_R)(\omega - \omega_R)}{n(\omega_R)} + \cdots\right]E_{k_x,k_y}(\omega, z)\\ &+ \frac{\mathrm{i}\omega}{2\epsilon_0 cn(\omega_R)}\left[1 - \frac{n'(\omega_R)(\omega - \omega_R)}{n(\omega_R)} + \cdots\right]P_{k_x,k_y}(\omega, z)\,.\end{aligned} \quad (9)$$

Now we invoke a restriction on the medium and spectral bandwidth of pulses admitted to our model

$$|n'(\omega_R)(\omega - \omega_R)| \ll n(\omega_R)\,.$$

This is akin to the slowly-evolving-wave approximation (SEWA)

$$|n'(\omega_R)\omega_R| \ll n(\omega_R) ,$$

which is used in envelope models to accurately describe light propagation down to the single-cycle regime [8]. Our approximation is somewhat more demanding as it puts additional restrictions on the medium when propagating high harmonic generating pulses; as the harmonic order increases, ω will deviate more from ω_R and hence $n'(\omega_R)$ will need to be small enough to account for this. After approximations, the resulting propagation equation is

$$\partial_z E_{k_x,k_y}(\omega, z) = \mathrm{i}k(\omega)E_{k_x,k_y}(\omega, z) - \frac{\mathrm{i}ck_\perp^2}{2n(\omega_R)\omega}E_{k_x,k_y}(\omega, z) + \frac{\mathrm{i}\omega}{2\epsilon_0 cn(\omega_R)}P_{k_x,k_y}(\omega, z) . \quad (10)$$

The following dispersion relation approximation is utilized

$$k(\omega) = a\omega^3 - \frac{b}{\omega} + q\omega . \quad (11)$$

While no explicit ω_R is present in the dispersion relation, Eq. (11) is a local approximation as the parameters a, b, and q are chosen to accurately reflect the full dispersion profile (e.g., Sellmeier, etc.) near a specific frequency ω_R. Each example in this work considers a single laser pulse launched with a carrier frequency ω_0 corresponding to a wavelength λ_0. Therefore, we take the reference frequency as $\omega_R = \omega_0$ and let $n_0 = n(\omega_R)$. Upon substitution into Eq. (10) and transforming to time-domain real-space, the propagation model simplifies to

$$\partial_\tau\left(\partial_z E - a\partial_\tau^3 E + \frac{1}{2\epsilon_0 cn_0}\partial_\tau P\right) + bE = \frac{c}{2n_0}\Delta_\perp E , \quad (12)$$

where $\tau = t - qz$ and the fields are functions of (x, y, τ, z). At this point it is easy to separate the nonlinear current density from the nonlinear polarization by replacing $\partial_\tau P$ with $\partial_\tau P + J$. Furthermore, we split J into a lossless plasma defocusing term and a nonlinear absorption term

$$J = J_{\mathrm{plas}} + J_{\mathrm{abs}}, \quad (13)$$

$$\partial_\tau J_{\mathrm{plas}} = \frac{e^2}{m_e}\rho E, \quad (14)$$

$$\frac{J_{\mathrm{abs}}}{\epsilon_0 cn_0} = \frac{W(I)U_i}{I}(\rho_{\mathrm{nt}} - \rho)E , \quad (15)$$

$$\partial_\tau \rho = W(I)(\rho_{\mathrm{nt}} - \rho) , \quad (16)$$

where $I = \epsilon_0 c n_0 \langle E^2 \rangle$ is the intensity of electric field, U_i is the potential, ρ_{nt} is the neutral plasma density, and W is the ionization rate of free charge ρ. We also use an instantaneous Kerr nonlinear response such that

$$P = \epsilon_0 \chi^{(3)} E^3 = \tfrac{4}{3}\epsilon_0(\epsilon_0 c n_0^2 n_2) E^3 \, .$$

This produces a propagation model with the electric field in units of V/m,

$$\partial_\tau \left(\partial_z E - a\partial_\tau^3 E + \frac{\chi^{(3)}}{2cn_0}\partial_\tau E^3 + \frac{1}{2\epsilon_0 c n_0} J_{\mathrm{abs}} \right) + \left(b + \frac{e^2}{2\epsilon_0 n_0 c m_e}\rho \right) E = \frac{c}{2n_0}\Delta_\perp E \, . \tag{17}$$

Rescaling E by

$$\tilde{E} = \left(\frac{\epsilon_0 c n_0}{2} \right)^{1/2} E \, ,$$

and removing tilde's after substitution, we obtain the GMKP optical filamentation equation

$$\partial_\tau \left(\partial_z E - a\partial_\tau^3 E + \frac{4n_2}{3c}\partial_\tau E^3 + \mathcal{N}_{\mathrm{abs}}(\rho, I) E \right) + \big(b + b_{\mathrm{plas}}(\rho) \big) E = \frac{c}{2n_0}\Delta_\perp E, \tag{18}$$

with

$$b_{\mathrm{plas}}(\rho) = \frac{e^2}{2\epsilon_0 n_0 c m_e}\rho, \tag{19}$$

$$\mathcal{N}_{\mathrm{abs}}(\rho, I) = \frac{W(I) U_i (\rho_{\mathrm{nt}} - \rho)}{2I} \, . \tag{20}$$

At long-wavelengths, a full Keldysh ionization response $W(I)$ is preferable to the multiphoton approximation [9]. These ionization rates are computed from formulas derived in [10].

The MKP equation is generated by dropping the nonlinear current density terms and ignoring anomalous dispersion ($b = 0$). With these simplifications, we have

$$\partial_\tau \left(\partial_z E - a\partial_\tau^3 E + \frac{4n_2}{3c}\partial_\tau E^3 \right) = \frac{c}{2n_0}\Delta_\perp E \, . \tag{21}$$

It's instructive to normalize and nondimensionalize this equation with

$$U = \frac{E}{\sqrt{I_0}}, \quad \tilde{\tau} = \omega_0 \tau, \quad \tilde{r} = \frac{r}{w_0}, \quad \tilde{z} = \frac{z}{L},$$

where I_0 is the initial intensity, w_0 is the beam diameter, and L is an arbitrary length scale. After substitutions we obtain

$$\partial_{\tilde{\tau}}\left(\partial_{\tilde{z}}U - \frac{L}{12L_{\mathrm{CSD}}}\partial_{\tilde{\tau}}^3 U + \frac{4L}{3L_{\mathrm{NL}}}\partial_{\tilde{\tau}} U^3\right) = \frac{L}{4L_{\mathrm{DF}}}\Delta_{\tilde{\perp}} U, \tag{22}$$

with length scales

$$L_{\mathrm{DF}} = \frac{k(\omega_0)w_0^2}{2}, \quad L_{\mathrm{NL}} = \frac{c}{\omega_0 n_2 I_0}, \quad L_{\mathrm{CSD}} = \frac{1}{12a\omega_0^3}. \tag{23}$$

The diffractive length scale L_{DF} and nonlinear length scale L_{NL} are standard length scales in nonlinear optics that describe the propagation of an electric field envelope. On the other hand, the carrier shock dispersion length scale L_{CSD} describes the propagation of the electric field, not the envelope; in particular, L_{CSD} measures the distance over which carrier wave steepening dissipates. With the MKP dispersion relation (11), L_{CSD} can alternatively be written as

$$L_{\mathrm{CSD}} = \frac{2}{k(3\omega_0) - 3k(\omega_0)},$$

which clearly relates the carrier shock dispersion to the coherence between the fundamental and the 3rd harmonic. This important observation connects carrier wave shock to harmonic generation and harmonic walk-off.

1.3 The Nonlinear Envelope Equation

The extended nonlinear Schrödinger equation (NLS) in the main model describing the nonlinear propagation of the envelope of the electric field of a laser pulse in transparent nonlinear dispersive media.

The nonlinear propagation of ultra-short laser pulses in transparent materials is modeled using an extended NLS, which describes the evolution of the slowly varying electric field envelope $\mathcal{E}$ along z:

$$K\frac{\partial \mathcal{E}}{\partial z} = \frac{\mathrm{i}}{2}[\nabla_{\perp}^2 \mathcal{E} + D\mathcal{E}] + k_0[T^2 N_{\mathrm{Kerr}}(\mathcal{E}) + T N_{\mathrm{NLL}}(\mathcal{E}) + N_{\mathrm{Plasma}}(\mathcal{E}, \rho)]. \tag{24}$$

Equation (24) takes into account both linear (first term on the right-hand side) and nonlinear effects (second term on the right-hand side) that are acting on the electric field envelope $\mathcal{E}$. $\nabla_\perp^2 = \Delta_\perp = (\partial^2/\partial x^2) + (\partial^2/\partial y^2)$ is the transverse Laplacian operator which is used to model diffraction, while D is the dispersive operator used to model the material chromatic dispersion. The operators $K = k_0 + \mathrm{i}k_0'(\partial/\partial t)$ and $T = 1 + \mathrm{i}\omega_0^{-1}(\partial/\partial t)$ account for space-time focusing and self-steepening of the laser pulse envelope. The nonlinear source terms on the right-hand side of (24) are associated with the optical Kerr effect, nonlinear losses from multi-photon absorption (MPA) and plasma related effects like plasma defocusing and absorption. They are given by:

$$N_{\mathrm{Kerr}}(\mathcal{E}) = \mathrm{i}\frac{\omega_0}{c}n_2\left((1-a)|\mathcal{E}|^2 + a\int_{-\infty}^{t} R(t-\tau)|\mathcal{E}|^2\,\mathrm{d}\tau\right)\mathcal{E}\ ; \tag{25}$$

$$N_{\mathrm{NLL}}(\mathcal{E}) = -\frac{\beta_K}{2}\left[1 - \frac{\rho}{\rho_{\mathrm{nt}}}\right]|\mathcal{E}|^{2K-2}\mathcal{E}\ ; \tag{26}$$

$$N_{\mathrm{Plasma}}(\mathcal{E},\rho) = -\frac{\sigma}{2}(1+\mathrm{i}\omega\tau_c)\rho\mathcal{E}\ . \tag{27}$$

The optical Kerr effect described by (25) consists out of an instantaneous and a delayed component with fraction alpha. The instantaneous part is associated with the electronic response of the medium and is described by the term $\mathrm{i}(\omega_0/c)n_2(1-a)\times|\mathcal{E}|^2\mathcal{E}$, while the delayed component associated with stimulated molecular Raman scattering can be written as $\mathrm{i}(\omega_0/c)n_2 a\int_{-\infty}^{t} R(t-\tau)|\mathcal{E}|^2\,\mathrm{d}\tau\mathcal{E}$. $R(t)$ is the function describing the delayed molecular response:

$$R(t) = R_0 \exp[-\Gamma t]\sin(\omega_R t) \tag{28}$$

where Γ and ω_R are the characteristic time and frequency of the molecular response. Equation (25) can be re-written for instantaneous material response only in the form:

$$N_{\mathrm{Kerr}}(\mathcal{E}) = \mathrm{i}\frac{\omega_0}{c}n_2|\mathcal{E}|^2\mathcal{E} \tag{29}$$

The nonlinear losses term of (26) describes the losses to the field envelope due to MPA of order K. The plasma contribution term described by (28) consists of an imaginary part describing plasma defocusing and a real part describing plasma absorption. Equation (24) is coupled with an electron evolution equation in order to follow the evolution of the generated plasma in time:

$$\frac{\partial\rho}{\partial t} = \sigma_K|E(t)|^{2K}(\rho_{\mathrm{nt}} - \rho), \tag{30}$$

Equation (30) is taking into account MPI (first term on the right-hand side) with rate $W_{\mathrm{MPI}} = \sigma_K I^K$, and avalanche ionization (second term on the right-hand side). The last term of the right-hand side accounts for the electron to ion recombination with

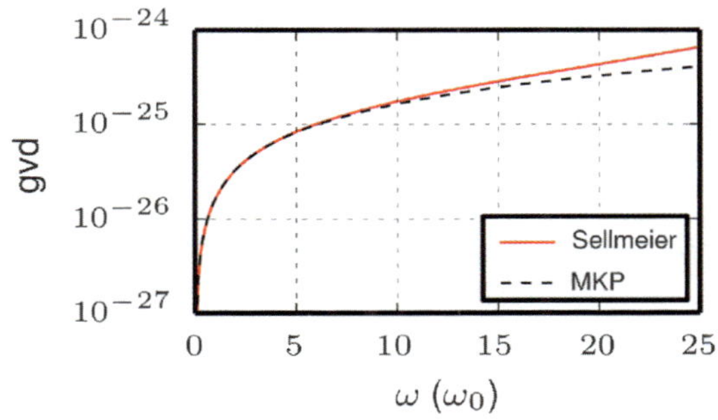

Fig. 1 Group velocity dispersion as modeled by GMKP and UPPE for single-crystal diamond vs. normalized frequency $\omega_0 = 2\pi/(\lambda_0(= 8\,\mu\text{m}))$

coefficient:

$$a_{\text{rec}} = \frac{1}{\tau_{\text{rec}} \times \rho_{\text{breakdown}}(\approx 3 \times 10^{19}\,\text{cm}^{-3})} \tag{31}$$

since recombination time is much longer than the typical femtosecond laser pulse duration, the last term of (30) can safely be neglected in most cases (gases). The nonlinear envelope equation (NEE) as defined here is broadly used for the simulation of nonlinear pulse propagation when electric field related effects can be safely neglected [1].

1.4 UPPE vs. GMKP

The essential difference between the UPPE and GMKP models is in the way dispersion is modeled; the UPPE accurately represents dispersion over a global frequency range using the Sellmeier relation, whereas the MKP dispersion, as shown in (11), is a local approximation whose accuracy depends on the carrier frequency. A comparison of the dispersion curves used in the two models can be seen in Fig. 1. In addition, the UPPE model takes into account vibrational Raman, the full complex χ relation including transmittance, and plasma losses due to a finite electron collision time. As will be shown later on, even though UPPE is a more complex and complete model for nonlinear wave propagation, the physics that govern carrier shock formation can be adequately described by the simpler GMKP model.

2 Numerical Simulations

In this section the UPPE, GMKP, and NLS numerical models will be used to simulate nonlinear pulse propagation in various transparent media.

2.1 Carrier Wave Shock Formation

In this section carrier wave shock formation in realistically modeled dispersive materials will be investigated.

Since both UPPE and GMKP models are full field resolved electromagnetic propagators they can be used to simulated electric field related effects such as steepening and shock formation, which by definition is impossible by NLS type envelope models.

Carrier wave shock formation, although predicted in ideal dispersion-less media in [11], is considered extremely difficult in realistically modeled dispersive media using typical near infrared (near-IR) femtosecond lasers (800 nm chirped pulse amplification Ti:Sapphire systems). This problem was circumvented by the use of mid-infrared (mid-IR) laser pulses in the spectral region of 4–8 μm, where the dispersive landscape of many transparent materials is relatively flat. Materials that fit this criteria are atmospheric air, most noble gases, and some solid materials such as single crystal diamond.

Utilizing the GMKP model one can simulate the nonlinear propagation of an 8 μm–40 fs laser pulse in Xenon gas. Throughout this chapter other materials with similar optical properties, such as other noble gases, air and single crystal diamond, will be utilized to study carrier wave field shocks. Figure 2 shows the electric field of the 8 μm pulse before and after the propagation of 5.63 cm inside Xenon. We can clearly see that the electric field of the carrier wave is undergoing self-steepening, reforming into a smooth "shark-fin" like shape. It is evident that the steepening of the electric field is proportional to its amplitude, since the center cycle is affected the most. This impressive field shock formation is the first to our knowledge predicted for a realistic dispersive medium.

In Fig. 3a we can see the dependence of the carrier shock on the initial wavelength. The temporal axis is scaled by the respective pulse durations (τ_p) so that the electric fields can be compared directly. As the wavelength is increased from 2 to 4 μm, and finally 8 μm, we can see that the optical shock becomes significantly more pronounced.

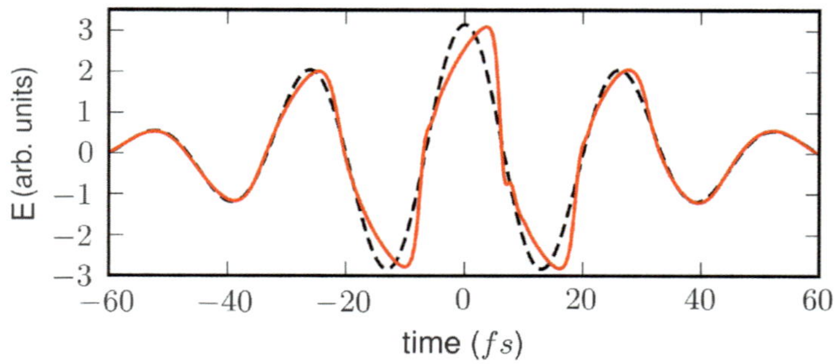

Fig. 2 Electric field of a 8 μm, 40 fs laser pulse (*black dashed line*) undergoing self-steepening after 5.63 cm propagation in Xenon gas (*red continuous line*). Taken from[12]

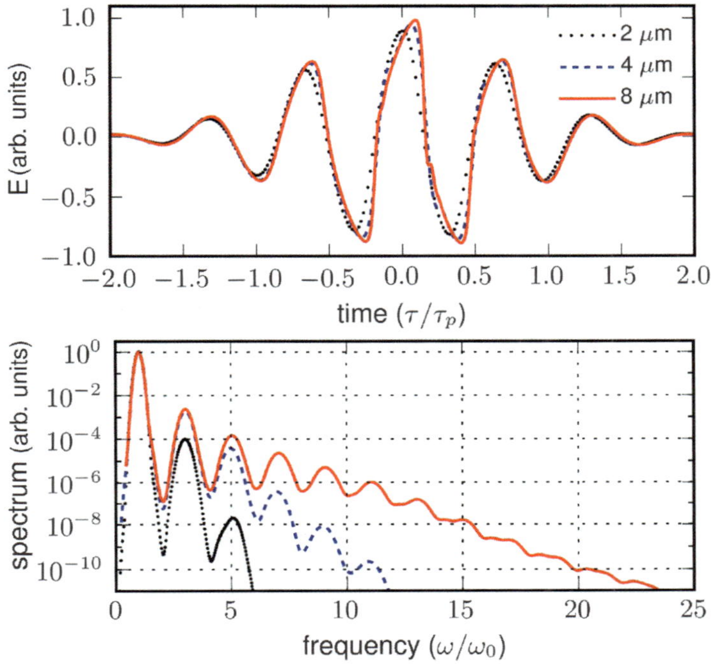

Fig. 3 (**a**) Electric fields (at optimal distance). (**b**) Corresponding spectral intensities of three laser pulses after propagation in Xenon gas. *Black dotted line*: $\lambda_0 = 2\,\mu$m, *blue dashed line*: $\lambda_0 = 4\,\mu$m, *red continuous line*: $\lambda_0 = 8\,\mu$m. Horizontal axes are normalized to pulse duration τ_p and fundamental frequency ω_0, respectively. Taken from [12]

This behavior is reflected in the spectral domain, which is shown in Fig. 3b. As before, the horizontal axis, here frequency, is scaled by the fundamental frequency ω_0 for each pulse, respectively. We can see that as the wavelength increases the number of generated harmonics and their amplitudes increase dramatically. In the case of the 8 μm beam, up to ten odd harmonics are generated.

We identify this higher harmonic generation process as the driving force for carrier wave shock formation. The interaction of the generated higher harmonics with the fundamental carrier wave has the potential to generate a field shock. However, in order for the field shock to manifest itself and be observable, the different harmonics must co-propagate with the fundamental over a minimum distance, which can only be possible in weakly dispersive materials. The increase in field shock at longer wavelengths is therefore expected, since more harmonics reside in the flat dispersion regime and are able to co-propagate with the fundamental over a longer distance.

Note that although the electric field profiles for the 4 and 8 μm beams are almost identical, their respective spectral intensities differ significantly in amplitude after the 3rd harmonic. From this observation we can infer that most of the steepening originates from the interaction between the 3rd harmonic and fundamental, since the additional higher harmonics have only a small impact on the field shape.

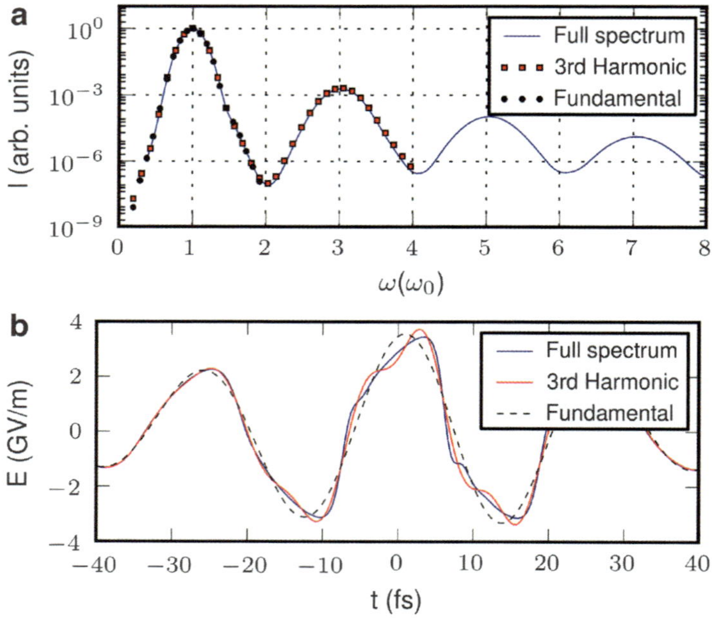

Fig. 4 (**a**) Broadened spectrum of the 8 μm wavepacket after propagation in single-crystal diamond (*cyan continuous line*). *Dashed lines*: artificially truncated spectra, and (**b**) corresponding electric fields. Taken from [13]

Additional simulations showed that when the spectrum is artificially truncated after the 3rd harmonic, most of the steepening seen in the full spectral box case is preserved. To show this important observation we will repeat the numerical experiment in single crystal diamond where the slightly higher dispersion makes ++.

In Fig. 4a we can see the generated higher harmonics of the 8 μm beam in diamond, which is represented by the cyan continuous line. The corresponding field shape is shown in Fig. 4b again with the cyan continuous line. To isolate the effect each of the harmonics has on field shape, we artificially truncated the spectrum up to the 3rd harmonic, shown in Fig. 4a with the red squares. The corresponding field shape, shown in Fig. 4b (red continuous line), is much more jagged when compared to the case where all the harmonics are taken into account. However, the electric field is still steepening, and the shock is still observable. In addition, the field shape is only slightly changing as the number of generated higher harmonics is varied from 1 to the full spectrum. On the other hand, if we keep only the fundamental by truncating all higher harmonics (including the third), the field shape remains virtually unmodified as shown in Fig. 4b (black dashed line). This clearly demonstrates that the actual steepening of the field is mainly coming from the 3rd harmonic, while the rest of the harmonics smooth out the jagged oscillations observed when only the 3rd harmonic is taken into account. The oscillations on the field profile are interpreted as temporal interference fringes between multiple

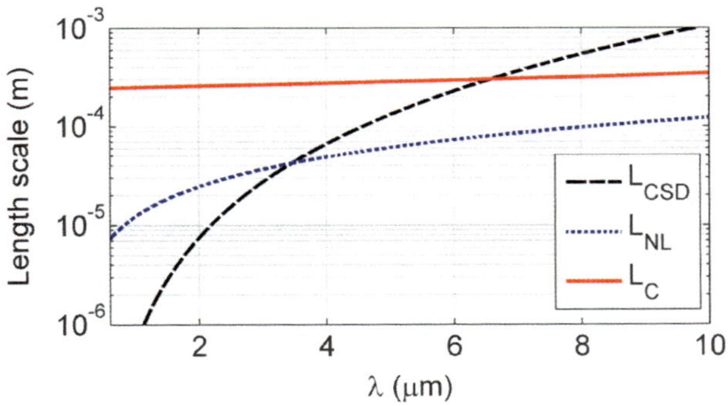

Fig. 5 Characteristic length scales for the nonlinear Kerr effect (*blue dotted line*), carrier shock dispersion (*black dashed line*), and collapse distance (*red continuous line*) in single-crystal diamond as a function of wavelength. The input intensity is 1×10^{13} W/cm^2. Taken from [13]

harmonics. In the case of the full spectrum, the large number of harmonics decrease the period of the fringes, effectively smoothing out the field profile.

It is now possible to directly compare the significance of each physical effect for a given intensity value by their characteristic lengths. In Fig. 5 we can see all three characteristic lengths plotted for a peak intensity of $I_0 = 1 \times 10^{13}$ W/cm^2 (a typical value in the filamentation regime) as a function of wavelength in single-crystal diamond. The propagation of the wave packet is governed by the physical effect with the shortest characteristic length. The general trend is that at shorter wavelengths carrier shock dispersion is strong and the generated harmonics walk-off before field steepening can occur. At longer wavelengths, where dispersion is much weaker, the walk-off of the higher harmonics is slow, and a carrier wave shock is expected to occur.

Note that the carrier shock dispersion length scale is independent of the intensity. On the other hand, L_{NL} depends on intensity, which for practical reasons makes a direct comparison between the two difficult due to the ever shifting peak intensity found in the filamentation regime. For this reason it is helpful to scan over a range of intensity values and produce a surface for each L_{NL}, and L_{CSD}, the latter being invariant along the intensity-axis. Figure 6a shows the overlapping surfaces for L_{NL} (color coded surface) and L_{CSD} (black grid) for single-crystal diamond. Since the shortest length scale dominates, carrier shock is expected to manifest itself in areas where the surface corresponding to L_{NL} is under the surface corresponding to L_{CSD}. To simplify the 3D surfaces shown in Fig. 6a we project them onto the $I - \lambda$ plane. In Fig. 6b we can see that the $I - \lambda$ plane is now separated in two regions, a region where carrier shock is expected to occur (blue) and a region where it is not (black), depending on which of the relevant length scales is shorter. The boundary between the two regions is given by the relation

$$I(\omega) = \frac{c\omega|k''(3\omega) - 3k''(\omega)|}{n_2} \tag{32}$$

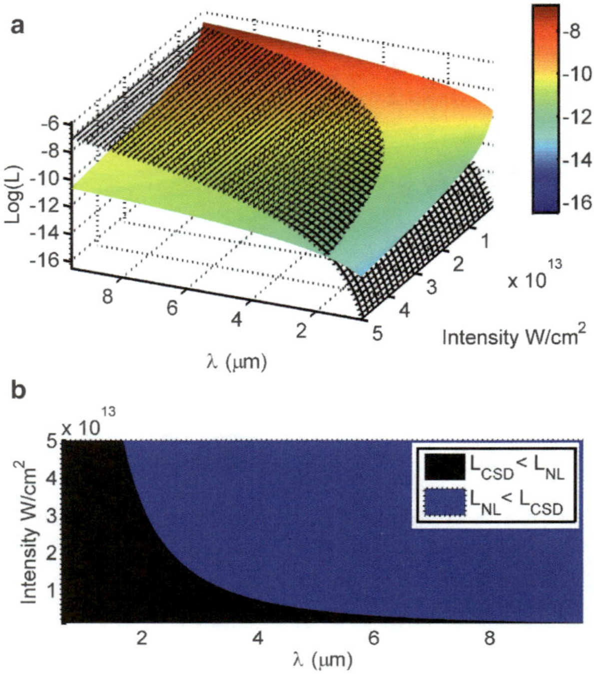

Fig. 6 (**a**) Characteristic length surfaces of single-crystal diamond plotted in 3D. *Black grid*: $L_{\rm CSD}$, color-coded surface: $L_{\rm NL}$. (**b**) Minimum length scale over the $I - \lambda$ plane. *Black region*: $L_{\rm CSD} < L_{\rm NL}$ and *blue region*: $L_{\rm NL} < L_{\rm CSD}$ indicative of shock formation. Taken from [13]

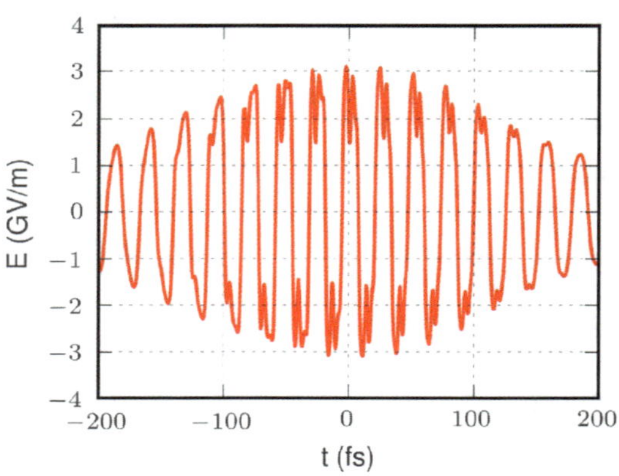

Fig. 7 Electric field of a $\tau_p = 200$ fs, $\lambda_0 = 8\,\mu$m pulse, propagated in single-crystal diamond for $170\,\mu$m. Taken from [13]

when $L_{\rm CSD} = L_{\rm NL}$. Figure 6 provides a practical estimate for whether or not carrier shock formation is expected to happen in diamond. Similar boundaries can be found for most transparent media in [13].

A very important observation is that the carrier shock formation is unaffected by the pulse duration since neither $L_{\rm CSD}$ nor $L_{\rm NL}$ depend on it. In Fig. 7 we can see the electric field of a 200 fs pulse undergoing self-steepening in diamond after $170\,\mu$m of propagation. The carrier wave shock forms over multiple oscillations of the electric field. Interestingly, the reshaping of the electric field is not constant over the whole pulse, but rather the steepening follows the field amplitude. Oscillations at the front and back of the pulse show very little steepening, while as one moves closer to the central time slice, "shark-fin" like features become evident. At the center of the pulse the electric field is strongly reshaped into a "top-hat," effectively having gone through all stages of the carrier shock formation. As stated before, the carrier shock formation is intensity (or else field amplitude) related and is acting on each point of the field separately through the co-propagation of the generated harmonics. Therefore, the overall pulse duration is of little importance concerning carrier shock formation. However, the use of longer duration high power pulses will be limited by optical damage in solid state media like single-crystal diamond.

2.1.1 Envelope vs. Carrier Wave Resolved Models

The choice between using envelope code such as NEE or a carrier wave resolved propagator such as UPPE is still a subject for debate. While envelope based propagation models properly describe the majority of physical effects of nonlinear pulse propagation they fail by definition in describing electric field related effects. In addition the NEE models are in general faster than field propagators, since the later demand proper discretization of the rapidly oscillating electric field rather than a slowly varying envelope.

However, envelope based model can in some cases be problematic even when modeling linear properties that act on high power wavepackets, such as chromatic dispersion for example. The restricted frequency interval in which the NEE dispersion operator approximates the medium becomes very evident in some cases in simulations, especially when there is a strong and narrow absorption band inside the computational spectral window. This is shown in Fig. 8, which compares a UPPE-based supercontinuum generation (black) in the water with an added "agent" that exhibits a narrow absorption band around 500 nm wavelength with that calculated using the dispersion operator approach (red) used by NEE. The UPPE spectral solver correctly exhibits a gap in the supercontinuum spectrum caused by the absorber, but the rest of the supercontinuum radiation remains unaffected. In the case of the NEE, the dispersion operator cuts off the spectrum close to the absorption band, and thus restricts the applicability of the model to a single "transparency window."

It is evident that the UPPE model as a Maxwell-like carrier-resolved propagator does not suffer from this restriction. The chromatic properties of the medium enter through a complex-valued function, say $\epsilon(\omega)$ which captures both the index of refraction and losses over an arbitrarily broad spectral range. The "only" real restriction here is in fact the availability of reliable and sufficiently accurate experimental data that span the relevant spectral window to serve as input to the model $\epsilon(\omega)$. Once we have a sufficiently accurate, smooth (i.e., free of numerical interpolation artifacts) representation of the permittivity, this is "applied" directly during the solution of the UPPE. This is made possible by the fact that the native computational variables are actually the spectral amplitudes of the electric field.

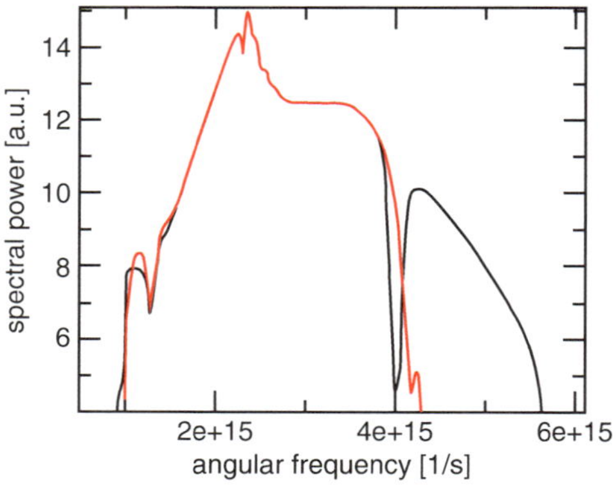

Fig. 8 Effect of the finite radius of convergence in the dispersion operator of the NEE. The simulated spectra for a water sample with an absorption band. The *black curve* shows the UPPE-based results that take into account index and loss over the whole spectra range. The *red curve* corresponds to an ideal implementation of the Brabec and Krausz NEE, this model fails at frequencies near and above the absorption band. Taken from [14]

2.2 *High Power Airy Wavepackets*

In this section we will explore the uses of a new type of wavepacket recently introduced to optics, the Airy.

2.2.1 The Airy Beam

The Airy function was first introduced in the context of quantum mechanics as a non-spreading solution of the wave equation by Berry and Balazs in 1979 [15]. In optics, Airy beams with finite energy were recently proposed as genuine non spreading optical waves [16, 17]. Airy beams are intrinsically one-dimensional structures but several combinations were demonstrated since 2007 for realizing a two- or three-dimensional non-spreading optical wave packet by linear superposition of subdimensional non-spreading structures [18–20]. In 2010, Chong et al. generated Bessel–Airy beams consisting of a Bessel beam in the transverse plane and an Airy distribution along the longitudinal dimension [21]. Abdollahpour et al. generated 3D Airy light-bullets consisting of an Airy beam profile in each dimension [22]. Nonlinear generation techniques were also used, e.g., by Ellenbogen et al. who generated Airy beams at the second harmonic of a pump beam by using nonlinear three-wave mixing in nonlinear crystals [23]. The one- and two-dimensional representations of the Airy function can be seen in Figs. 9 and 10.

Ideal Airy beams are infinite in size, and therefore carry infinite energy. In practice, both in experiments and simulations, Airy beams are finite in size and energy, and for this purpose an exponential decay function is utilized. For one transverse dimension x we can write:

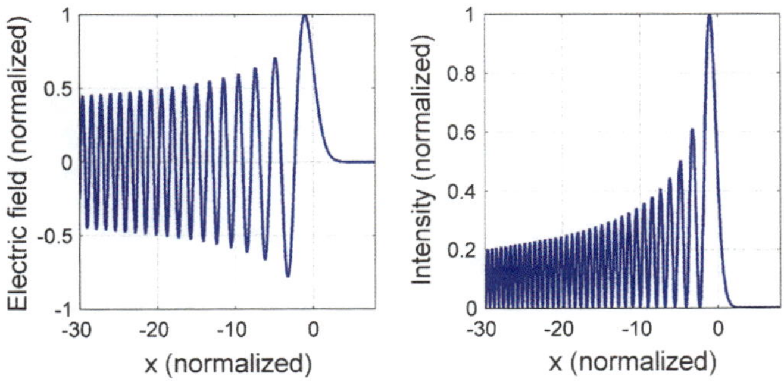

Fig. 9 Normalized (**a**) electric field, and (**b**) intensity distribution of a 1D Airy beam

Fig. 10 Normalized 2D Airy intensity distribution

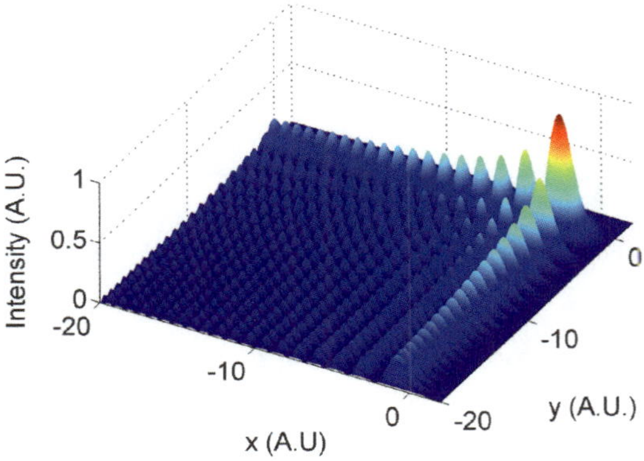

$$E(z=0,x) = E_0\,\mathrm{Ai}(x/w_0)\exp(ax/w_0)\;, \tag{33}$$

where Ai denotes the Airy function and w_0 the width of the central lobe. Typical value for the strength of the apodization factor a is between 0.1 (less number of lobes) and 0.05 (high number of lobes).

The generation of an Airy beam relies on the fact that a finite energy Airy distribution is in fact the Fourier transform of a Gaussian with the addition of a spatial cubic phase. This means that they can readily be generated from typical Gaussian laser beam profiles with use of spatial light modulator (used for inducing the cubic phase) and a focusing lens [16]. The Airy beam will thus be generated at the Fourier plane of the focusing lens, as can be seen in Fig. 11. An alternative technique that is suitable for high power applications uses cylindrical lenses and precise control of their aberrations instead of an SLM [18].

In the linear regime the propagation dynamics of Airy beams has been studied extensively. The two most exotic features of Airy beams, namely, self-healing and transverse acceleration, have been investigated by Broky et al. [25] and Siviloglou et al. [26]. The dependence of Airy wave packets on wavelength, spatial coherence, and dispersion has been investigated in [27, 28].

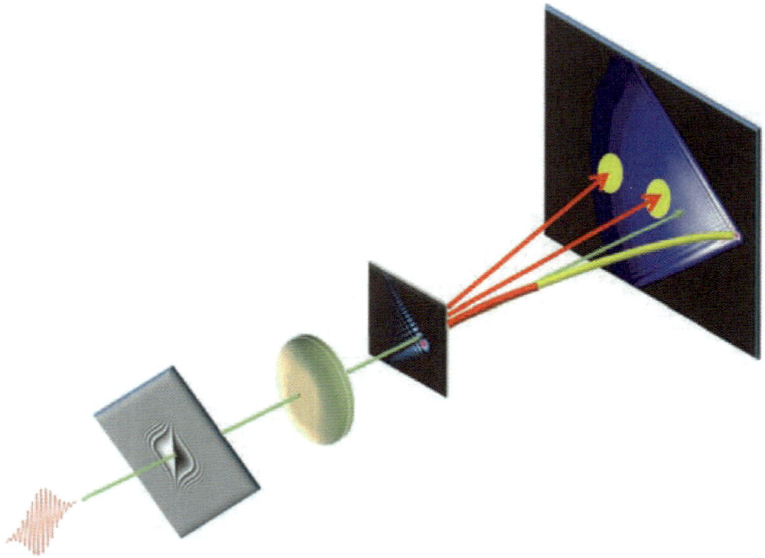

Fig. 11 Experimental setup for the generation two dimensional Airy beams. Taken from [24]

Transverse acceleration is namely the ability of the Airy beam intensity lobes to turn as the beam propagates in space. The transverse acceleration of the peak intensity lobe of Airy beams obeys $s - \zeta^2 = s_0$, i.e., in physical units:

$$x = x_0 + \frac{z^2}{4k^2 w_0^3}, \tag{34}$$

where x_0 denotes the initial position of the peak at $z = 0$. Equation (34) shows that the Airy beam follows a parabolic trajectory $x = x_0 + z^2/2r_c$ with curvature radius $r_c = 2k^2 w_0^3$. This can seen in Fig. 12, where the peak intensity of an Airy beam is shown versus propagation distance, close to the focus position at $z = 20$ cm. Note that in all paraxial Airy beams, the spatial displacement due to transverse acceleration is much smaller than the distance the beam needs to propagate, i.e., the turning speed is relatively slow. This is evident if one takes a closer look at the axis units of Fig. 12.

Self-reconstruction, or else self-healing, is the remarkably ability of the Airy to reconstruct its intensity profile even after it has been strongly perturbed or even obscured. Self-reconstruction was demonstrated in [25] as can be seen in Fig. 13.

Both the transverse acceleration and self-reconstruction of Airy beams are linear properties and can be found even when propagation is taking place in vacuum. They arise from the unique transverse energy flux inherent to the Airy profile that pushes energy along the diagonal, much like in the case of the Bessel beam [29] generated by axicons [30] and other conical wavepackets.

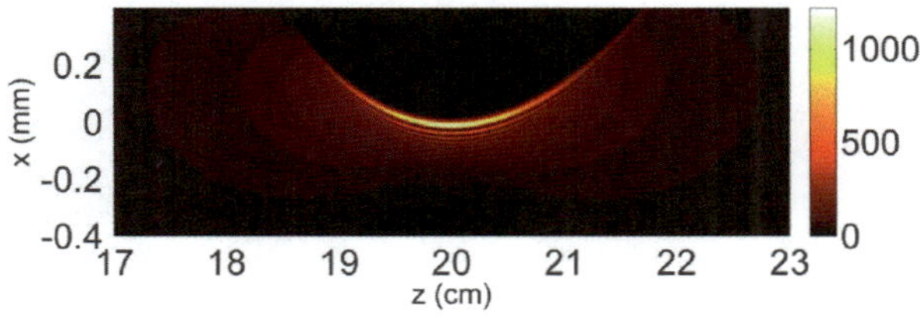

Fig. 12 Simulated peak intensity along propagation distance of a 1D finite energy Airy beam. The generation technique mimics the experimental one, using a phase mask (at $z = 0$ cm) which is applied to a Gaussian and a focusing lens $f = 10$ cm located at $z = 10$ cm. Focal plane at $z = 20$ cm

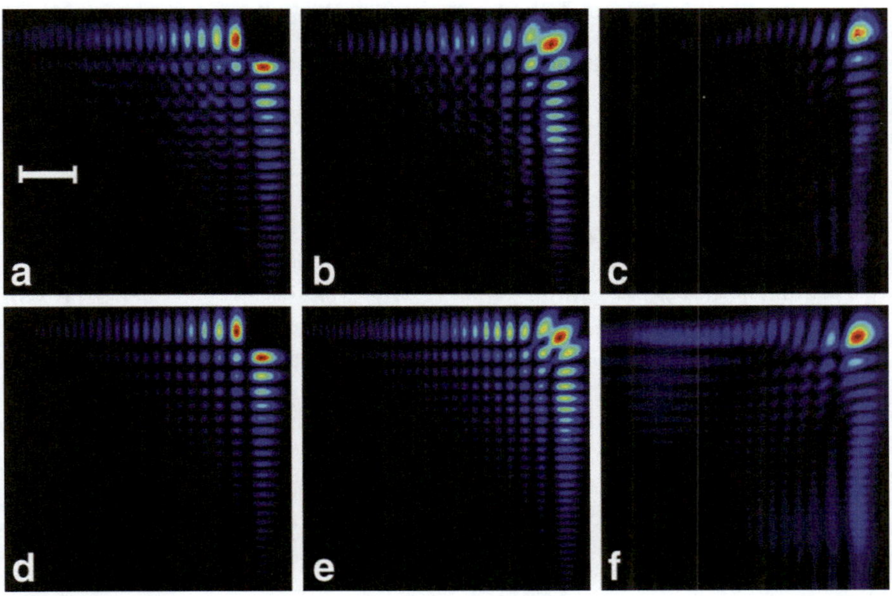

Fig. 13 Self-healing of an Airy beam when its main lobe is blocked. Observed intensity profile at (**a**) the input $z = 0$, (**b**) $z = 11$ cm, and (**c**) $z = 30$ cm. The corresponding numerical simulations are shown in (**d**–**f**). Taken from [25]

Airy beams in the nonlinear regime were discussed in [22, 24, 31–33] from the observation of spectacular manifestations common with the nonlinear dynamics of light filaments, such as supercontinuum generation and conical emission. Lotti et al. recently discovered the existence of stationary nonlinear Airy beams in one transverse dimension [34]. An open question concerns the properties of multidimensional Airy beams in the nonlinear regime.

In the following sections we will take a short look at the propagation of high intensity Airy beams in atmospheric air. As we will see, optical Airy beams can be strongly affected by nonlinear effects that play a key role in the dynamics of ultra-short laser pulse propagation, i.e., the optical Kerr effect, multiphoton absorption and ionization to cite only a few [1, 35, 36].

2.2.2 Nonlinear Airy Beams

High power pulsed Airy beams and in general all non-diffracting wavepackets are of particular interest in nonlinear optics since diffraction, one of the basic effects regulating optical filaments, is of limited importance here. The scientific interest around the propagation regime of the Airy, that is by definition different than classical Gaussian beam filamentation, is further more enhanced by the exotic linear characteristics discussed in the previous section.

It was shown in [33], by the use of the NEE model, that a competition between the linear and nonlinear effects takes place when high power two (XY) dimensional Airy beams are propagating in the atmosphere. The outcome of this competition may be modified by tuning the width of the main lobe of the Airy beam and the size of the truncating diaphragm, which in both cases is equivalent of regulating the linear transverse flux that give rises to the Airy's exotic characteristics.

Figure 14 shows the intensity cross sections of the high power Airy beams at different propagation distances during their nonlinear propagation in atmospheric air. For the three cases displayed in column, $w_0 = 100, 200,$ and $300\,\mu$m, the radius of the circular diaphragm is $r_d = 4.5$ mm. Propagation distances increase from top to bottom from $z = 0$ up to $z = 80$ cm with 20 cm steps. The last row of Fig. 14 shows the beam profiles that would be obtained for a linear propagation over 80 cm of each input Airy beam with the same w_0, for comparison. The intense peak of the Airy beam accelerates transversely faster when the initial size of the central lobe is small. At a propagation distance of 80 cm, the peak of the Airy beam with $w_0 = 100\,\mu$m (first column in Fig. 14) reached the position $x = y \sim 2.5$ mm, whereas the peaks of the Airy beams with $w_0 = 200$ and $w_0 = 300\,\mu$m (second and third column in Fig. 14) show a slower transverse acceleration by a factor of 8 and 27, respectively. The quadratic relation between transverse acceleration and w_0 is a purely linear property that carries over to the nonlinear regime.

It is readily seen in the third column in Fig. 14 that the Airy beam with the largest lobe remains nearly in place but undergoes self-focusing. The effect of self-focusing on the Airy lobes is also clear from the comparison of the beam cross-sections after nonlinear and linear propagation over 80 cm, shown in the last two rows of Fig. 14. For the same input intensity, the main lobe of the Airy beam carrying the highest power ($w_0 = 300\,\mu$m, third column of Fig. 14) shrinks faster than that of the less powerful Airy beams. Even the secondary lobes of the most powerful Airy beam are significantly reshaped. This trend to self-focus competes with the acceleration of the Airy beam inherited from the properties of linearly propagating Airy beams. Both processes can be viewed as a transfer of power: the acceleration corresponds to a transfer from the secondary lobes to the main lobe and is responsible for the self-healing property of Airy beams [25], whereas self-focusing concentrates the power of a given lobe upon itself.

Even though high power Airy beams are intense enough to undergo Kerr self-focusing and induce multiphoton absorption, the trajectory of the main lobe obtained in the linear regime is preserved in the three cases shown in Fig. 14. The parabolic trajectory of the Airy beam with the smallest lobe width (Fig. 14a) exhibits the

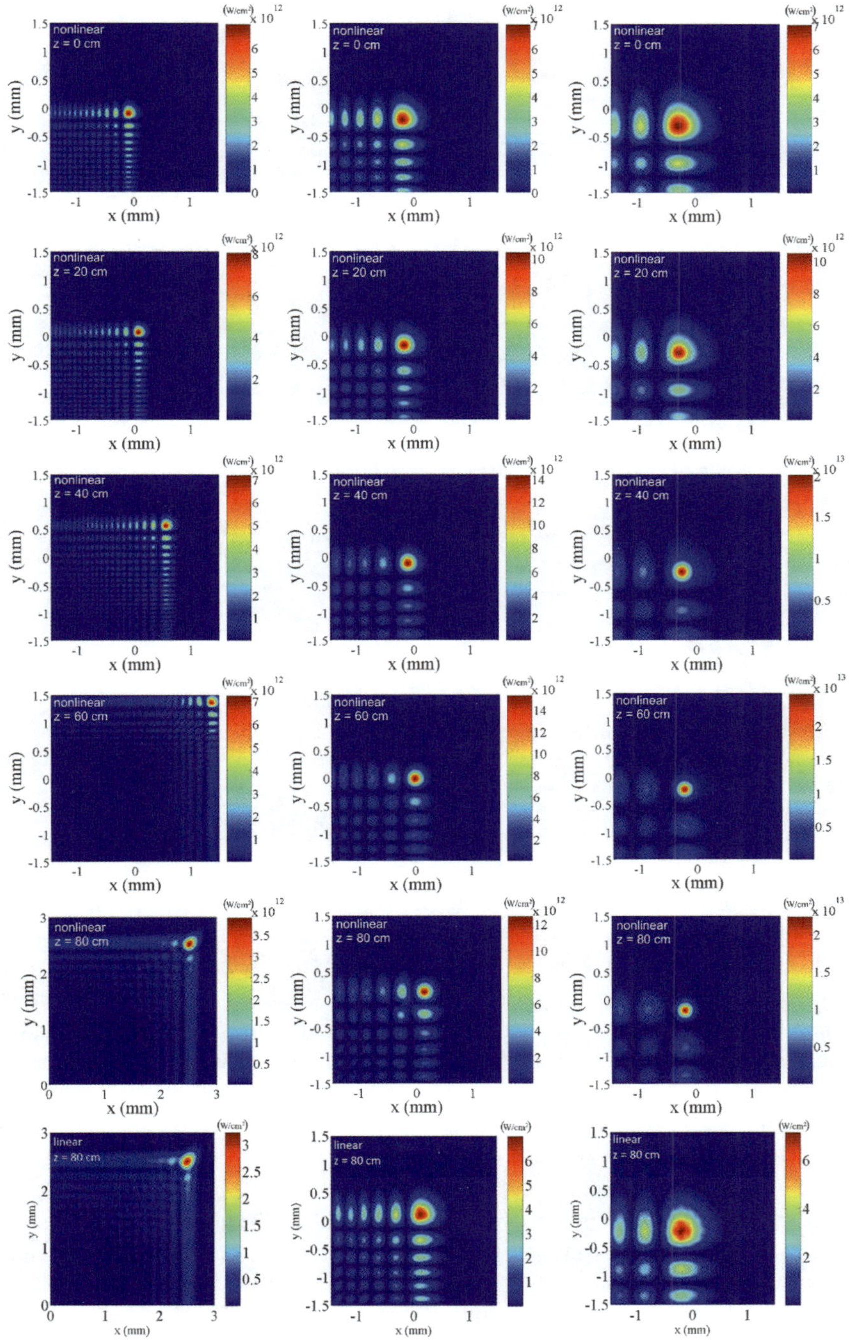

Fig. 14 (x, y) intensity plots of nonlinear Airy beams with initial peak intensity 7×10^{12} W/cm^2 and different central lobe size. *First column*: $w_0 = 100\,\mu$m; *Second column*: $w_0 = 200\,\mu$m; *Third Column* $w_0 = 300\,\mu$m. The propagation distance varies from $z = 0$ (*first line*) to $z = 80$ cm (*fifth line*) by steps of 20 cm. The diaphragm radius for all cases is $r_d = 4.5$ mm. The *last line* shows the beam profiles after linear propagation of the three Airy beams over $z = 80$ cm. Note the larger window used for plotting the cross section in the *bottom left corner*. Taken from [33]

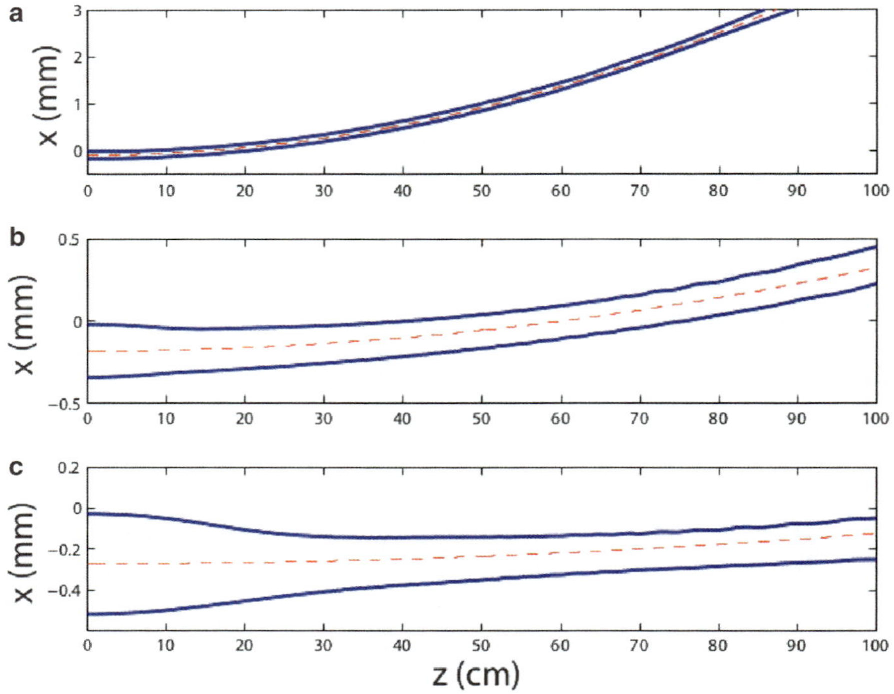

Fig. 15 Projection of the main lobe trajectory and FWHM of the Airy beam along the x-direction. (**a**) Input Airy beam with $w_0 = 100\,\mu\text{m}$. (**b**) Input Airy beam with $w_0 = 200\,\mu\text{m}$. (**c**) Input Airy beam with $w_0 = 300\,\mu\text{m}$. Taken from [33]

smallest curvature radius. Figure 15 shows the projection of the main lobe trajectory in the (x, z) plane and its FWHM diameter. The peak clearly turns as a linear Airy beam would do, following a parabolic trajectory that coincides with that given by (34) (dashed curves in Fig. 15). For the wider input Airy beams (Fig. 15b, c), the nonlinear propagation starts by a Kerr induced self-focusing stage with a decrease of the FWHM diameter from $z = 0\,\text{cm}$ to $z = 30\,\text{cm}$. This indicates a competition between Kerr self-focusing and the acceleration of the Airy beam. It is therefore possible to retain a high intensity curved light channel given that the transverse energy flux of the Airy is strong enough (i.e., transverse size of the beam is big enough) to overcome the optical Kerr effect.

Utilizing this extraordinary property, Polynkin et al. in [24] were able to produce curved plasma channels in air using ultraintense Airy beams, as can be seen in Fig. 16. In this setup the region of nonlinear interaction follows the Airy beam's intensity curved trajectory, and is strong enough to generate a free electron plasma through MPA, and the associate with it supercontinuum. While a curved plasma channel is extraordinary and interesting in it own right, the supercontinuum, which is a spectral imprint of the Kerr and plasma related self-phase modulation of every point of the plasma channel, is of fundamental interest in nonlinear optics. Since

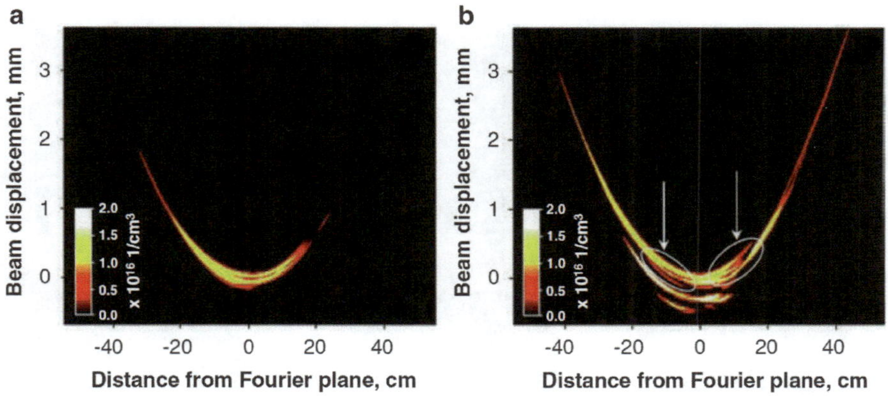

Fig. 16 (**a**) Results of numerical simulations for the peak free-electron density in plane of curvature of the Airy beam for 5 mJ pulses. The generated plasma channel is continuous and smooth. (**b**) Same as in (**a**) but for 10 mJ pulses. Secondary lobes of the Airy beam start generating their own plasma channels, and the main channel develops bifurcations along the propagation path. Locations of the two most pronounced bifurcations are indicated with *arrows*. Taken from [24]

the generated frequencies are not sharing the curved trajectory of the wavepacket that generated them, they will propagate in the forward direction on a straight line. Exploiting this difference in behavior between generating and generated radiation, it is possible to measure the generated spectrum of each point along the curved plasma channel separately. An achievement that is rather difficult in the coaxial geometry of standard non-curved plasma channels.

Self-reconstruction of high power wavepackets is another promising application of the nonlinear Airy beam, since nonlinear instabilities can be considered to act as perturbation on the linear beam profile. Again, as in the case of transverse acceleration the deciding factor is the competition between the linear transverse flux of the Airy and the optical Kerr effect which tends to lead to the nucleation of individual filaments on the intensity lobes of the Airy. This competition was investigated in [33] for different input powers and Airy beam geometries.

Figure 17 shows a comparison of the propagation of three Airy beams which have identical main lobe width (200 μm) but which are truncated differently: the diaphragm radius is $r_{\mathrm{d}} = 4.5$ mm in the first column, $r_{\mathrm{d}} = 2.5$ mm in the second column and $r_{\mathrm{d}} = 1$ mm in the third column of Fig. 17. The propagation is taking place in air with an increased nonlinearity by one order of magnitude. This keeps unchanged the curvature radius of the trajectories of the linearly propagating Airy beams but increases $P_{\mathrm{A}}/P_{\mathrm{cr}}$ by a factor of 10 and allows us to isolate the effect of self-focusing on the beam shape from other nonlinearities, like multiphoton absorption and plasma defocussing.

As we can see the in the process is highly dynamic. In the case of the large apodizer, the linear energy flux of the Airy beam is sustained over a larger propagation distance. The beam truncated by the small diaphragm containing only

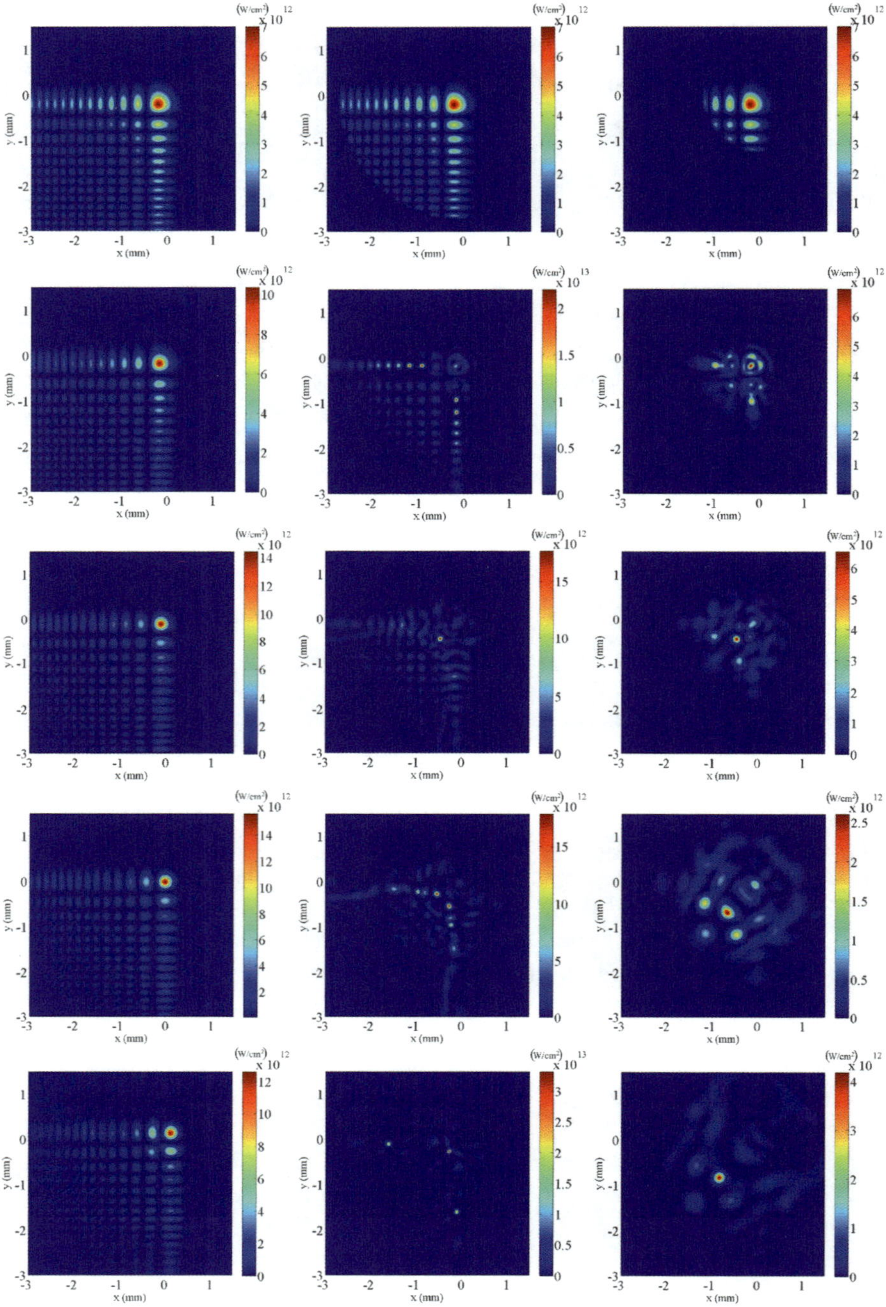

Fig. 17 Effect of truncation in the high-power regime. The three columns correspond to the propagation of of nonlinear Airy beams with initial peak intensity $7 \times 10^{12}\,\mathrm{W/cm^2}$, central lobe size $w_0 = 200\,\mu\mathrm{m}$, diaphragmed by circular apertures of radius $r_\mathrm{d} = 4.5\,\mathrm{mm}$ (*first column*), $r_\mathrm{d} = 2.5\,\mathrm{mm}$ (*second column*), $r_\mathrm{d} = 1\,\mathrm{mm}$ (*third column*), with nonlinear index coefficient $n_2 = 3.2 \times 10^{-18}\,\mathrm{cm^2/W}$. The propagation distance varies from $z = 0$ (*first line*) to $z = 80\,\mathrm{cm}$ (*fifth line*) by steps of 20 cm

three lobes of the input Airy beam propagates under the action of nonlinear effects over a much shorter distance. The linear energy flux responsible for the curved trajectory is in this case not strong enough to bring energy from the tail to the peak of the Airy beam over the whole Airy zone. The small diaphragm quenches the power refilling of the secondary lobes and therefore simultaneously tends to quench the turning ability of the Airy beam. Secondary lobes still undergo self-focusing and form filaments but they subsequently follow a standard filament interaction process featured by energy exchange with the neighborhood. Their intensity increases until multiphoton absorption becomes efficient and plasma is generated that eventually participates in beam spreading and nucleation of other filaments from the reservoir. Figure 17 shows that although these multiple filaments are initiated at the same locations regardless of the diaphragm size, the smaller the diaphragm radius the earlier in their propagation they depart from each other. In this case the nonlinear dynamics dominate and seemingly prevent the main lobe to turn.

2.2.3 Linear Auto-Focusing Waves

By exploiting the transverse acceleration and the 1D nature of the Airy function, Efremidis et al. recently introduced the cylindrically symmetric Airy beam, or ring-Airy wavepacket, which is able to abruptly autofocus in the linear regime [37]. As shown in a recent work [38], abruptly autofocusing waves are experimentally feasible, and can be used to deliver high energy pulses inside thick transparent samples without damaging the material prior to the focus. The abrupt intensity increase at the focus of the cylindrical symmetric Airy makes it easy to combine a long working distance with a small focal volume.

The electric field envelope of such a ring-Airy wavepacket follows the function:

$$\mathcal{E}_0(r,0) = \mathrm{Ai}\left(\frac{r_0 - r}{w_0}\right)\exp\left[a\left(\frac{r_0 - r}{w_0}\right)\right] \tag{35}$$

where Ai corresponds to the Airy function from (35), r is the radial coordinate, r_0 represents the radius of the primary ring, a is a constant, and w_0 is a scaling factor. The radial fluence distribution of a ring-Airy beam can be seen in Fig. 18.

Figure 19 shows the normalized radial electric field amplitude distribution of a ring-Airy wavepacket along propagation distance z. The normalized coefficients of (35) are $a = 0.05$, $r_0 = 10$, and $w_0 = 1$. As we can see at z 6 the beam exhibits an intense autofocus. The maximum intensity inside the auto-focus depends on the beam initial beam characteristics and a and r_0. The maximum contrast $I_{\max}/I_0$ was found in [37] to be 156 for $a = 0.05$, $r_0 = 15$, and $w_0 = 1$ (dimensionless units), which will be the values used in Fig. 18 and in the next section. The experimental realization of ring-Airy wavepackets is utilizing the same approach as is used in the 1D and 2D (x, y) Airy beams, a phase mask in the form of an SLM and a Fourier transformation lens [38].

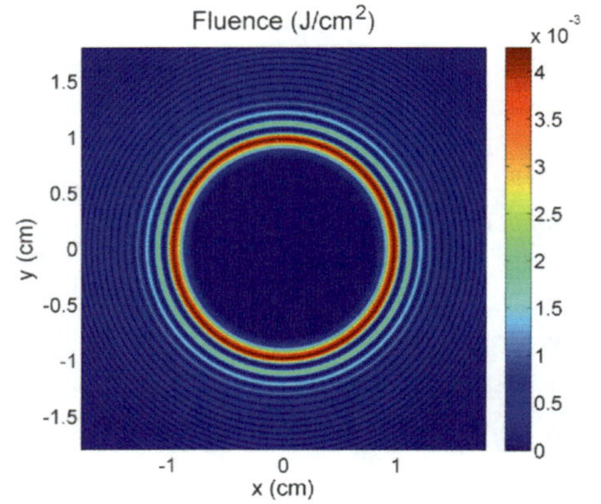

Fig. 18 Radial fluence distribution of a ring-Airy beam with $a = 0.05$, $r_0 = 921\,\mu\text{m}$ and $w_0 = 61.4\,\mu\text{m}$

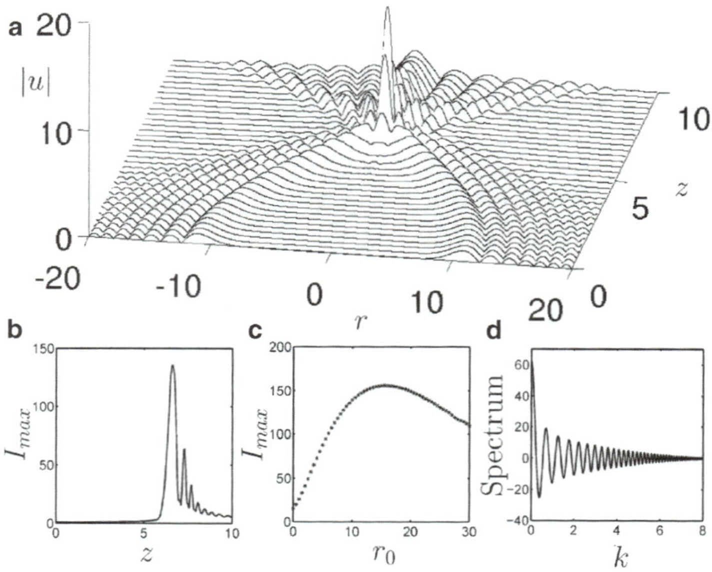

Fig. 19 Dynamics of radially symmetric Airy beams for $a = 0.05$, $r_0 = 10$, and $I_{\max}(z = 0) = 1$. (**a**) Detailed plot of the central part of the propagation dynamics. (**b**) Maximum intensity as a function of z. (**c**) Maximum intensity that the Airy beam reaches during propagation for different values of the initial radius r_0. (**d**) Spectrum obtained from Hankel transformation of the initial condition. Taken from [37]

2.2.4 Nonlinear Ring-Airy Beams

In this section, we will numerically explore the nonlinear dynamics of high power ring-Airy wavepackets generated with Gaussian laser pulses. We compare the attributes of the high power ring-Airy beams to those of Gaussian beams carrying the same power and identify a number of very useful and interesting features for practical applications.

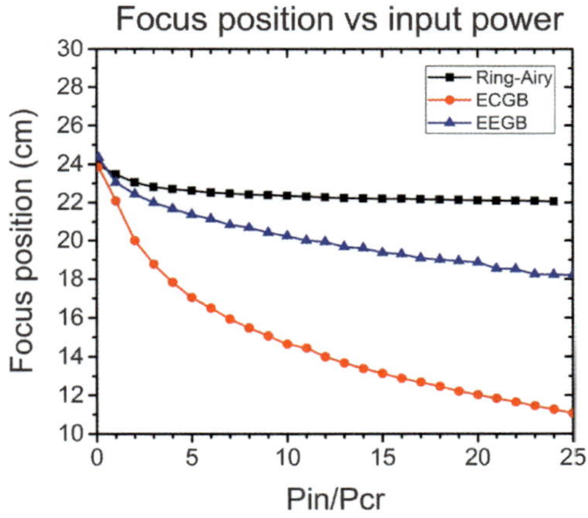

Fig. 20 Focus position as a function of the input power for the HPRAW (*black squares*), and the two Gaussian beams with equivalent contrast (*red dots*), and equivalent envelope (*blue triangles*).Taken from [39]

It was shown in [39], both numerically and experimentally, that ring-Airy beams behave very differently than their Gaussian equivalents in the nonlinear regime. The well-known shift of the linear focus position in the nonlinear regime, which was one of the first nonlinear phenomena studied by Marburger et al. in 1975 [40], is much smaller in case of ring-Airy beams when compared to Gaussians.

As we can see in Fig. 20, where the focus position versus total input power of a ring-Airy and two equivalent Gaussians [18] are shown, the ring-Airy focal position (black squares) shifting in a very different way than the two Gaussians. For this case the ring-Airy beam is identical to the one depicted in Fig. 18, while the two Gaussians have widths at $1/e^2$ radius of 1738 and 855 μm. The Gaussian are defined as having equivalent beam envelope (EEGB) and equivalent contrast (ECGB) in [18]. Regardless if the spatial profile of the beams used the temporal profile is in all cases Gaussian with an FWHM of 200 fs at a central wavelength of 800 nm.

Interestingly, in the nonlinear regime, the high power ring-Airy beam focus shifts a significantly smaller amount compared to the two equivalent Gaussian beams as the beam power increases. For the ring-Airy, the focus position moves from 24 cm to roughly 22 cm, while the input power is gradually increased up to 25 Pcr. Further increase in power up to seems to have negligible effect on the focus position. For both Gaussian beams, the focus shift is much larger than in the high power ring-Airy beam case. Especially for the ECGB, the focus position is shifted down to 11 cm for 24, which is more than half the initial distance. For the EEGB, the focus shift is not as strong, but still reaches 6 cm, about three times more than for the high power ring-Airy beam case.

Acknowledgements The work reported on in this review paper was carried out with support from an AFOSR MURI grant FA9550-10-1-0561. A grant from the National Science Foundation (NSF) DMS-1324288 provided partial travel support for U.S participants to the workshop.

References

1. A. Couairon, A. Mysyrowicz, Phys. Rep. **441**, 47 (2007)
2. M. Kolesik, J.V. Moloney, Phys. Rev. E **70**(3), 036604 (2004)
3. P. Kinsler, S.B.P. Radnor, G.H.C. New, Phys. Rev. A **72**(6), 063807 (2005)
4. P. Kinsler, J. Opt. Soc. Am. B **24**(9), 2363 (2007)
5. P. Kinsler, Phys. Rev. A **81**(1), 013819 (2010)
6. P. Kinsler, Phys. Rev. A **81**(2), 023808 (2010)
7. M. Kolesik, P. Jakobsen, J.V. Moloney, Phys. Rev. A **86**(3), 035801 (2012)
8. T. Brabec, F. Krausz, Phys. Rev. Lett. **78**(17), 3282 (1997)
9. L. Bergé, J. Rolle, C. Köhler, Phys. Rev. A **88**(2), 023816 (2013)
10. K. Mishima, M. Hayashi, J. Yi, S.H. Lin, H.L. Selzle, E.W. Schlag, Phys. Rev. A **66**(3), 033401 (2002)
11. R.G. Flesch, A. Pushkarev, J.V. Moloney, Phys. Rev. Lett. **76**(14), 2488 (1996)
12. P. Whalen, P. Panagiotopoulos, M. Kolesik, J.V. Moloney, Phys. Rev. A **89**(2), 023850 (2014)
13. P. Panagiotopoulos, P. Whalen, M. Kolesik, and J. V. Moloney, Journal of the Optical Society of America B **32**, 1718–1730 (2015)
14. M. Kolesik, P.T. Whalen, J.V. Moloney, IEEE J. Sel. Top. Quantum Electron. **18**(1), 494 (2012)
15. M.V. Berry, N.L. Balazs, Am. J. Phys. **47**(3), 264 (1979)
16. G.A. Siviloglou, D. Christodoulides, Opt. Lett. **32**(8), 979 (2007)
17. G.A. Siviloglou, J. Broky, A. Dogariu, D. Christodoulides, Phys. Rev. Lett. **99**(21), 213901 (2007)
18. D.G. Papazoglou, S. Suntsov, D. Abdollahpour, S. Tzortzakis, Phys. Rev. A **81**(6), 061807 (2010)
19. D.M. Cottrell, J.A. Davis, T.M. Hazard, Opt. Lett. **34**(17), 2634 (2009)
20. J.A. Davis, M.J. Mitry, M.A. Bandres, I. Ruiz, K.P. McAuley, D.M. Cottrell, Appl. Opt. **48**(17), 3170 (2009)
21. A. Chong, W.H. Renninger, D.N. Christodoulides, F.W. Wise, Nat. Photonics **4**(2), 103 (2010)
22. D. Abdollahpour, S. Suntsov, D. Papazoglou, S. Tzortzakis, Phys. Rev. Lett. **105**(25), 253901 (2010)
23. T. Ellenbogen, N. Voloch-Bloch, A. Ganany-Padowicz, A. Arie, Nat. Photonics **3**(7), 395 (2009)
24. P. Polynkin, M. Kolesik, J.V. Moloney, G.A. Siviloglou, D.N. Christodoulides, Science **324**(5924), 229 (2009)
25. J. Broky, G.A. Siviloglou, A. Dogariu, D.N. Christodoulides, Opt. Express **16**(17), 12880 (2008)
26. G.A. Siviloglou, J. Broky, A. Dogariu, D.N. Christodoulides, Opt. Lett. **33**(3), 207 (2008)
27. J.E. Morris, M. Mazilu, J. Baumgartl, T. Čižmár, K. Dholakia, Opt. Express **17**(15), 13236 (2009)
28. I.M. Besieris, A.M. Shaarawi, Phys. Rev. E **78**(4), 046605 (2008)
29. M. Abramowitz, I.A. Stegun (eds.), *Handbook of Mathematical Functions: With Formulas, Graphs, and Mathematical Tables*. Dover Books on Mathematics (Dover, New York, 1965)
30. J.H. McLeod, J. Opt. Soc. Am. **44**(8), 592 (1954)
31. J. Parravicini, P. Minzioni, V. Degiorgio, E. DelRe, Opt. Lett. **34**(24), 3908 (2009)
32. P. Polynkin, M. Kolesik, J. Moloney, Phys. Rev. Let. **103**(12), 123902 (2009)
33. P. Panagiotopoulos, D. Abdollahpour, A. Lotti, A. Couairon, D. Faccio, D.G. Papazoglou, S. Tzortzakis, Phys. Rev. A **86**(1), 013842 (2012)
34. A. Lotti, D. Faccio, A. Couairon, D.G. Papazoglou, P. Panagiotopoulos, D. Abdollahpour, S. Tzortzakis, Phys. Rev. A **84**(2), 021807 (2011)
35. A. Couairon, E. Brambilla, T. Corti, D. Majus, O. Ramírez-Góngora, M. Kolesik, Eur. Phys. J. Spec. Top. **199**(1), 5 (2011)
36. I. Kaminer, M. Segev, D.N. Christodoulides, Phys. Rev. Let. **106**(21), 213903 (2011)
37. N.K. Efremidis, D.N. Christodoulides, Opt. Lett. **35**(23), 4045 (2010)

38. D.G. Papazoglou, N.K. Efremidis, D.N. Christodoulides, S. Tzortzakis, Opt. Lett. **36**(10), 1842 (2011)
39. P. Panagiotopoulos, D.G. Papazoglou, A. Couairon, S. Tzortzakis, Nat. Commun. **4**, 2622 (2013)
40. J.H. Marburger, Prog. Quantum Electron. **4**(1), 35 (1975)

Index

A.D. Bandrauk et al. (eds.), *Laser Filamentation*, CRM Series in Mathematical Physics, DOI 10.1007/978-3-319-23084-9

Printed by Printforce, the Netherlands